LONDON MATHEMATICAL SOCIETY LECTURE NOTE SERIES

Managing Editor: Professor J.W.S. Cassels, Department of Pure Mathematics and Mathematical Statistics, University of Cambridge, 16 Mill Lane, Cambridge CB2 1SB, England

The titles below are available from booksellers, or, in case of difficulty, from Cambridge University Press.

London Mathematical Society Lecture Note Series. 252

Geometry and Cohomology in Group Theory

Edited by

Peter H. Kropholler
Queen Mary and Westfield College, London

Graham A. Niblo
University of Southampton

Ralph Stöhr
UMIST

CAMBRIDGE UNIVERSITY PRESS

CAMBRIDGE UNIVERSITY PRESS
Cambridge, New York, Melbourne, Madrid, Cape Town, Singapore, São Paulo

Cambridge University Press
The Edinburgh Building, Cambridge CB2 8RU, UK

Published in the United States of America by Cambridge University Press, New York

www.cambridge.org
Information on this title: www.cambridge.org/9780521635561

First published 1998

A catalogue record for this publication is available from the British Library

ISBN 978-0-521-63556-1 paperback

Transferred to digital printing 2008

Table of Contents

vi

Preface

Refereed Conference Proceedings
The London Mathemtical Society Symposium† on
Geometry and Cohomology in Group Theory
12–22 July 1994

The cross-fertilization of ideas from abstract group theory with those from geometry, the use of topological and cohomological techniques have contributed
over recent years to revitalising group theory and representation theory in
many exciting ways. Not only has this brought impressive mathematical advances, but again it has drawn in different kinds of specialists. At this London
Mathematical Society Symposium we aimed to draw together experts in algebra, geometry, representation theory and cohomology to make a contribution
to this interchange of ideas. It is an area where there is marked strength in
the United Kingdom, traditionally the home of many group theorists, but
nowadays also active in cohomological and geometric group theory.

The closing report to the Science and Engineering Research Council* began
thus:

"The success of this meeting owed much to the strong list of participants, the
high standard of lectures by invited speakers and to the good environment
at Grey College which encouraged many collaborative research projects. The
success can be measured by the strength of the articles which we are assembling for the Conference Proceedings."

We thank all those authors whose articles were submitted for this volume, and
also the equal number of referees. We emphasise that every effort was made

† Supported by SERC Grant GR/H92159.
* Now replaced by the Engineering and Physical Sciences and the Particle
Physics and Astronomy Research Councils.

to have papers refereed to the same standard as academic and learned mathematical journals. There are a number of survey articles here including those by Carlson, Cornick, Grigorchuk, Linnell, Mikhajlovskii and Ol'shanskii, and Wilson. These cover ground from cohomology and representation theory, analytic methods and the use of von Neumann algebras, the application of hyperbolic groups and the structure of soluble groups. Davis gave a series of lectures on buildings. Further contributions were made following collaborations at the meeting, for example the paper by Neumann and Rowley.

Every participant contributed to the success of this meeting. For the smooth organization we are indebted to the hard work of Sue Nesbitt, Rachel Duke and Ruth Silverstone. The support of the London Mathematical Society and Tony Scholl were invaluable, and the accommodation at Grey College provided a comfortable and congenial atmosphere. The first named editor would like to thank Richard Platten for his assistance with authors' proofs.

Peter H. Kropholler
School of Mathematical Sciences
Queen Mary & Westfield College
Mile End Road
London E1 4NS

Graham A. Niblo
Faculty of Mathematical Studies
University of Southampton
Highfield
Southampton SO17 1BJ

Ralph Stöhr
Mathematics Department
UMIST
P.O. Box 88
Manchester M60 1QD

LMS Durham Symposium on: Geometry and Cohomology in Group Theory

Tuesday 12th July to Friday 22nd July 1994

Back row – left to right: W. Dicks, D. Johnson, R. Gilman, Y. Ginosar, G. Noskov, B. Borovik, H-W. Henn, J. Alonso, R. M. Thomas, N. Gilbert, P. J. Rowley, R. Sandling, R. Bieri, J. R. J. Groves, I. J. Leary, J. Rickard, J. Greenlees, M. A. Roller, B. Hartley, G. Mislin.

Third row: D. E. Cohen, A. Juhász, K. W. Gruenberg, B. Eckmann, K. Ruane, C. F. Miller, C. J. B. Brookes, I. M. Chiswell, H. Hoare, D. J. Collins, U. Stammbach, E. Aljadeff, M. W. Davis, K. Vogtmann, M. J. Dunwoody, G. A. Swarup, C. B. Thomas, H. Glover, E. Ventura.

Second row: E. C. Turner, N. Benakli, H. Meinert, Yu. Kuz'min, C. Champetier, R. I. Grigorchuk, H. Bass, M. Bridson, J. S. Wilson, A. Ol'shanskii, R. Stöhr, G. Baumslag, P. Kropholler, J. Carlson, C. T. C. Wall, D. J. Benson, F. E. A. Johnson, S. Pride, A. Adem, A. Mann.

Front row: R. Bryant, R. Geoghegan, G. Connor, X. Wang, P. Papasoglu, J. Cornick, W. Neumann, D. B. A. Epstein, A. Duncan, J. Harlander, M. Hartl, B. Funke, S. Rees, O. Talelli, Mrs. Bestvina, P. A. Linnell, M. Edjvet.

List of Participants

A. Adem, Wisconsin, Madison	adem@math.wisc.edu
E. Aljadeff* , Haifa	mar3300@technion.ac.il
J. Alonso, Stockholm	alonso@matematik.su.se
H. Bass, Columbia, NY	hb@math.columbia.edu
M. Batty* , Warwick	batty@maths.warwick.ac.uk
G. Baumslag, City College, NY	gilbert@groups.sci.ccny.cuny.edu
N. Benakli* , Columbia, NY	benakli@math.columbia.edu
D. J. Benson, Athens, Georgia	djb@byrd.math.uga.edu
M. Bestvina, Utah	bestvina@math.utah.edu
R. Bieri, Franfurt	bieri@math.uni-frankfurt.de
B. Borovik, UMIST	sasha@lanczos.ma.umist.ac.uk
B. Bowditch, Southampton	bhb@maths.soton.ac.uk
M. Bridson, Oxford	bridson@maths.ox.ac.uk
C. J. B. Brookes, Cambridge	C.J.B.Brookes@pmms.cam.ac.uk
R. Bryant, UMIST	bryant@umist.ac.uk
C. M. Campbell, St. Andrews	cmc@st-and.ac.uk
J. Carlson, Athens, Georgia	jfc@math.uga.edu
C. Champetier, Grenoble	champet@fourier.grenet.fr
I. M. Chiswell, QMW, London	I.M.Chiswell@qmw.ac.uk
D. E. Cohen, QMW, London	D.E.Cohen@qmw.ac.uk
D. J. Collins, QMW, London	D.J.Collins@qmw.ac.uk
G. Conner* , BYU	conner@math.byu.edu
J. Cornick, Columbus, Ohio	cornick@math.ohio-state.edu
M. W. Davis, Columbus, Ohio	mdavis@mail.math.ohio-state.edu
W. Dicks, Barcelona	dicks@manwe.mat.uab.es
M. du Sautoy, Oxford	marcus.dusautoy@all-souls.ox.ac.uk
A. Duncan, Newcastle	A.Duncan@ncl.ac.uk
M. J.Dunwoody, Southampton	mjd@maths.soton.ac.uk
B. Eckmann, Zürich	beno.eckmann@math.ethz.ch

---

* Not supported by SERC

M. Edjvet, Nottingham	Martin.Edjvet@maths.nott.ac.uk
D. B. A. Epstein, Warwick	dbae@maths.warwick.ac.uk
R. A. Fenn, Sussex	R.A.Fenn@sussex.ac.uk
R. Geoghegan, Binghampton, NY	ross@math.binghamton.edu
N. Gilbert, Heriot-Watt	n.d.gilbert@hw.ac.uk
R. Gilman* , New Jersey	rgilman@stevens-tech.edu
Y. Ginosar* , Haifa	mat2810@technion.ac.il
H. Glover, Columbus, Ohio	glover@math.ohio-state.edu
J. P. C. Greenlees, Sheffield	J.Greenlees@sheffield.ac.uk
R. I. Grigorchuk, Moscow	grigorch@alesia.ips.ras.ru
J. R. J. Groves, Melbourne	jrjg@maths.mu.oz.au
K. W. Gruenberg, QMW, London	K.W.Gruenberg@qmw.ac.uk
J. Harlander, Frankfurt	harl@math.uni-frankfurt.de
M. Hartl* , Strasbourg	hartl@math.u-strasbg.fr
B. Hartley, Manchester	
H-W. Henn, Heidelberg	henn@mathi.uni-heidelberg.de
J. Hillman, Sydney	jonh@maths.usyd.edu.au
H. Hoare, Birmingham	A.H.M.Hoare@bham.ac.uk
J. Humphreys, Liverpool	su01@liv.ac.uk
D. L. Johnson, Nottingham	David.Johnson@maths.nott.ac.uk
F. E. A. Johnson, UC, London	feaj@math.ucl.ac.uk
A. Juhàsz* , Haifa	
P. H. Kropholler, QMW, London	P.H.Kropholler@qmw.ac.uk
Yu. Kuz'min, Moscow	post@miit.msk.su
I. J. Leary, Southampton	ijl@maths.soton.ac.uk
P. A. Linnell, Blacksburg, Virginia	linnell@calvin.math.vt.edu
A. Mann, Jerusalem	Mann@vms.huji.ac.il
H. Meinert, Frankfurt	meinert@math.uni-frankfurt.de
C. F. Miller, Melbourne	cfm@mundoe.maths.mu.oz.au
G. Mislin, Zürich	guido.mislin@math.ethz.ch
P. M. Neumann, Oxford	neumann@vax.ox.ac.uk
W. Neumann, Melbourne	Neumann@mundoe.maths.mu.oz.au
G. A. Niblo, Southampton	gan@maths.soton.ac.uk
G. Noskov, Omsk	noskov@univer.omsk.su
B. E. A. Nucinkis* , Southampton	bean@maths.soton.ac.uk
A. Ol'shanskii, Moscow	ay@olshan.msk.ru
P. Papasoglu, Paris Sud	panos@matups.matups.fr
S. Pride, Glasgow	sjp@maths.gla.ac.uk
S. Rees, Newcastle	sarah.rees@ncl.ac.uk
J. Rickard, Bristol	j.rickard@bristol.ac.uk
M. A. Roller, Regensburg	
M. A. Ronan, Illinois	ronan@mathematik.uni-bielefeld.de

xii

P. J. Rowley, UMIST   peter.rowley@umist.ac.uk
K. E. Ruane* , Vanderbilt   ruane@math.Vanderbilt.Edu
R. Sandling, Manchester   rsandling@manchester.ac.uk
V. Sergiescu, France   sergiesc@fourier.grenet.fr
L. Soicher, QMW, London   L.H.Soicher@qmw.ac.uk
U. Stammbach, Zürich   urs.stammbach@math.ethz.ch
R. Stöhr, UMIST   Ralph.Stohr@umist.ac.uk
G. A. Swarup, Melbourne   G.Swarup@ms.unimelb.edu.au
O. Talelli* , Athens   otalelli@atlas.uoa.gr
R. Thomas, Leicester   R.Thomas@mcs.le.ac.uk
C. B. Thomas, Cambridge   C.B.Thomas@dpmms.cam.ac.uk
E. C. Turner, Albany   ted@math.albany.edu
E. Ventura* , Barcelona   ventura@dumbo.ups.es
K. Vogtmann, Cornell   vogtmann@math.cornell.edu
P. Wakefield, Newcastle   paul.wakefield@ncl.ac.uk
C. T. C. Wall, Liverpool   C.T.C.Wall@liverpool.ac.uk
X. Wang* , Glasgow   xw@maths.gla.ac.uk
J. S. Wilson, Birmingham   J.S.Wilson@bham.ac.uk

On the Cohomology of $SL_2(\mathbb{Z}[1/p])$

Alejandro Adem and Nadim Naffah

Mathematics Department, University of Wisconsin, Madison, WI 53706 USA.
Department of Mathematics, ETH–Zürich, Zürich CH–8092.

0. Introduction

In this note we compute the integral cohomology of the discrete groups $SL_2(\mathbb{Z}[1/p])$, p a prime. According to Serre [6] these are groups of virtual cohomological dimension 2. The method we use is to exploit the fact that these groups can be expressed as an amalgamation of two copies of $SL_2(\mathbb{Z})$ along the subgroup $\Gamma_0(p)$ of 2×2 matrices with lower left hand entry divisible by p. We first compute the cohomology of this virtually free group (using a tree on which it acts with finite isotropy and compact quotient), and then use the well–known Mayer–Vietoris sequence in cohomology to obtain our result. We assume that p is an odd prime larger than 3. The cases $p = 2, 3$ must be treated separately; we discuss them at the end of the paper. We are grateful to the referee for his extremely useful remarks, and to J.-P. Serre for pointing out Proposition 3.1.

1. Double cosets and permutation modules

In this section we calculate certain double coset decompositions which will play a key rôle in our approach. Let $G = SL_2(\mathbb{F}_p)$, and $B \subset G$ the subgroup consisting of all matrices with lower left hand entry equal to zero. It is easy to see that $B \cong \mathbb{Z}/p \times_T \mathbb{Z}/p - 1$, a semidirect product. We will denote by C_2, C_4 and C_6 the cyclic subgroups generated by the following three respective matrices of orders 2, 4 and 6:

$$a_2 = \begin{pmatrix} -1 & 0 \\ 0 & -1 \end{pmatrix}, \; a_4 = \begin{pmatrix} 0 & -1 \\ 1 & 0 \end{pmatrix}, \; a_6 = \begin{pmatrix} 0 & -1 \\ 1 & 1 \end{pmatrix}.$$

The first author was partly supported by the NSF.

It is easy to check that the set of right cosets $B\backslash G$ decomposes as follows:

$$B\backslash G = Ba_4 \sqcup (\bigsqcup_{x \in \mathbb{F}_p} B \begin{pmatrix} 1 & 0 \\ x & 1 \end{pmatrix}).$$

Now as a_2 is central, we obtain a double coset decomposition for G using B and C_2 as follows:

$$G = Ba_4 C_2 \sqcup (\bigsqcup_{x \in \mathbb{F}_p} B \begin{pmatrix} 1 & 0 \\ x & 1 \end{pmatrix} C_2).$$

Applying the usual induction restriction formula we obtain

$$\mathbb{Z}[G/C_2] \mid_B \cong (\mathbb{Z}[B/C_2])^{p+1}. \tag{1.1}$$

Next we consider the double cosets using C_4. Note that we have

$$\begin{pmatrix} 1 & 0 \\ x & 1 \end{pmatrix} \begin{pmatrix} 0 & -1 \\ 1 & 0 \end{pmatrix} = \begin{pmatrix} 0 & -1 \\ 1 & -x \end{pmatrix}$$

and

$$\begin{pmatrix} 0 & -1 \\ 1 & -x \end{pmatrix} \begin{pmatrix} 1 & 0 \\ 1/x & 1 \end{pmatrix} = \begin{pmatrix} -1/x & -1 \\ 0 & -x \end{pmatrix}.$$

From this we conclude that if $x \ne -1/x$ or 0, then

$$B \begin{pmatrix} 1 & 0 \\ x & 1 \end{pmatrix} \begin{pmatrix} 0 & -1 \\ 1 & 0 \end{pmatrix} = B \begin{pmatrix} 1 & 0 \\ -1/x & 1 \end{pmatrix}$$

and so

$$B \begin{pmatrix} 1 & 0 \\ x & 1 \end{pmatrix} C_4 = B \begin{pmatrix} 1 & 0 \\ x & 1 \end{pmatrix} \sqcup B \begin{pmatrix} 1 & 0 \\ -1/x & 1 \end{pmatrix}, \quad \text{and } Ba_4 C_4 = B \sqcup Ba_4.$$

In each case the two cosets are permuted by the matrix a_4. In the case when $x^2 + 1 = 0$, then the coset is fixed under this action, and hence the associated coset is equal to the double coset. It is an elementary fact that the polynomial $t^2 + 1$ will have roots in $\mathbb{F}_p$ (necessarily two distinct ones) if and only if $p \equiv 1 \mod(4)$. Using this and the induction restriction formula yields

$$\mathbb{Z}[G/C_4] \mid_B \cong \mathbb{Z}[B/s_1 C_4 s_1^{-1}] \oplus \mathbb{Z}[B/s_2 C_4 s_2^{-1}] \oplus (\mathbb{Z}[B/C_2])^{(p-1)/2}$$

if $p \equiv 1 \mod(4)$, and

$$\mathbb{Z}[G/C_4] \mid_B \cong (\mathbb{Z}[B/C_2])^{(p+1)/2} \quad \text{otherwise.} \tag{1.2}$$

The elements s_1, s_2 correspond to the two roots of the polynomial. For C_6, we must look at the action of the matrix of order 3, $\begin{pmatrix} 0 & 1 \\ -1 & -1 \end{pmatrix}$ on the double cosets. In this case the orbit of the action will be of the form

$$ B \begin{pmatrix} 1 & 0 \\ x & 1 \end{pmatrix}, \ B \begin{pmatrix} 1 & 0 \\ 1/(1-x) & 1 \end{pmatrix}, \ B \begin{pmatrix} 1 & 0 \\ (x-1)/x & 1 \end{pmatrix} $$

provided $x \neq 1$. A coset will be fixed if and only if $x^2 - x + 1 = 0$; given that $p > 3$, this will have roots in $\mathbb{F}_p$ (necessarily two distinct ones) if and only if $p \equiv 1 \bmod (3)$. As for $x = 1$, the corresponding coset gives rise to the singular orbit

$$ B, \ B \begin{pmatrix} 1 & 0 \\ 1 & 1 \end{pmatrix}, \ Ba_4. $$

From this we can deduce the following decomposition (notation as before):

$$ \mathbb{Z}[G/C_6] \, |_B \cong \mathbb{Z}[B/s_1 C_6 s_1^{-1}] \oplus \mathbb{Z}[B/s_2 C_6 s_2^{-1}] \oplus (\mathbb{Z}[B/C_2])^{(p-1)/3} $$

if $p \equiv 1 \bmod (3)$ and

$$ \mathbb{Z}[G/C_6] \, |_B \cong (\mathbb{Z}[B/C_2])^{(p+1)/3} \quad \text{otherwise.} \tag{1.3} $$

2. The cohomology of $\Gamma_0(p)$

The subgroup $\Gamma_0(p) \subset SL_2(\mathbb{Z})$ is defined by

$$ \Gamma_0(p) = \{ \begin{pmatrix} a & b \\ c & d \end{pmatrix} \in SL_2(\mathbb{Z}) \mid c \equiv 0 \bmod(p) \}. $$

If $\Gamma(p)$ denotes the level p congruence subgroup, then clearly $\Gamma_0(p)$ can be expressed as an extension

$$ 1 \to \Gamma(p) \to \Gamma_0(p) \to B \to 1. $$

Recall [6] that $SL_2(\mathbb{Z})$ acts on a tree T with finite isotropy, and quotient a single edge. As $\Gamma(p)$ is torsion free, it acts freely on this tree, and so $G = SL_2(\mathbb{F}_p)$ acts on the finite graph $T/\Gamma(p)$. The isotropy subgroups of this action are precisely C_4 and C_6 for the vertices and C_2 for the edge. Let EB denote the universal B–space, then clearly using the projection $\pi : \Gamma_0(p) \to B$, $\Gamma_0(p)$ can be made to act diagonally on $EB \times T$, with trivial isotropy. As this space is contractible, its quotient under this action has the same homotopy type as the classifying space $B\Gamma_0(p)$, and so we have $B\Gamma_0(p) \cong EB \times_B T/\Gamma(p)$.

Let C^* denote the cellular cochains on the B–CW complex $T/\Gamma(p)$; then it is clear from the above that

$$C^0 \cong \mathbb{Z}[G/C_4] \mid_B \oplus \mathbb{Z}[G/C_6] \mid_B \quad \text{and that} \quad C^1 \cong \mathbb{Z}[G/C_2] \mid_B.$$

In this situation, there is a spectral sequence converging to $H^*(\Gamma_0(p), \mathbb{Z})$ (see [2]) with $E_1^{p,q} \cong H^q(B, C^p)$, which degenerates into a long exact sequence:

$$\ldots \to H^i(\Gamma_0(p), \mathbb{Z}) \to H^i(B, C^0) \to H^i(B, C^1) \to H^{i+1}(\Gamma_0(p), \mathbb{Z}) \to \ldots$$
(2.1)

where the middle arrow is induced by the coboundary map δ on C^*.

As a first application of the long exact sequence (2.1) we obtain

Proposition 2.2. *Under the above conditions, $H^1(\Gamma_0(p), \mathbb{Z}) \cong (\mathbb{Z})^{N(p)}$ where*

$$N(p) = \begin{cases} (p-7)/6, & \text{if } p \equiv 1 \ mod(12); \\ (p+1)/6, & \text{if } p \equiv 5 \ mod(12); \\ (p-1)/6, & \text{if } p \equiv 7 \ mod(12); \\ (p+7)/6, & \text{if } p \equiv 11 \ mod(12). \end{cases}$$

Proof The sequence (2.1) starts as

$$0 \to \mathbb{Z} \to (C^0)^B \to (C^1)^B \to H^1(\Gamma_0(p), \mathbb{Z}) \to H^1(B, C^0).$$

Recall that $H^1(B, C^0) = 0$, as C^0 is a permutation module. Hence calculating ranks completes the proof.

To compute the remaining cohomology groups we first switch to the associated projective group $P\Gamma_0(p)$. Note that there will be a situation analogous to that for the original group, except that throughout we must divide out by the central $\mathbb{Z}/2$. Note that if PB is the associated group for B, then C^1 will now be a free PB-module. Hence the corresponding long exact sequence degenerates to yield the isomorphism $H^{2i}(P\Gamma_0(p), \mathbb{Z}) \cong H^2(PB, C^0)$ for all $i \geq 0$, and the fact that all its odd dimensional cohomology (except in dimension 1) is zero. This is summarized in

Proposition 2.3. *For any integer $i \geq 1$, we have that*

$$H^{2i}(P\Gamma_0(p), \mathbb{Z}) \cong \begin{cases} \mathbb{Z}/6 \oplus \mathbb{Z}/6, & \text{if } p \equiv 1 \ mod(12); \\ \mathbb{Z}/2 \oplus \mathbb{Z}/2, & \text{if } p \equiv 5 \ mod(12); \\ \mathbb{Z}/3 \oplus \mathbb{Z}/3, & \text{if } p \equiv 7 \ mod(12); \\ 0, & \text{if } p \equiv 11 \ mod(12). \end{cases}$$

and $H^{2i+1}(P\Gamma_0(p), \mathbb{Z}) = 0$.

$\square$

Note that $H^1(P\Gamma_0(p), \mathbb{Z}) \cong H^1(\Gamma_0(p), \mathbb{Z})$.

Next we apply the spectral sequence over $\mathbb{Z}$ associated to the central extension

$$1 \to C_2 \to \Gamma_0(p) \to P\Gamma_0(p) \to 1.$$

Note that the interesting cases are if $p \equiv 1$, $5 \mod (12)$, and that the 3–torsion plays no role. As the group has periodic cohomology and is virtually free, it suffices to compute H^2 and H^3. In total degree 2 we simply have the contributions from $H^2(C_2, \mathbb{Z}) \cong \mathbb{Z}/2$ and $H^2(P\Gamma_0(p), \mathbb{Z})_{(2)} \cong \mathbb{Z}/2 \oplus \mathbb{Z}/2$. As 4–torsion must appear (there will be a subgroup of that order), we conclude that $H^2(\Gamma_0(p), \mathbb{Z})_{(2)} \cong \mathbb{Z}/4 \oplus \mathbb{Z}/2$.

In total degree 3, we only have one term

$$H^1(P\Gamma_0(p), H^2(C_2, \mathbb{Z})) \cong (\mathbb{Z}/2)^{N(p)+2}.$$

However, it is not hard to see that the map induced by the quotient in co-homology, $H^4(P\Gamma_0(p), \mathbb{Z}) \to H^4(\Gamma_0(p), \mathbb{Z})$ must be zero. This can be proved by comparing the two long exact sequences described above and using the corresponding fact for the map induced by the quotient $\mathbb{Z}/4 \to \mathbb{Z}/2$. The only possible differential on this horizontal edge group is $d_3 : E_3^{1,2} \to E_3^{4,0}$, hence it must have an image of 2–primary rank 2. We conclude that $E_\infty^{1,2} \cong (\mathbb{Z}/2)^{N(p)} \cong H^3(\Gamma_0(p), \mathbb{Z})$, and we obtain

Theorem 2.4. *For any integer $i \geq 1$, we have*

$$H^{2i}(\Gamma_0(p), \mathbb{Z}) \cong \begin{cases} \mathbb{Z}/12 \oplus \mathbb{Z}/6, & \text{if } p \equiv 1 \bmod(12); \\ \mathbb{Z}/4 \oplus \mathbb{Z}/2, & \text{if } p \equiv 5 \bmod(12); \\ \mathbb{Z}/3 \oplus \mathbb{Z}/6, & \text{if } p \equiv 7 \bmod(12); \\ \mathbb{Z}/2, & \text{if } p \equiv 11 \bmod(12) \end{cases}$$

and

$$H^{2i+1}(\Gamma_0(p), \mathbb{Z}) \cong (\mathbb{Z}/2)^{N(p)}.$$

$\square$

3. Calculation of the cohomology

To begin we recall that aside from the natural inclusion we also have an injection $\rho : \Gamma_0(p) \to SL_2(\mathbb{Z})$, given by

$$\begin{pmatrix} a & b \\ c & d \end{pmatrix} \mapsto \begin{pmatrix} a & pb \\ p^{-1}c & d \end{pmatrix}.$$

Using these two imbeddings, we can construct the amalgamated product [6]

$$SL_2(\mathbb{Z}[1/p]) \cong SL_2(\mathbb{Z}) *_{\Gamma_0(p)} SL_2(\mathbb{Z}).$$

In addition we have that

$$\overline{H}^r(SL_2(\mathbb{Z}), \mathbb{Z}) \cong \begin{cases} \mathbb{Z}/12 & \text{if r is even;} \\ 0 & \text{if r is odd.} \end{cases}$$

We can identify ρ^* with the ordinary restriction map.

Using the Mayer–Vietoris sequence associated to an amalgamated product we see that $H^1(SL_2(\mathbb{Z}[1/p]), \mathbb{Z}) = 0$ and that we have exact sequences

$$0 \to \mathbb{Z}^{N(p)} \to H^2(SL_2(\mathbb{Z}[1/p]), \mathbb{Z}) \to \mathbb{Z}/12 \oplus \mathbb{Z}/12 \to H^2(\Gamma_0(p), \mathbb{Z}) \to H^3(SL_2(\mathbb{Z}[1/p]), \mathbb{Z}) \to 0$$

and

$$0 \to (\mathbb{Z}/2)^{N(p)} \to H^{2i}(SL_2(\mathbb{Z}[1/p]), \mathbb{Z}) \to \mathbb{Z}/12 \oplus \mathbb{Z}/12 \to H^{2i}(\Gamma_0(p), \mathbb{Z}) \to H^{2i+1}(SL_2(\mathbb{Z}[1/p]), \mathbb{Z}) \to 0.$$

The cohomology will evidently be 2-fold periodic above dimension 2, which is in fact the virtual cohomological dimension of $SL_2(\mathbb{Z}[1/p])$.

We will need the following result, which is due to J.-P. Serre [7]. It can also be proved using an explicit presentation for the group, described in [1].

Proposition 3.1.

$$H_1(SL_2(\mathbb{Z}[1/p]), \mathbb{Z}) \cong \begin{cases} \mathbb{Z}/3 & \text{if } p=2; \\ \mathbb{Z}/4 & \text{if } p=3; \\ \mathbb{Z}/12 & \text{otherwise.} \end{cases}$$

□

Hence we have that for $p > 3$, $H^2(SL_2(\mathbb{Z}[1/p]), \mathbb{Z}) \cong (\mathbb{Z})^{N(p)} \oplus \mathbb{Z}/12$.

Let $A(p)$ denote the number $12/|Q(p)|$, where $Q(p)$ is the largest cyclic subgroup in $H^2(\Gamma_0(p), \mathbb{Z})$. Then, from the fact that the restriction from the cohomology of $SL_2(\mathbb{Z})$ to that of its cyclic subgroups factors through $\Gamma_0(p)$, we deduce that the sequence above simplifies to yield

$$0 \to (\mathbb{Z}/2)^{N(p)} \to H^{2i}(SL_2(\mathbb{Z}[1/p]), \mathbb{Z}) \to \mathbb{Z}/12 \oplus \mathbb{Z}/A(p) \to 0 \qquad (3.2)$$

and

$$H^{2i+1}(SL_2(\mathbb{Z}[1/p]), \mathbb{Z}) \cong H^{2i}(\Gamma_0(p), \mathbb{Z})/Q(p).$$

Moreover, from our previous calculation for $\Gamma_0(p)$, we have that

$$Q(p) = \begin{cases} \mathbb{Z}/12 & \text{if } p \equiv 1 \bmod(12); \\ \mathbb{Z}/4 & \text{if } p \equiv 5 \bmod(12); \\ \mathbb{Z}/6 & \text{if } p \equiv 7 \bmod(12); \\ \mathbb{Z}/2 & \text{if } p \equiv 11 \bmod(12). \end{cases}$$

It remains only to determine precisely what this extension (3.2) looks like. We need only be concerned with the 2–primary component. Recall that $H^{2i}(SL_2(\mathbb{Z}[1/p]), \mathbb{Z})$ is a quotient of $H^2(SL_2(\mathbb{Z}[1/p]), \mathbb{Z})$, as $SL_2(\mathbb{Z}[1/p])$ is a group of virtual cohomological dimension 2, which has 2-fold periodic cohomology. This can also be explained by saying that the sequence in high even dimensions can be identified with the corresponding sequence in 2-dimensional Farrell cohomology (see [2]). Mapping one sequence into the other, we see that a $\mathbb{Z}/4$ summand must split off for all values of $p > 3$. This means that the sequence *will split* for $p \equiv 1, 5 \bmod(12)$. For the remaining cases $p \equiv 7, 11$ $\bmod(12)$, it remains to solve the extension problem after splitting off the $\mathbb{Z}/4$ summand. However, from our knowledge of H^2, we know that the dimension of $H^{2i} \otimes \mathbb{Z}/2$ can be at most $N(p) + 1$. We infer that the reduced extension *does not split*, and must necessarily have a $\mathbb{Z}/4$ summand present. Hence we have the following complete calculation:

Theorem 3.3. *Let p be an odd prime larger than 3, then*

$$H^1(SL_2(\mathbb{Z}[1/p]), \mathbb{Z}) \cong 0$$

and

$$H^2(SL_2(\mathbb{Z}[1/p]), \mathbb{Z}) \cong \begin{cases} \mathbb{Z}^{(p-7)/6} \oplus \mathbb{Z}/12 & \text{if } p \equiv 1 \bmod(12); \\ \mathbb{Z}^{(p+1)/6} \oplus \mathbb{Z}/12 & \text{if } p \equiv 5 \bmod(12); \\ \mathbb{Z}^{(p-1)/6} \oplus \mathbb{Z}/12 & \text{if } p \equiv 7 \bmod(12); \\ \mathbb{Z}^{(p+7)/6} \oplus \mathbb{Z}/12 & \text{if } p \equiv 11 \bmod(12). \end{cases}$$

For $i \geq 2$, we have

$$H^{2i}(SL_2(\mathbb{Z}[1/p]), \mathbb{Z}) \cong \begin{cases} (\mathbb{Z}/2)^{(p-7)/6} \oplus \mathbb{Z}/12 & \text{if } p \equiv 1 \bmod(12); \\ (\mathbb{Z}/2)^{(p+1)/6} \oplus \mathbb{Z}/12 \oplus \mathbb{Z}/3 & \text{if } p \equiv 5 \bmod(12); \\ (\mathbb{Z}/2)^{(\nu-7)/0} \oplus \mathbb{Z}/12 \oplus \mathbb{Z}/4 & \text{if } p \equiv 7 \bmod(12); \\ (\mathbb{Z}/2)^{(p+1)/6} \oplus \mathbb{Z}/12 \oplus \mathbb{Z}/12 & \text{if } p \equiv 11 \bmod(12), \end{cases}$$

and

$$H^{2i-1}(SL_2(\mathbb{Z}[1/p]), \mathbb{Z}) \cong \begin{cases} \mathbb{Z}/6 & \text{if } p \equiv 1 \ mod(12); \\ \mathbb{Z}/2 & \text{if } p \equiv 5 \ mod(12); \\ \mathbb{Z}/3 & \text{if } p \equiv 7 \ mod(12); \\ 0 & \text{if } p \equiv 11 \ mod(12). \end{cases}$$

□

Of the two remaining cases ($p = 2, 3$) the second one can be done in a manner totally analogous to what we have presented. Specifically we have that

$$H^i(\Gamma_0(3), \mathbb{Z}) \cong \begin{cases} \mathbb{Z} & \text{if } i = 0, 1; \\ \mathbb{Z}/6 & \text{if } i \text{ is even}; \\ \mathbb{Z}/2 & \text{if } i > 1 \text{ is odd}. \end{cases}$$

We obtain that $H^2(SL_2(\mathbb{Z}[1/3]), \mathbb{Z}) \cong \mathbb{Z} \oplus \mathbb{Z}/4$, $H^{2i+1}(SL_2(\mathbb{Z}[1/3]), \mathbb{Z}) \cong 0$, and it only remains to deal with the extension

$$0 \to \mathbb{Z}/2 \to H^{2i}(SL_2(\mathbb{Z}[1/p]), \mathbb{Z}) \to \mathbb{Z}/12 \oplus \mathbb{Z}/2 \to 0.$$

As before the $\mathbb{Z}/12$ must split off, and by rank considerations an extra $\mathbb{Z}/4$ summand must appear. To summarize, we have

$$H^i(SL_2(\mathbb{Z}[1/3]), \mathbb{Z}) \cong \begin{cases} 0 & \text{if } i \text{ is odd}; \\ \mathbb{Z} \oplus \mathbb{Z}/4 & \text{if } i = 2; \\ \mathbb{Z}/12 \oplus \mathbb{Z}/4 & \text{if } i = 2j, \ j > 1. \end{cases}$$

The case $p = 2$ is complicated by the fact that $\Gamma(2)$ is not torsion–free. However, one can still make use of the associated projective groups; $P\Gamma(2)$ is free of rank 2, and there is an extension

$$1 \to P\Gamma(2) \to P\Gamma_0(2) \to \mathbb{Z}/2 \to 1.$$

Analyzing the action on the corresponding graph, it is not hard to show that

$$H^*(P\Gamma_0(2), \mathbb{Z}) \cong \begin{cases} \mathbb{Z} & \text{if } i \text{ is 0 or 1}; \\ \mathbb{Z}/2 & \text{if } i \text{ is even}; \\ 0 & \text{otherwise}. \end{cases}$$

Then, using the central extension, it is direct to show that

$$H^i(\Gamma_0(2), \mathbb{Z}) \cong \begin{cases} \mathbb{Z} & \text{if } i \text{ is 0 or 1}; \\ \mathbb{Z}/4 & \text{if } i \text{ is even}; \\ \mathbb{Z}/2 & \text{if } i \text{ is odd}, i > 1. \end{cases}$$

Using the Mayer–Vietoris sequence as before, we obtain that

$$H^2(SL_2(\mathbb{Z}[1/2]), \mathbb{Z}) \cong \mathbb{Z} \oplus \mathbb{Z}/3, \, H^i(SL_2(\mathbb{Z}[1/2]), \mathbb{Z}) = 0$$

if i is odd, and a short exact sequence

$$0 \to \mathbb{Z}/2 \to H^{2i}(SL_2(\mathbb{Z}[1/2]), \mathbb{Z})) \to \mathbb{Z}/12 \oplus \mathbb{Z}/3 \to 0.$$

In this case, we do not know that the $\mathbb{Z}/12$ summand must split off. Looking at the 2-primary part, we see that it can have *at most* one cyclic summand. The only possibility is $\mathbb{Z}/8$, and we have

$$H^i(SL_2(\mathbb{Z}[1/2]), \mathbb{Z}) \cong \begin{cases} 0 & \text{if } i \text{ is odd;} \\ \mathbb{Z} \oplus \mathbb{Z}/3 & \text{if } i = 2; \\ \mathbb{Z}/24 \oplus \mathbb{Z}/3 & \text{if } i = 2j, \, j > 1. \end{cases}$$

Remarks. Note that this last group has no finite subgroups of order eight, which makes its cohomology rather interesting. We are grateful to Hans–Werner Henn for pointing out the correct cohomology of this group. Also we would like to point out that Naffah [4] has calculated the 3-adic component of the Farrell cohomology $\widehat{H}^*(SL_2(\mathbb{Z}[1/N]), \mathbb{Z})$ for *any* integer N. This of course can be used to recover our calculations at $p = 3$ in dimensions larger than 2. Moss [5] has computed the rational cohomology of $SL_2(\mathbb{Z}[1/N])$ which again can be used to recover part of our results. The general calculation of $H^*(SL_2(\mathbb{Z}[1/N]), \mathbb{Z})$ seems to be a rather complicated but interesting open problem. We refer to [3] for more on this.

References

[1] H. Behr and J. Mennicke, *A Presentation of the Groups $PSL(2,p)$*, Canadian Journal of Mathematics **20** No. 6, (1968), pp. 1432–1438.

[2] K. Brown, *Cohomology of Groups*, Graduate Texts in Mathematics, **87** Springer–Verlag (1982).

[3] S. Hesselmann, *Zur Torsion in der Kohomologie S-arithmetischer Gruppen*, Bonner Math. Schriften. To appear.

[4] N. Naffah, Thesis, ETH–Zürich.

[5] K. Moss, *Homology of $SL(n, \mathbb{Z}[1/p])$*, Duke Mathematical Journal **47** No. 4, (1980).

[6] J. -P. Serre, *Trees*, Springer–Verlag (1980).

[7] J -P. Serre, *Groupes de Congruence*, Annals of Mathematics **92** No. 3, (1970), pp. 489–527.

Cohomology of Sporadic Groups, Finite Loop Spaces, and the Dickson Invariants

David J. Benson

Department of Mathematics, University of Georgia, Athens GA 30602, USA
djb@byrd.math.uga.edu

1. Introduction

Whatever one may think of a proof that covers over ten thousand pages of journal articles, some of which have still not appeared in print, the classification of finite simple groups is a remarkable theorem. It says that (apart from the cyclic groups of prime order and the alternating groups) most finite simple groups are groups of Lie type. These are the analogues of the compact Lie groups, defined over a finite field. They admit a uniform description in terms of the fixed points of certain automorphisms related to the Frobenius map, on the corresponding algebraic groups in prime characteristic. The classifying space of a finite group of Lie type fibers over that of the corresponding Lie group with fibers which are cohomologically finite away from the defining characteristic. Apart from the groups of Lie type, there are the alternating groups A_n ($n \geq 5$), and twenty-six other groups called the sporadic simple groups.

The first five sporadic groups were discovered by Mathieu in the late nineteenth century. The remaining twenty-one were discovered in the nineteen sixties and seventies. The largest is the Fischer–Griess Monster, which has order roughly 8×10^{53}. For a wealth of information on the sporadic groups and other "small" finite simple groups, the reader is referred to the ATLAS of finite groups [Atlas]. A great deal of effort has gone into trying to understand these sporadic groups. For example, the "Monstrous Moonshine" [CN] (see also [CS]) is a remarkable series of observations, still not fully understood, connecting various of the sporadic groups, especially those involved in the Monster, with the theory of modular functions. This article is about another

Partly supported by a grant from NSF

attempt to understand some of the sporadic groups from a different point of view, namely that of the topology of their classifying spaces.

The story started with the construction by Bill Dwyer and Clarence Wilkerson [DW2] of a new finite loop space at the prime two, which they call DI(4). The mod two cohomology of its classifying space is equal to the rank four Dickson invariants, whence the name. Finite loop spaces are sufficiently like compact Lie groups that this prompted the question of what finite simple groups could be associated to them in the way that the finite groups of Lie type are associated to the compact Lie groups.

The first group that was considered in this regard was Conway's group Co_3, which has a 2-local structure closely associated with the 2-local structure of DI(4). It turns out that the classifying space of Co_3 fibers over the classifying space of DI(4) with fibers cohomologically finite at the prime two. However, this turns out to be more akin to the relation between M_{12} and G_2 than to the relation between $G_2(q)$ and G_2.

The finite simple groups which should be regarded as the groups "of Lie type DI(4)" over the field of q elements don't literally exist as groups. They were first considered by Ron Solomon [So] in an attempt to prove that Co_3 is determined by a Sylow 2-subgroup. He was led to consider simple groups in which the centralizer of an involution was $\mathrm{Spin}(q)$ (q odd), possibly with some decoration of odd order. In order to prove nonexistence, Solomon was obliged to write down what appeared to be a consistent fusion pattern for the Sylow 2-subgroup, and then pass to an odd prime to obtain a contradiction. It turns out that despite the nonexistence of these groups, they have perfectly good 2-completed classifying spaces, which map to the classifying space of DI(4) with fibers which are cohomologically finite at the prime two.

This prompts the speculation that there should exist a theory of "p-local groups", in which one only gives a Sylow p-subgroup and a fusion pattern. The fusion pattern should obey a set of axioms which are strong enough to be able to build a p-completed classifying space. One possible candidate for such a set of axioms has been written down by Puig [Pu] in connection with his attempts to formalise the local theory of blocks, but much work remains to be done in this area.

2. Classifying spaces

We begin by describing the theory of classifying spaces, and explain why they are rigid enough that new maps between them should be considered to be interesting.

For any topological group G, there exists a contractible space EG together with a free action of G on EG. A uniform construction was found by Milnor [Mi]. Roughly speaking, the idea is to form an infinite join $G * G * \cdots$ of copies of G as a topological space, and let G act diagonally. The join $X * Y$ of two spaces X and Y consists of the space obtained from $X \times I \times Y$ by identifying $(x, 0, y)$ with $(x', 0, y)$ for all $x, x' \in X$ and $y \in Y$, and $(x, 1, y)$ with $(x, 1, y')$ for all $x \in X$ and $y, y' \in Y$. The difficult point in Milnor's construction is the exact description of the topology on the infinite join. However, we shall mostly be interested in the case where G is either discrete or a compact Lie group. In this case, we can just give the infinite join the weak topology with respect to the finite joins. Then EG is a CW-complex, so to see that it is contractible, it suffices to see that any map from the n-sphere S^n to EG extends to a map from the $(n+1)$-disk D^{n+1}. Since S^n is compact, any such map has image contained in some finite join of copies of G, and then the next higher join contains the cone on this join, so that the map can be extended to D^{n+1} in this next join.

We write BG for the quotient of EG by the action of G. It turns out that the classifying space BG is determined up to homotopy by G, independently of the choice of contractible space EG. Thus if G is a subgroup of G', then using EG' instead of EG, we see that there is an induced fibration $BG \longrightarrow BG'$ with fiber the coset space G'/G. Similarly if N is a normal subgroup of G, we may use $EG \times E(G/N)$ instead of EG, with G acting on both factors, and obtain an induced fibration $BG \longrightarrow B(G/N)$ with fiber BN. In this way, for any homomorphism of topological groups $\alpha : G \longrightarrow G'$, we have an induced map of classifying spaces $B\alpha : BG \longrightarrow BG'$, unique up to homotopy.

If G is discrete, BG is an Eilenberg–Mac Lane space $K(G, 1)$, and EG is its universal cover. In any case, the loop space ΩBG is homotopy equivalent to G as a topological space.

Classifying spaces are homotopically rather rigid. For example, if G and G' are discrete groups, then the set of homotopy classes of maps $[BG; BG']$ is in natural bijection with $\mathrm{Hom}_{grp}(G, G')/\mathrm{Inn}(G')$. Here, the group $\mathrm{Inn}(G')$ of inner automorphisms of G' acts in the obvious way by composition on the set $\mathrm{Hom}_{grp}(G, G')$ of group homomorphisms from G to G'. The bijection sends a map $\alpha : G \longrightarrow G'$ to $B\alpha : BG \longrightarrow BG'$. The inverse of this bijection sends a map $BG \longrightarrow BG'$ to its effect on π_1. This, of course, depends on a choice of basepoints, but the orbit under the action of $\mathrm{Inn}(G')$ does not.

On the other hand, if G is a connected simple compact Lie group, then a theorem of Jackowski, McClure and Oliver (Theorem 3 in [JMO]) says that there is a natural bijection

$$\mathrm{Hom}_{Liegrp}(G, G)/\mathrm{Inn}(G) \wedge \{k \geq 0 : \ k = 0 \ \mathrm{or}(k, |W|) = 1\} \longrightarrow [BG; BG]$$

sending (α, k) to $\Psi^k \circ B\alpha$. Here, the smash product denotes the quotient of the direct product given by identifying all elements in which at least one coordinate is zero. For $k > 0$, the operations Ψ^k are the unstable Adams psi operations, which are known to exist for any connected simple compact Lie group and any integer k coprime to the order of the Weyl group W. They commute up to homotopy with maps coming from Lie group homomorphisms. If $k = 0$, Ψ^k is taken to be the identity map. The map Ψ^k is characterised by the property that its effect on degree $2i$ rational cohomology is multiplication by k^i for each $i \geq 0$.

These theorems indicate that it is of interest to construct maps of classifying spaces which do not come from group homomorphisms and Adams operations. Some examples for connected Lie groups can be found in [JMO]. For maps from the classifying space of a finite group to that of a compact Lie group, the situation is rather unclear, and it is possible that one should try to build a branch of representation theory based on such maps.

A theorem of Dwyer and Wilkerson [DW] indicates that we may work one prime at a time in this situation. Writing $X_p^\wedge$ for the Bousfield–Kan p-completion of a space X, their theorem states the following.

Theorem 2.1. *If π is a finite group and G is a connected compact Lie group then*

$$\mathrm{Map}(B\pi, BG) \simeq \prod_p \mathrm{Map}((B\pi)_p^\wedge, (BG)_p^\wedge).$$

Here, and from now on in this article, one must work with simplicial sets rather than topological spaces, and "Map" denotes the simplicial mapping space. The homotopy category of topological spaces is equivalent to the homotopy category of simplicial sets (these homotopy categories are formed by inverting weak equivalences), via singular simplices and geometric realization. So we use the term "space" from now on to mean "simplicial set", and we replace a classifying space by its simplicial set of singular simplices.

The effect of Bousfield–Kan p-completion on the classifying space of a finite group G is as follows. Without changing the mod p cohomology, it gets rid of the mod l cohomology for every prime $l \neq p$. Its effect on the fundamental group G is to replace it by $G/O^p(G)$. So for example if G is a nonabelian finite simple group then the p-completion of BG is simply connected. The group G may not be recovered from the p-completion of BG, though the Sylow p-subgroup and the strong fusion of its subgroups may be recovered.

3. Dickson invariants

Let $\mathbb{F}_q$ be the finite field of q elements, with $q = p^a$ a prime power. Let $V = (\mathbb{F}_q)^n$ be an n-dimensional vector space over $\mathbb{F}_q$. Then the finite general linear group $G = GL_n(\mathbb{F}_q)$ acts on V in the obvious way, and hence on the coordinate ring

$$\mathbb{F}_q[V] = \mathbb{F}_q \oplus V^* \oplus S^2(V^*) \oplus \cdots$$

We regard this as a graded ring, with $\mathbb{F}_q$ in degree zero, $V^* = \text{Hom}_{\mathbb{F}_q}(V, \mathbb{F}_q)$ in degree one, $S^2(V^*)$ in degree two, and so on. It is a polynomial ring over $\mathbb{F}_q$ on generators forming an basis for V^* in degree one.

L. E. Dickson's theorem [Di] states that the invariants of this action also form a polynomial ring:

$$\mathbb{F}_q[V]^G = \mathbb{F}_q[c_{n,0}, \ldots, c_{n,n-1}]$$

in generators $c_{n,i}$ (the Dickson invariants) of degree $q^n - q^i$. In fact, there is an explicit formula for $c_{n,i}$ as follows:

$$c_{n,i} = (-1)^{n-i} \sum_{\substack{W \subsetneq V \\ \dim_{\mathbb{F}_q}(W) = i}} \prod_{\substack{\phi \in V^* \\ \phi|_W \neq 0}} \phi.$$

If you are a topologist, you may wish to double all the degrees in the above discussion, because if p is odd, elements of odd degree are supposed to anti-commute rather than commuting. Having done that, the polynomial ring on generators in degree two over $\mathbb{F}_p$ acquires an action of the Steenrod operations, since it is the mod p cohomology of a torus $T^n = \mathbb{R}^n/\mathbb{Z}^n = B\mathbb{Z}^n$ of rank n. The group $GL(n, \mathbb{Z})$ acts on T^n, and induces an action of the quotient group $GL(n, \mathbb{F}_p)$ on its mod p cohomology. This action commutes with the Steenrod operations, and so the invariants inherit an action of the Steenrod algebra.

Even if you are a topologist, in the case where $q = p = 2$, you may also wish to keep the polynomial generators in degree one. In this case, the polynomial ring is the cohomology of $(\mathbb{R}P^\infty)^n = B(\mathbb{Z}/2)^n$, and so the mod two Steenrod algebra acts on this polynomial ring. Again, the invariants inherit an action of the Steenrod algebra. In this case, let us look at the question of when the algebra of Dickson invariants can be the cohomology of a topological space, as an algebra over the Steenrod algebra. When $n = 1$, the group $GL(1, \mathbb{F}_2)$ is the trivial group, and the invariants form a polynomial ring on a single generator in degree one:

$$\mathbb{F}_2[c_{1,0}] = H^*(\mathbb{R}P^\infty, \mathbb{F}_2) = H^*(B\mathbb{Z}/2, \mathbb{F}_2).$$

The situation gets slightly more interesting in rank two. $GL(2, \mathbb{F}_2)$ is isomorphic to the symmetric group of degree three, and the invariants form a polynomial ring on generators in degrees two and three:

$$\mathbb{F}_2[c_{2,0}, c_{2,1}] \cong H^*(BSO(3), \mathbb{F}_2).$$

The group $SO(3)$ contains a subgroup E_2 isomorphic to $(\mathbb{Z}/2)^2$, and the normalizer quotient is $N_{SO(3)}(E_2)/E_2 \cong GL(2, \mathbb{F}_2)$. The restriction map

$$H^*(BSO(3), \mathbb{F}_2) \longrightarrow H^*(B(\mathbb{Z}/2)^2, \mathbb{F}_2)^{GL(2,\mathbb{F}_2)}$$

identifies the cohomology with the Dickson invariants.

In rank three, a similar thing happens, but with the exceptional compact Lie group G_2. The Dickson invariants of rank three form a polynomial ring on generators in degrees four, six and seven:

$$\mathbb{F}_2[c_{3,0}, c_{3,1}, c_{3,2}] \cong H^*(BG_2, \mathbb{F}_2).$$

The group G_2 contains a subgroup E_3 isomorphic to $(\mathbb{Z}/2)^3$, and the normalizer quotient is $N_{G_2}(E_3)/E_3 \cong GL(3, \mathbb{F}_2)$. The restriction map

$$H^*(BG_2, \mathbb{F}_2) \longrightarrow H^*(B(\mathbb{Z}/2)^3, \mathbb{F}_2)^{GL(3,\mathbb{F}_2)}$$

again identifies the cohomology with the Dickson invariants.

In rank four, there is no longer a compact Lie group with the right cohomology. However, Dwyer and Wilkerson [DW2] recently constructed a space $BDI(4)$ with the Dickson invariants of rank four as its cohomology. The Dickson invariants in this case form a polynomial ring on generators in degrees eight, twelve, fourteen and fifteen:

$$\mathbb{F}_2[c_{4,0}, c_{4,1}, c_{4,2}, c_{4,3}] \cong H^*(BDI(4), \mathbb{F}_2).$$

The space $BDI(4)$ is not the classifying space of a group, but rather of a finite loop space $DI(4)$ at the prime two. We shall have more to say about this in Section 6. This finite loop space may be thought of as containing a subgroup E_4 isomorphic to $(\mathbb{Z}/2)^4$, with normalizer quotient $N_{DI(4)}(E_4)/E_4 \cong GL(4, \mathbb{F}_2)$. The "restriction map"

$$H^*(BDI(4), \mathbb{F}_2) \longrightarrow H^*(B(\mathbb{Z}/2)^4, \mathbb{F}_2)^{GL(4,\mathbb{F}_2)}$$

again identifies the cohomology with the Dickson invariants.

The game stops here. A theorem of Lin and Williams [LW] states that there is no topological space whose mod two cohomology is the ring of Dickson invariants of rank five or more (as an algebra over the Steenrod algebra).

4. Finite groups of Lie type

As we pointed out in the introduction, most finite simple groups are of Lie type. Actually, just as in the representation theory, from the point of view of the classifying space, it is better to work with certain "almost simple" groups. For example, instead of working with $PSL(n, \mathbb{F}_q)$, it is usually easier to work with $SL(n, \mathbb{F}_q)$, or even $GL(n, \mathbb{F}_q)$.

With this proviso, the data needed to specify a finite group of Lie type are a compact Lie group G, a finite field $\mathbb{F}_q$, and a diagram automorphism α of the corresponding Dynkin diagram. For a large portion of the groups of Lie type (the *untwisted* groups), α is the identity map. For example, both the finite linear groups and the finite unitary groups come from the Dynkin diagram A_n, with the identity automorphism in the first case, and a diagram automorphism of order two in the second case. The finite group associated with the above data is denoted ${}^{\alpha}G(q)$, or just $G(q)$ if α is the identity.

The following theorem of Quillen and Friedlander (see Quillen [Nice], and Friedlander [Fr], Chapter 12) relates the classifying space of the finite group ${}^{\alpha}G(q)$ to the classifying space of G.

Theorem 4.1. *There is a commutative diagram*

$$
\begin{array}{ccc}
B\,{}^{\alpha}G(q) & \longrightarrow & BG \\
\downarrow & & \downarrow{\scriptstyle\Delta} \\
BG & \xrightarrow{\mathrm{id}\times(\Psi^q\circ B\alpha)} & BG \times BG
\end{array}
$$

which becomes a homotopy pullback square after Bousfield–Kan completing at any prime l not dividing q.

In particular, there is a fibration $B\,{}^{\alpha}G(q) \longrightarrow BG$ whose fibers have finite homology at primes l not dividing q.

Here, Δ denotes the diagonal map from BG to $BG \times BG$. Construction of this diagram involves replacing the compact Lie group by the corresponding algebraic group defined over $\mathbb{Z}$, and lifting from characteristic p using étale homotopy theory. The details may be found in Friedlander [Fr], Chapter 12. This diagram gives a method of calculating the cohomology of the finite groups of Lie type away from their defining characteristic. First, one uses the Eilenberg–Moore spectral sequence on this homotopy pullback to calculate the cohomology modulo a certain filtration, then one uses the subgroup structure to ungrade. This was carried out by Quillen [Qu] in the case of $GL(n, \mathbb{F}_q)$, Fiedorowicz and Priddy [FP] in the case of the other classical groups, and Kleinerman [Kl] for the exceptional Lie types.

To give an example, consider the fibration $BG_2(q) \longrightarrow BG_2$ for q an odd prime power. The homology of the fiber at the prime two is an exterior algebra on generators in degrees three, five and six. So we get an embedding

$$H^*(BG_2, \mathbb{F}_2) = \mathbb{F}_2[c_{3,0}, c_{3,1}, c_{3,2}] \subseteq H^*(BG_2(q), \mathbb{F}_2).$$

The corresponding Eilenberg–Moore spectral sequence calculation gives a filtration of $H^*(BG_2(q), \mathbb{F}_2)$ whose associated graded is a tensor product of a polynomial ring on the Dickson generators in degrees four, six and seven, with an exterior algebra on generators in degrees three, five and six. After ungrading, we see that $H^*(BG_2(q), \mathbb{F}_2)$ is a free module of rank eight over the Dickson algebra. In particular, it is a Cohen–Macaulay ring, and the Poincaré series for the cohomology is

$$\sum_{i=0}^{\infty} t^i \dim_{\mathbb{F}_2} H^i(BG_2(q), \mathbb{F}_2) = \frac{(1+t^3)(1+t^5)(1+t^6)}{(1-t^4)(1-t^6)(1-t^7)}.$$

5. The Mathieu group M_{12}

A few years ago, Adem, Maginnis and Milgram computed $H^*(BM_{12}, \mathbb{F}_2)$ (see [AMM]). Their answer was rather complicated, but they observed that the answer was a finitely generated free module over a subring isomorphic to the Dickson invariants of rank three:

$$H^*(BG_2, \mathbb{F}_2) = \mathbb{F}_2[c_{3,0}, c_{3,1}, c_{3,2}] \subseteq H^*(BM_{12}, \mathbb{F}_2).$$

It is clear that there can be no homomorphism of groups from M_{12} to G_2 inducing the above map in mod two cohomology, because G_2 has a seven dimensional complex representation, while the smallest nontrivial complex representation of M_{12} is eleven dimensional. However, it turns out that there is a fibration $BM_{12} \longrightarrow BG_2$ whose fibers have finite mod two cohomology. There are (at least) two possible strategies for constructing such a map. One (Milgram [Milg]) is to construct maps from BM_{12} to $BG_2(q)$ for suitable q, and then use the Quillen–Friedlander theorem. A more direct approach (Benson and Wilkerson [BW]) goes as follows. Let P be a Sylow 2-subgroup of M_{12}, of order 2^6. Let W be the centralizer of the center of P, and W' be the normalizer of a normal fours group in P. Then $|W : P| = |W' : P| = 3$, but $W \not\cong W'$. The obvious map from the amalgamated free product $W *_P W'$ to M_{12} is a cohomology isomorphism, and so after completing at the prime two, we obtain a homotopy equivalence $(B(W *_P W'))^\wedge_2 \longrightarrow (BM_{12})^\wedge_2$. It turns out that there is a group homomorphism from $W *_P W'$ to G_2, whose image is dense, and having the desired effect in cohomology. Now using Theorem 2.1, we may pass from a map between 2-completions to a map between the original classifying spaces $BM_{12} \longrightarrow BG_2$ by specifying any map we want at odd primes.

The classifying space $B(W *_P W')$ can be obtained by taking the homotopy colimit (see Chapter XII of Bousfield and Kan [BK]) of the diagram

which may be thought of in this case as a "double mapping cylinder".

6. Finite loop spaces

From the point of view of a homotopy theorist, the two properties of a compact Lie group G which stand out are:

(i) G is a finite CW-complex, and

(ii) G has the homotopy type of a loop space (namely $G \simeq \Omega BG$).

We say that a space X is a *finite loop space* if it satisfies properties (i) and (ii). It turns out that finite loop spaces enjoy many of the properties of compact Lie groups. There is an algebraically defined maximal torus, and a Weyl group which is a finite reflection group [AW].

Since homotopy theorists like to work one prime at a time using Bousfield–Kan p-completion, it is appropriate also to introduce the p-complete version of this definition. We say that X is a finite loop space at the prime p if it satisfies the following two properties:

$(i)_p$ X is p-complete and $H^*(X, \mathbb{F}_p)$ is finite, and

$(ii)_p$ X has the homotopy type of a loop space ΩBX.

Note that BX is now just the name of the space whose loops are homotopy equivalent to X. To give a finite loop space, we must really give both X and BX, but X is determined by BX so it is only necessary to construct BX.

Again, a finite loop space at the prime p has an algebraically defined maximal torus, and a Weyl group which is a finite p-adic reflection group. The finite reflection groups over the complex numbers were classified by Shephard and Todd [ST]. Clark and Ewing [CE] determined which of these were defined over the p-adics for p not dividing the group order, and demonstrated how to produce a finite loop space at the prime p in this "non-modular" case. There remains a well defined list of questions, one for each prime dividing the order of each of the groups in the list of Shephard and Todd.

There is essentially only one known example of a connected finite loop space at the prime two, which is not the 2-completion of a compact Lie group. All other known examples are given by taking products. The example in question is the space $DI(4) \simeq \Omega BDI(4)$ constructed by Dwyer and Wilkerson [DW2]. The Weyl group of $DI(4)$ is $\mathbb{Z}/2 \times GL(3, \mathbb{F}_2)$, acting as a three dimensional 2-adic reflection group. It is group number 24 in the list of Shephard and Todd. The corresponding "root system" has 21 pairs of opposite roots, so the dimension of DI(4) is $3 + 2 \times 21 = 45$. The 2-adic integral cohomology of $BDI(4)$ is a polynomial algebra on three variables in degrees 8, 12 and 28, corresponding to (twice) the degrees of the fundamental invariants of the Weyl group. The mod two cohomology is the Dickson algebra on four generators in degrees 8, 12, 14 and 15.

Dwyer and Wilkerson constructed the space $BDI(4)$ as a homotopy colimit as follows. Consider the category whose objects are the groups $(\mathbb{Z}/2)^n$ for $1 \le n \le 4$, and whose arrows are the injective homomorphisms between these groups. They constructed a functor to the homotopy category taking the following values:

$$\mathbb{Z}/2 \mapsto B\mathrm{Spin}(7)_2^\wedge \qquad (\mathbb{Z}/2)^2 \mapsto B(SU(2)^3/ \pm 1)_2^\wedge$$
$$(\mathbb{Z}/2)^3 \mapsto B(T^3 \rtimes \mathbb{Z}/2)_2^\wedge \qquad (\mathbb{Z}/2)^4 \mapsto B(\mathbb{Z}/2)^4.$$

They then proved that this could be lifted to a functor to the category of spaces, so that they could then take the homotopy colimit. Some of the maps involved do not exist until the spaces have been 2-completed.

The homotopy colimit of the above functor over a full subcategory consisting of a single object plays the role of the classifying space of the "normalizer" of the corresponding elementary abelian 2-subgroup in DI(4). For example, the normalizer in this sense of the subgroup $(\mathbb{Z}/2)^4$ is a non-split extension of this elementary abelian subgroup by its automorphism group $GL(4, \mathbb{F}_2)$.

It is interesting to speculate on the existence of a suitable algebraic system of dimension 45 which should play the role of a tangent space at the identity for DI(4). It should be almost, but not quite, a Lie algebra over the 2-adics. One tempting candidate is the set of 3×3 skew-hermitian matrices over a suitable 2-adic version of the Cayley numbers. This would have the right dimension, but on the face of it, it would seem to contain the wrong 21-dimensional Lie subalgebra (C_3 instead of B_3). But there may be some twisted version of this which works.

7. Conway's group Co_3

Conway's sporadic simple group Co_3 was discovered as part of a series of groups arising from Leech's close packing of spheres in twenty-four dimensions [Co,CS]. It is defined as the stabilizer of the origin and a type three vector in the group of automorphisms of the Leech lattice, and has order $2^{10}.3^7.5^3.7.11.23$. The following theorem was proved in [Co3]:

Theorem 7.1. *There is a map from BCo_3 to $BDI(4)$, which induces an embedding*

$$H^*(BDI(4), \mathbb{F}_2) \subseteq H^*(BCo_3, \mathbb{F}_2)$$

in such a way that the latter is a finitely generated module over the former.

Prior to the proof of this theorem, nothing was known about the mod two cohomology ring of BCo_3 (apart from in degrees one and two). It is still not known whether the mod two cohomology of BCo_3 is a free module over the rank four Dickson invariants, which amounts to the question of whether it is a Cohen–Macaulay ring. It is known, however, that the cohomology of a Sylow 2-subgroup of Co_3 is not Cohen–Macaulay, because there are maximal elementary abelian 2-subgroups of different ranks, so that it is not even equidimensional. But all maximal elementary abelian 2-subgroups of Co_3 have rank four.

The idea of the proof of the above theorem is as follows. Let P be a Sylow 2-subgroup of Co_3, of order 2^{10}. Then $Z(P) = \langle t_1 \rangle$ has order two, and $C = C_{Co_3}(t_1)$ is a double cover of the simple group $S_6(2)$. Now the Weyl group of type E_7 is $\mathbb{Z}/2 \times S_6(2)$, and so there is an obvious map from $S_6(2)$ to $SO(7)$, which lifts to a map from C to $\mathrm{Spin}(7)$.

There is a normal fours group $\langle t_1, t_2 \rangle$ in P whose normalizer N is a solvable maximal 2-local subgroup of Co_3 of order $2^{10}.3^3$. There is also a normal subgroup $V = \langle t_1, t_2, t_3, t_4 \rangle \cong (\mathbb{Z}/2)^4$ whose normalizer $X = N_{Co_3}(V)$ is a non-split extension of $(\mathbb{Z}/2)^4$ by $GL(4, \mathbb{F}_2)$. It turns out that the natural map from the homotopy colimit of the diagram formed by the classifying spaces of C, N, X and their intersections to the classifying space of Co_3 is a mod two cohomology equivalence. This is proved essentially by the methods of Jackowski and McClure [JM].

Now we have already chosen a suitable map from C to $\mathrm{Spin}(7)$, and hence from $BC_2^\wedge$ to $B\mathrm{Spin}(7)_2^\wedge$. So we choose a map from $BN_2^\wedge$ to $B(SU(2)^3/\pm1 \rtimes GL(2, \mathbb{F}_2))_2^\wedge$ (the classifying space of the "normalizer" of $(\mathbb{Z}/2)^2$ in DI(4)) and a homotopy equivalence from $BX_2^\wedge$ to the classifying space of the "normalizer" of $(\mathbb{Z}/2)^4$ in DI(4), in such a way that the diagrams obtained by restricting to intersections commutes up to homotopy.

Finally, in order to obtain a map of classifying spaces, it is necessary to lift to strictly commuting maps. There is an obstruction theory for doing this, and it turns out that the obstruction lies in a group isomorphic to $\mathbb{Z}/2$. Fortunately, there is just enough room for maneuver to change this obstruction to zero, and so a map of classifying spaces is obtained.

8. Some nonexistent finite simple groups

In his 1974 paper [So], Ron Solomon considered the problem of classifying finite simple groups in which the Sylow 2-subgroups are isomorphic to those of Conway's group Co_3. In this paper, he proved that Co_3 is the only such group. In the process of proving this, he was forced to examine a configuration in which the centralizer of an involution is isomorphic to $\mathrm{Spin}(7, \mathbb{F}_q)$ (q an odd prime power), possibly with some decoration of odd order at the top and bottom. In fact, only the case where q is congruent to 3 or 5 (mod 8) was relevant to the problem he was trying to solve, because for other values of q the Sylow 2-subgroup is larger than that of Co_3. However, we shall be interested in all odd prime powers q.

The final result of Solomon's analysis was that if G is a finite group with an involution z satisfying

$$O^{2'}C_G(z)/(O_{2'}C_G(z) \cap O^{2'}C_G(z)) \cong \mathrm{Spin}(7, \mathbb{F}_q)$$

then $z \in Z^*(G)$ (that is, $zO_{2'}(G) \in Z(G/O_{2'}(G))$). In particular, G is not simple. In order to reach this conclusion, he was obliged to examine a configuration which appeared to be consistent at the prime two, but which gave rise to a contradiction upon examining a Sylow p-subgroup, where p is the prime of which q is a power.

Despite the nonexistence of Solomon's groups, they have 2-completed classifying spaces. Namely, Dwyer and Wilkerson [DW3] show that for each 2-adic unit q, there is a (unique up to homotopy) self homotopy equivalence $\Psi^q : BDI(4) \longrightarrow BDI(4)$ whose effect on degree $2i$ integral cohomology is multiplication by q^i. In fact, they prove that the group of homotopy classes of self equivalences inducing the identity on mod two cohomology is precisely the group of such operations Ψ^q; since minus the identity is in the Weyl group, Ψ^q is indistinguishable from Ψ^{-q}, so the group is isomorphic to the 2-adic units modulo ± 1. For an odd integer q, write $BSol(q)$ (Sol for Solomon) for the space defined by the following homotopy pullback diagram:

$$
\begin{array}{ccc}
BSol(q) & \longrightarrow & BDI(4) \\
\downarrow & & \downarrow{\scriptstyle \Delta} \\
BDI(4) & \xrightarrow{\mathrm{id} \times \Psi^q} & BDI(4) \times BDI(4).
\end{array}
$$

The following diagram of maps of classifying spaces commutes up to homotopy:

$$
\begin{array}{ccccc}
B(\mathbb{Z}/2 \times M_{12}) & \longrightarrow & & & BCo_3 \\
\downarrow & & & & \downarrow \\
B(\mathbb{Z}/2 \times G_2(q)) & \longrightarrow & B\mathrm{Spin}(7, \mathbb{F}_q) & \longrightarrow & BSol(q) \\
\downarrow & & \downarrow & & \downarrow \\
B(\mathbb{Z}/2 \times G_2) & \longrightarrow & B\mathrm{Spin}(7) & \longrightarrow & BDI(4).
\end{array}
$$

Using the Eilenberg–Moore spectral sequence on the pullback diagram defining $BSol(q)$, one can easily deduce that the Poincaré series for the cohomology is given by

$$\sum_{i=0}^{\infty} \dim_{\mathbb{F}_2} H^i(BSol(q),\mathbb{F}_2) = \frac{(1+t^7)(1+t^{11})(1+t^{13})(1+t^{14})}{(1-t^8)(1-t^{12})(1-t^{14})(1-t^{15})}.$$

We have maps

$$H^*(BDI(4),\mathbb{F}_2) \longrightarrow H^*(BSol(q),\mathbb{F}_2) \longrightarrow H^*(B(\mathbb{Z}/2)^4,\mathbb{F}_2)^{GL(4,\mathbb{F}_2)}$$

whose composite is an isomorphism. Using this, one can show that, as an algebra over the Steenrod algebra, $H^*(BSol(q),\mathbb{F}_2)$ splits as a tensor product of $H^*(BDI(4),\mathbb{F}_2)$ with the cohomology of the fiber of $BSol(q) \longrightarrow BDI(4)$; namely with $H^*(DI(4),\mathbb{F}_2)$. The latter is generated by an element λ in degree 7 whose fourth power is zero, and two elements μ and ν of degrees 11 and 13 which square to zero. The action of the Steenrod algebra is determined by $\mathrm{Sq}^4(\lambda)=\mu$, $\mathrm{Sq}^2(\mu)=\nu$, and $\mathrm{Sq}^1(\nu)=\lambda^2$.

References

[AW]   J. F. Adams and C. W. Wilkerson, *Finite H-spaces and algebras over the Steenrod algebra*, Ann. of Math. **111** (1980), pp. 95–143.

[AMM]   A. Adem, J. Maginnis and R. J. Milgram, *The geometry and cohomology of the Mathieu group M_{12}*, J. Algebra **139** (1991), pp. 90–133.

[Co3]   D. J. Benson, *Conway's group Co_3 and the Dickson invariants*, Manuscripta Mathematica **85** (1994), pp. 177–193.

[BW]   D. J. Benson and C. W. Wilkerson, *Finite simple groups and Dickson invariants*, Contemp. Math. **188** (1995), pp. 31–42.

[BK]   A. Bousfield and D. Kan, *Homotopy limits, completions and localizations*, Springer Lecture Notes in Mathematics 304, Springer-Verlag, Berlin/New York, 1972.

[CE]   A. Clark and J. Ewing, *The realization of polynomial algebras as cohomology rings*, Pacific J. of Math. **50** (1974), pp. 425–434.

[Co]   J. H. Conway, *A group of order* $8,315,553,613,086,720,000$, Bull. London Math. Soc. **1** (1969), pp. 79–88.

[CN]   J. H. Conway and S.P. Norton, *Monstrous moonshine*, Bull. London Math. Soc. **11** (1979), pp. 308–339.

[CS]   J. H. Conway and N. Sloane, *Sphere packings, lattices and groups*, Grundlehren der mathematischen Wissenschaften, **290** Springer-Verlag (1988).

[Atlas] J. H. Conway, R. T. Curtis, S. P. Norton, R. A. Parker and R. A. Wilson, *Atlas of Finite Groups*, Oxford University Press (1985).

[Di]   L. E. Dickson, *A fundamental system of invariants of the general modular linear group with a solution of the form problem*, Trans. Amer. Math. Soc. **12** (1911), pp. 75–98.

[DW] W. G. Dwyer and C. W. Wilkerson, *Maps of B**Z**/p**Z** to BG,* in: Algebraic Topology, Rational Homotopy (Y. Felix, ed.), Springer Lecture Notes in Mathematics, **1318** (1988), pp. 92–98.

[DW2] W. G. Dwyer and C. W. Wilkerson, *A new finite loop space at the prime two,* J. Amer. Math. Soc. **6** (1993), pp. 37–64.

[DW3] W. G. Dwyer and C. W. Wilkerson, *The uniqueness of BDI(4),* In preparation.

[FP] Z. Fiedorowicz and S. Priddy, *Homology of Classical Groups over Finite Fields and their Associated Infinite Loop Spaces,* Springer Lecture Notes in Mathematics, **674** Springer-Verlag (1978).

[Fr] E. Friedlander, *Etale homotopy of simplicial schemes,* Annals of Math. Studies, **104** Princeton Univ. Press (1982).

[JM] S. Jackowski and J. McClure, *Homotopy decomposition of classifying spaces via elementary abelian subgroups,* Topology **31** (1992), pp. 113–132.

[JMO] S. Jackowski, J. McClure and R. Oliver, *Homotopy classes of self-maps of BG via G-actions,* Ann. of Math. **135** (1992), pp. 227–270.

[Kl] S. Kleinerman, *The cohomology of Chevalley groups of exceptional Lie type,* Memoirs of the A.M.S., **268** (1982).

[LW] J. P. Lin and F. Williams, *On 14-connected finite H-spaces,* Israel J. Math. **66** (1989), pp. 274–288.

[Milg] R. J. Milgram, Unpublished.

[Mi] J. W. Milnor, *Construction of universal bundles, I, II.,* Ann. of Math. **63** (1956), pp. 272–284, 430–436.

[Pu] L. Puig, Handwritten notes.

[Nice] D. G. Quillen, *Cohomology of groups,* in: International Congress of Mathematicians, Nice (1970).

[Qu] D. G. Quillen, *On the cohomology and K-theory of the general linear groups over a finite field,* Ann. of Math. **96** (1972), pp. 552–586.

[ST] G. C. Shephard and J. A. Todd, *Finite unitary reflection groups,* Canad. J. Math. **6** (1954), pp. 274–304.

[So] R. Solomon, *Finite groups with Sylow 2-subgroups of type .3 ,* J. Algebra **28** (1974), pp. 182–198.

[Wi] C. W. Wilkerson, *A primer on the Dickson invariants,* in: Proc. of the Northwestern Homotopy Theory Conference Contemp. Math., **19** American Mathematical Society (1983), pp. 421–434.

Kernels of Actions on Non-Positively Curved Spaces

Robert Bieri and Ross Geoghegan[1]

Fachbereich Mathematik,Department of Mathematical Sciences, Universität Frankfurt, Robert-Mayer-Str. 6, D 6000, Frankfurt 1, Germany.

State University of New York, Binghamton, NY 13902-6000, USA.

email:bieri@math.uni-frankfurt.de

email:ross@math.binghamton.edu

1. Introduction

1.1. Summary

This is an outline of recent work aimed at understanding the finiteness properties F_n, FP_n, FD and FP of the group N in a short exact sequence

$$N \rightarrowtail G \overset{\pi}{\twoheadrightarrow} Q$$

where Q acts properly discontinuously by isometries on a space having, in some rather general sense, non-positive curvature. Our results apply, in particular when Q is a free group, or a discrete cocompact subgroup of a virtually connected linear semi-simple Lie group, or an S-arithmetic subgroup of a reductive algebraic group of global rank zero.

---

[1] This work was begun during a 1992 Semester on Geometric Methods in Group Theory at the Centra de Recerca Mathematica of the Universitat Autonoma de Barcelona. It was continued at the Durham Symposium, partially supported by the National Science Foundation and by means of travel grants from the Deutsche Forschungsgemeinschaft. We are grateful for all of these sources of support.

1.2. The finiteness properties

Recall that a group G is *of type F_n* (respectively FP_n) if there exists a $K(G, 1)$-complex (respectively a free G-resolution $F \twoheadrightarrow \mathbb{Z}$) with finite n-skeleton. Both F_1 and FP_1 are equivalent to G being finitely generated. Condition F_2 is equivalent to G being finitely presented, and the weaker (but possibly equivalent[2]) condition FP_2 is equivalent to $G = H/P$ with H finitely presented and $P \triangleleft H$ a perfect normal subgroup. G is *of type F* (respectively FD) if there exists a finite (respectively finitely dominated) $K(G, 1)$-complex, and *of type FP* if there exists a finite projective G-resolution $P \twoheadrightarrow \mathbb{Z}$. It is not known if FD implies F.

1.3. The Problem

It is easily observed that if both N and G in §1.1 are of type F_n (respectively FP_n), so is Q. And if both N and Q are of type F_n (respectively FP_n), so is G. Thus we are left with the problem of studying the behaviour of N under the assumption that both G and Q are of type F_n (respectively FP_n), and we shall from now on make this assumption without specially mentioning it. The case when G is free of rank 2 and $Q \cong \mathbb{Z}$ shows drastically that N will not, in general, be of type F_n (respectively FP_n). Hence the Problem is *to find computable parameters which control the exceptional case that N succeeds in being of type F_n (respectively FP_n).* In case G is of type F we will also be interested in when N is of type FD (respectively FP).

1.4. The case when Q is Abelian

Our Problem has been previously investigated in case Q is Abelian. In the papers cited below, subsets $\Sigma^n(G)$ and $\Sigma^n(G; \mathbb{Z})$ of the $\mathbb{R}$-vector space $\mathrm{Hom}(G, \mathbb{R})$ are defined which should be thought of as "geometric invariants", for they parametrize the Problem in the following sense:

Theorem. ([BNS], [BR], [R₁]) *Consider the exact sequence in §1.1 with Q Abelian and G of type F_n (respectively FP_n). Then N is of type F_n (respectively FP_n) if and only if the image of $\pi^* : \mathrm{Hom}(Q, \mathbb{R}) \rightarrowtail \mathrm{Hom}(G, \mathbb{R})$ is contained in $\Sigma^n(G)$ (respectively $\Sigma^n(G; \mathbb{Z})$).*

The geometric invariant $\Sigma^1(G)$ was first studied in the special context of metabelian groups G in [BS₁] as a tool to characterize finite presentability of G. For arbitrary groups G, $\Sigma^1(G)$ was introduced in [BNS] and subsequently extended to $n \geq 2$ in [BR] and [R₂] (see also [R₁]). The desire to prove the above theorem has always served as a guideline. $\Sigma^1(G)$ has been computed in many cases, whereas the higher invariants are more difficult to compute.

---

[2] Bestvina and Brady have recently shown that $F_2 \neq FP_2$

For further information see the manuscript [BS$_2$] or Holger Meinert's survey article in this volume.

1.5. The set up

A group Q acts *properly discontinuously* on a space M if every point $x \in M$ has a neighbourhood U such that $\{q \in Q \mid qU \cap U \neq \emptyset\}$ is finite.

In the proof of the above theorem one starts by choosing a basis for the torsion-free part of the (finitely generated) Abelian group Q. This provides a properly discontinuous and cocompact translation action of Q on the Euclidean space $\mathbb{E}^m$, and this action is crucial throughout the proof. We will refer to this as the "flat case".

For our generalization, (the "non-positively curved case") we assume the group Q in §1.1 acts properly discontinuously and cocompactly by isometries on a "convex geodesic space" M. This is defined in §2, but one should think of prominent examples such as Euclidean and hyperbolic spaces, universal covers of closed manifolds with non-positive sectional curvature, locally finite trees and, more generally, locally finite Euclidean buildings. In particular, Q could be any discrete cocompact subgroup of a virtually connected linear semi-simple Lie group $\mathbb{G}$ acting on the homogenous space $\mathbb{G}$ modulo a maximal compact subgroup; or Q could be an S-arithmetic subgroup of a reductive algebraic group of global rank 0, acting on the product of the Bruhat-Tits buildings associated to the places in S. We note that our most fundamental result, Theorem 5, does not explicitly require cocompactness.

It will be convenient to fix a group of isometries $\mathbb{G}$ of M with the property that the action of Q on M is given by a monomorphism $\theta : Q \to \mathbb{G}$. We will prove that N is of type F_n if and only if the action $\theta\pi : G \to \mathbb{G}$ is "$(n-1)$-connected" in a sense to be defined. Now, other actions $\rho : G \to \mathbb{G}$, with $\rho(G)$ not necessarily properly discontinuous or cocompact, can be "$(n-1)$-connected"; so, as in the flat case, we use this property to define a subset $S\Sigma^n(G)$ of the space $\mathcal{R}(G, \mathbb{G}) = \mathrm{Hom}(G, \mathbb{G})$ of all G-actions (i.e. representations) into $\mathbb{G}$. The group $\mathbb{G}$ is given the compact-open topology, and since G is finitely generated $\mathcal{R}(G, \mathbb{G})$ can be viewed as a subspace of a finite product of copies of $\mathbb{G}$.

There is also an "$(n-1)$-acyclicity" property having a similar relationship to FP_n and leading to a corresponding subset $S\Sigma^n(G; \mathbb{Z})$ of $\mathcal{R}(G, \mathbb{G})$.

1.6. The main results

Theorem A. *Both $S\Sigma^n(G)$ and $S\Sigma^n(G; \mathbb{Z})$ are open subsets of $\mathcal{R}(G, \mathbb{G})$.*

Theorem B. *Consider the short exact sequence in §1.1 and assume we are given a properly discontinuous and cocompact action $\theta : Q \twoheadrightarrow \mathbb{G}$. Then N is of type F_n (respectively FP_n) if and only if $\theta\pi \in S\Sigma^n(G)$ (respectively $\theta\pi \in S\Sigma^n(G;\mathbb{Z})$).*

Theorem A generalizes Theorems A in [BNS], [BR] and [R_2], and Theorem B generalizes [BNS; Theorem B1], [BR; Theorem B] and [R_2, Satz C], all of which treat the flat case. As in those papers, by combining Theorems A and B we get an openness result for the finiteness properties F_n and FP_n – something which is intelligible without any knowledge of $S\Sigma^n$. Let $\mathcal{R}_0(G,\mathbb{G})$ denote the subspace of $\mathcal{R}(G,\mathbb{G})$ consisting of all cocompact actions $\rho : G \to \mathbb{G}$ such that $\rho(G)$ acts properly discontinuously on M. Then we have

Theorem C. *The set of all $\rho \in \mathcal{R}_0(G,\mathbb{G})$ with $\ker(\rho)$ of type F_n (respectively FP_n) is open in $\mathcal{R}_0(G,\mathbb{G})$.*

Remark. Weil [W] has proved the following: If $\mathbb{G}$ is a Lie group and $\rho \in \mathcal{R}(G,\mathbb{G})$ has the property that $H^1(G,\mathfrak{g}) = 0$, then every ρ' sufficiently close to ρ is obtained from ρ by conjugation in $\mathbb{G}$ (and hence has the same kernel). Thus Theorem C is only of interest when Weil's result or some other "local rigidity" theorem does not apply.

If, in §1, G is of type F then there exists a finite-dimensional $K(N,1)$-complex; so N has type FD (respectively FP) if and only if N has type F_n (respectively FP_n), where n is the dimension of some $K(G,1)$-complex (respectively projective G-resolution $P \twoheadrightarrow \mathbb{Z}$). Hence we have:

Corollary D. *Assume that the group G in §1.1 is of type F. The set of all $\rho \in \mathcal{R}_0(G,\mathbb{G})$ with $\ker(\rho)$ of type FD (respectively FP) is open in $\mathcal{R}_0(G,\mathbb{G})$.*

In §6.5 we give a description of $S\Sigma^n(G;\mathbb{Z})$ in terms of the vanishing of homology groups of G in dimensions $\leq n$ with coefficients in the Novikov modules $\mathbb{Z}G^e$ (these are generated by completions of the group ring $\mathbb{Z}G$ with respect to certain filtrations on G):

Theorem E. *With hypotheses as in Theorem B, N is of type FP_n if and only if $H_p(G;\mathbb{Z}G^e) = 0$ for all $p \leq n$ and all boundary points e of M.*

1.7. The invariants $\Sigma^n(G)$, $\Sigma^n(G;\mathbb{Z})$

These are true generalizations of the Bieri-Neumann-Strebel-Renz geometric invariants, and are introduced in §3.2 They are more subtle than the previously mentioned "symmetrized" versions $S\Sigma^n(G)$, $S\Sigma^n(G;\mathbb{Z})$, but are not relevant to our main theorems. However the example discussed in §3.4, of $\mathbb{G} = \mathrm{PSL}_2(\mathbb{R})$ and $n = 0$, suggests that they are of independent interest.

2. Convex geodesic spaces

2.1. The spaces

Let (M, d) be a metric space. If $a, b \in M$, a *geodesic segment* from a to b is an isometric embedding $[0, d(a, b)] \rightarrowtail M$. A *geodesic ray* based at $a \in M$ is an isometric embedding $[0, \infty) \rightarrowtail M$ taking 0 to a. The metric space (M, d) is a *length space* if there is a geodesic segment from any point to any other point. The metric d is *proper* if for each $a \in M$, the map $d(a, \cdot) : M \to \mathbb{R}$ is a proper map (i.e., preimages of compact sets are compact); proper metric spaces are locally compact, and all closed-and-bounded subsets are compact. The length space (M, d) is *convex* if whenever ω_1 and ω_2 are geodesic segments linearly reparametrized by $[0, 1]$, $d(\omega_1(t), \omega_2(t))$ is a convex function of t.

A *convex geodesic space* is a convex length space M with a proper metric such that for every $a \neq b \in M$, the geodesic segment from a to b can be prolonged to a (not necessarily unique) geodesic ray based at a.

2.2. Examples

Examples of convex geodesic spaces are (1) $\mathbb{E}^m$; (2) Hyperbolic space $\mathbb{H}^m$; (3) locally finite affine buildings; (4) complete simply connected open Riemannian manifolds of non-positive sectional curvature; (5) piecewise Euclidean $\mathrm{CAT}(0)$ complexes, in the sense of [Br], in which all closed-and-bounded sets are compact and no link is contractible. Every convex geodesic space is a "Busemann space" in the sense of [Bo]; we remark that the term "Busemann space" has different meanings in other places. It follows (see [Bo]) that if Q acts properly discontinuously and cocompactly on a convex geodesic space M (as is usually the case in this paper) then either Q is hyperbolic in the sense of Gromov or M contains a totally geodesic embedding of a Minkowskian plane.

2.3. Horoballs

Let M be a convex geodesic space. If $a \in M$, let E_a denote the set of all geodesic rays in M based at $a \in M$, topologized as a subset of $M^{[0,\infty)}$ with the compact open topology. It is not hard to show that E_a is compact and metrizable. For each $e \in E_a$ and $t \in [0, \infty)$ the *closed horoball* $HB_{(e,t)}$ is defined to be the closure of $\bigcup \{\mathring{B}_{u-t}(e(u)) \mid u > t\}$, where $\mathring{B}_r(x)$ denotes the open ball in M about x of radius r.

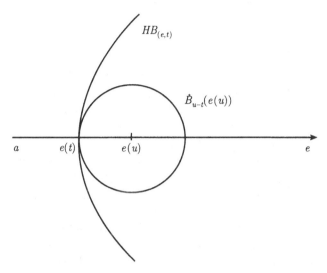

3. The Σ-invariants

3.1. Connectivity at infinity

Let (M, d) be a convex geodesic space and let G be a group of type F_n acting on M by isometries. Choose a $K(G, 1)$ complex Y having finite n-skeleton and let $X = \tilde{Y}$ be the universal cover on which G acts by covering transformations. Then we can construct a G-equivariant map $h : X^n \to M$ as follows. We pick a base point $* \in M$ and a finite set V of vertices of X containing exactly one vertex in each G-orbit of X^0. Define $h(v) = *$ for each $v \in V$ and extend equivariantly to X^0; then extend equivariantly skeleton by skeleton subject to the condition that for each cell σ of dimension $\leq n$, $h(\sigma)$ lies in the closed convex hull of $h(\dot{\sigma})$.

Each geodesic ray $e \in E_*$ based at $*$ gives rise to a decreasing filtration $(X^n_{(e,t)})_{t \geq 0}$ of X^n by subcomplexes: $X^n_{(e,t)}$ is defined to be the largest subcomplex of X^n lying in $h^{-1}(HB_{(e,t)})$. This filtration determines a sort of "end" of X^n defined by e, since different choices of $h : X^n \to M$ lead to equivalent filtrations. Thus we may talk of connectivity of X^n in the direction of this "end". For $p \geq -1$ we say that X^n is p-*connected in the direction* e if there is a constant $\lambda \geq 0$ such that for each $t \geq 0$ and each integer $-1 \leq q \leq p$ every map $f : S^q \to X^n_{(e,t+\lambda)}$ extends to a map $\tilde{f} : B^{q+1} \to X^n_{(e,t)}$. When $q = -1$ this says that each $X^n_{(e,t)}$ is non-empty.

Proposition 1. *For $p \leq n - 1$, whether or not X^n is p-connected in the direction e depends only on G, its action on M, and the ray e, not on the choice of X (i.e. of Y), nor on the choice of h.*

3.2. The invariants

Let $\mathbb{G}$ be a given group of isometries of M and $\mathcal{R}(G, \mathbb{G}) = \mathrm{Hom}(G, \mathbb{G})$ as in §1.5. Define $\Sigma^p(G) \subseteq \mathcal{R}(G, \mathbb{G}) \times E_*$ to be the set of all pairs (ρ, e) with the property that X^n is $(p - 1)$-connected in the direction e. By Proposition 1, $\Sigma^p(G)$ is an invariant of the group G when $p \leq n - 1$.

There is some redundancy in the product $\mathcal{R}(G, \mathbb{G}) \times E_*$. If H is a group of isometries of M which fixes the base point $*$ and normalizes $\mathbb{G}$ then H acts both on $\mathcal{R}(G, \mathbb{G})$ and on E_*, where the effect of the action of $\gamma \in H$ on $\rho \in \mathcal{R}(G, \mathbb{G})$ is given by $\gamma\rho(\cdot)\gamma^{-1}$. The subset $\Sigma^p(G) \subseteq \mathcal{R}(G, \mathbb{G}) \times E_*$ is invariant under the diagonal action of H. Hence one could also define $\Sigma^p(G)$ to be a subset of the space $\mathcal{R}(G, \mathbb{G}) \times_H E_*$ of all orbits under the diagonal H-action. This is particularly convenient if H acts transitively on E_*, for if so one has an isomorphism $\mathcal{R}(G, \mathbb{G}) \times_H E_* \cong \mathcal{R}(G, \mathbb{G})/H_{e_0}$, where $H_{e_0} \leq H$ is the stabilizer of some chosen "base ray" $e_0 \in E_*$.

3.3. Connection with the flat case

In the flat case we have $M = \mathbb{E}^1$ and $\mathbb{G} = $ Translation group of $\mathbb{E}^1$. We can take H to be the group $\{\pm\mathrm{id}_{\mathbb{E}^1}\}$. Then $H_{e_0} = \{1\}$ and we recover the original invariant $\Sigma^p(G) \subseteq \mathrm{Hom}(G, \mathbb{R})$ described in §1.4.

3.4. $\mathrm{PSL}_2(\mathbb{R})$-action on $\mathbb{H}^2$

Here we discuss the case where M is the hyperbolic plane $\mathbb{H}^2$ and $\mathbb{G} = \mathrm{PSL}_2(\mathbb{R})$ acting on the upper half plane model. For the base point $* \in \mathbb{H}^2$ we take the imaginary unit $i \in \mathbb{C}$ (which corresponds to the centre in the unit disk model). The stabilizer of $*$ in $\mathrm{PSL}_2(\mathbb{R})$ is then given by

$$H = \left\{ \pm \begin{pmatrix} a & b \\ -b & a \end{pmatrix} \,\middle|\, a^2 + b^2 = 1 \right\}.$$

Thus H is homeomorphic to a circle S^1 and acts transitively on the space of rays E_*. In the unit disk model H is the group of rotations around $*$. We identify each $e \in E_*$ with its endpoint on the boundary of $\mathbb{H}^2$. In the upper half plane model ∞ is a canonical endpoint, and its horoballs are given by $HB_{(\infty, t)} = \{z \mid \mathrm{Im}(z) \geq t\}$. If one thinks of the invariant $\Sigma^p(G)$ as a subset of $\mathcal{R}(G, \mathrm{PSL}_2(\mathbb{R}))$ then it suffices to restrict attention to this filtration. Thus $\rho : G \to \mathrm{PSL}_2(\mathbb{R})$ is in $\Sigma^p(G)$ if and only if X is $(p - 1)$-connected in the direction ∞. Given a fixed action ρ one can study the behaviour of ρ with respect to the various $e \in E_*$ by studying the orbit ρ^H of ρ in $\mathcal{R}(G, \mathrm{PSL}_2(\mathbb{R}))$ under H with respect to ∞.

Let us examine $\Sigma^0(G)$. Taking the $K(G, 1)$-complex Y with a single vertex makes it clear that a given representation $\rho : G \to \mathrm{PSL}_2(\mathbb{R})$ is in $\Sigma^0(G)$ if and only if $\mathrm{Im}(G*)$, i.e. the set of all imaginary components of the $g* \in \mathbb{C}$, is unbounded. In other words: the point $\infty \in \partial\mathbb{H}^2$ is an accumulation point of $G*$ in the *horocycle topology.* This should be compared with the much weaker condition that ∞ is an accumulation point of $G*$ in the Euclidean topology of the compactified unit disk model of $\mathbb{H}^2$, which simply means that the absolute values of points in $G*$ are unbounded. Note that if $\rho(G) = \mathrm{SL}_2(\mathbb{Z})$ then ∞ is an accumulation point of $G*$ in the Euclidean topology but not in the horocycle topology.

The situation can be analyzed further under the assumption that $\rho : G \to \mathrm{PSL}_2(\mathbb{R})$ is a *discrete* representation (i.e., $\rho(G)$ is a discrete subgroup of the Lie group $\mathrm{PSL}_2(\mathbb{R})$), for in that case one can say rather more about the behaviour of horocyclic accumulation points of $G*$ on $\partial\mathbb{H}^2$. Suppose ∞ is one of them and is, at the same time, fixed under some parabolic element of $\rho(G)$. Then $G*$ contains subsets $A \subseteq G*$ with $\mathrm{Im}(A)$ unbounded but $\mathrm{Re}(A)$ bounded; ∞ is then said to be a *point of approximation.* Points of approximation cannot lie on the boundary of any convex fundamental polygon but parabolic fixed points do (see [Be; p. 261]). Hence if $\rho \in \Sigma^0(G)$ then ∞ cannot be a parabolic fixed point. In fact, assembling results from Chapters 9 and 10 of [Be] one gets:

Proposition 2. *Let $\rho : G \to \mathrm{PSL}_2(\mathbb{R})$ be a discrete representation of the finitely generated group G. Then the following are equivalent:*

(i) $\rho \in \Sigma^0(G)$;

(ii) $\infty \in \partial\mathbb{H}^2$ is a point of approximation;

(iii) ∞ is not on the boundary of any convex locally finite fundamental domain.

Corollary 3. *Let $\rho : G \to \mathrm{PSL}_2(\mathbb{R})$ as in Proposition 2. Then we have:*

(a) $\rho^H \subseteq \Sigma^0(G)$ if and only if $G\backslash\mathbb{H}^2$ is compact.

(b) The complement of $\rho^H \cap \Sigma^0(G)$ in ρ^H is countable if and only if $G\backslash\mathbb{H}^2$ has finite area.

Remark. Let $\rho : G \to \mathrm{PSL}_2(\mathbb{R})$ be as in Proposition 2, and assume G has a fundamental domain with finite area. Then $\rho \in \Sigma^0(G)$ if and only if $\infty \in \partial\mathbb{H}^2$ is not a parabolic fixed point. The finite area assumption excludes the case that $\rho(G)$ is "elementary", so the G-orbit of each point of $\partial\mathbb{H}^2$ is dense in $\partial\mathbb{H}^2$. Hence, the set of parabolic fixed points is either empty or dense in $\partial\mathbb{H}^2$, and we find that if $\rho(G)$ is not cocompact then the complement of $\Sigma^0(G) \cap \rho^H$ in ρ^H is dense in ρ^H.

3.5. Independence of base point

It was a matter of convenience to define $\Sigma^p(G)$ to be a subset of $\mathcal{R}(G, \mathbb{G}) \times E_*$, or, if H fixes $*$ and normalizes $\mathbb{G}$, as a subset of $\mathcal{R}(G, \mathbb{G}) \times_H E_*$. This apparent dependence on $* \in M$ can be removed. Say that $e \in E_*$ and $e' \in E_{*'}$ are *parallel* if $\{d(e(t), e'(t)) \mid t \geq 0\}$ is bounded. It is shown in [H; 2.2] that given e and $*'$ there is a unique e' parallel to e (the proof in [H], ostensibly for CAT(0) spaces, works for convex geodesic spaces too). Since parallel rays e and e' lead to equivalent filtrations $(X^n_{(e,t)})$ and $(X^n_{(e',t)})$, X^n is $(p-1)$-connected in the direction e, using $*$, if and only if it is $(p-1)$-connected in the direction e', using $*'$. So, writing E for the set of "parallelism classes" of geodesic rays, we can equally well define $\Sigma^p(G)$ to be the corresponding subset of $\mathcal{R}(G, \mathbb{G}) \times E$, or of $\mathcal{R}(G, \mathbb{G}) \times_H E$ if H is the normalizer of $\mathbb{G}$ in the group of *all* isometries of M.

4. Discussion of the Theorems

4.1. The symmetrized invariants

As before we are given a group G, a convex geodesic space M with a base point $* \in M$, and a group of isometries $\mathbb{G}$ of M. In the flat case when $\mathbb{G} = T(\mathbb{E}^1)$, the 1-dimensional translation group, it was fairly easy (but crucial) to observe that the invariant $\Sigma^p(G)$ was an *open* subset of the space $\mathcal{R}(G, T(\mathbb{E}^1))$ for $p \leq n$. From the remark at the end of §3.4 we see that this will not, in general, be the case. For an openness result we have to replace $\Sigma^p(G) \subseteq \mathcal{R}(G, \mathbb{G}) \times E_*$ by a cruder version of it. This *symmetrized invariant* is the subset of $\mathcal{R}(G, \mathbb{G})$ defined by

$$S\Sigma^p(G) = \{\rho \mid (\rho, e) \in \Sigma^p(G), \quad \text{for all } e \in E_*\}.$$

Thus a homomorphism $\rho : G \to \mathbb{G}$ is in $S\Sigma^p(G)$ if X is $(p-1)$-connected in *all* directions $e \in E_*$.

In the flat case where $\mathbb{G} = T(\mathbb{E}^1)$, $S\Sigma^p(G)$ is just the intersection of $\Sigma^p(G)$ with its antipodal set $-\Sigma^p(G)$. In the example $\mathbb{G} = \mathrm{PSL}_2(\mathbb{R})$ of §3.4 it is the intersection $\bigcap_{h \in H} \Sigma^p(G)^h$.

4.2. The main results

Theorem 4. $S\Sigma^n(G)$ *is an open subset of* $\mathcal{R}(G, \mathbb{G})$.

As in the flat case, this openness result is based on a criterion for the action $\rho \in \mathcal{R}(G, \mathbb{G})$ to be in $\Sigma^n(G)$ in terms of a finite collection of small homotopies. In the flat case this was called the "Σ^n-criterion" (see [Bi] or [BS$_2$]) and its proof, by induction on n, was independent of the Theorem cited in §1.4. In the non-positively curved case, however, a proof of the analogous "$S\Sigma^n$-criterion" is intimately bound up with the proof of:

Theorem 5. *Let* $\rho \in \mathcal{R}(G, \mathbb{G})$ *be such that* $\rho(G)$ *acts properly discontinuously on* M. *If* $\rho \in S\Sigma^n(G)$ *then the kernel of* ρ *is of type* F_n.

We briefly discuss the proofs of Theorems 4 and 5. With notation as in §3.1, one wishes to find $R > 0$ such that the pair $(X^n, X^n(R))$ is n-connected; here $X^n(R)$ denotes the largest subcomplex of X^n lying in $h^{-1}(B_R(*))$. The complex $X^n(R)$ is acted on by $N \equiv \ker(\rho)$, and if the $\rho(G)$-action on M is properly discontinuous the map $N \backslash X^n \to M$ covered by h is proper. The last two sentences imply that $X^n(R)$ is a cocompact $(n-1)$-connected N-complex, so N has type F_n.

The proof that the pair $(X^n, X^n(R))$ is n-connected is delicate. By "elementary expansions" we attach finitely many G-orbits of cells to X to get a stronger version of $(n-1)$-connectedness in which the constant λ (see §3.1) is zero. The fact that E_* is compact is crucial in that, when we wish to perform homotopies of maps $f : (B^p, S^{p-1}) \to (X^p, X^p(R))$ in different directions e, compactness allows us to only use finitely many directions. We deform f, rel S^{p-1}, to a map into $X^p(R)$ by small homotopies in different directions, and then the use of only finitely many directions bounds certain important positive numbers away from 0. The construction of these small homotopies requires the "$\lambda = 0$" connectedness conditions; on the other hand, deriving the "$\lambda = 0$" connectedness actually requires the same kind of small homotopies in lower dimensions. Thus the proofs of Theorems 4 and 5 comprise one big induction. The details involve ensuring that the homotopies are suitably "Lipschitz over M" in order to have a sufficiently strong inductive hypothesis.

The simultaneous proof of Theorems 4 and 5 is long. It is based on ideas and techniques from the flat cases - more precisely: from the proof of the "Σ^n-criterion" (see [BS$_2$] or [Bi, Theorem A]) and the Theorem stated in §1.4 (and in [Bi] as Theorem E). The present generality entails new complications and technicalities - mostly coming from the fact that G does not act trivially on $\mathcal{R}(G, \mathbb{G})$ any more. We are preparing a detailed paper.

We now turn to the converse of Theorem 5. There is an obvious necessary condition for an action $\rho \in \mathcal{R}(G, \mathbb{G})$ to be in $S\Sigma^p(G)$: ρ has to be in $S\Sigma^0(G)$

– the point being that this latter condition depends only on the image $\rho(G)$ rather than on G itself: every horoball of M must contain a point of the orbit $G* = \rho(G)*$. In the case when M is the Euclidean m-space or the hyperbolic plane (see §3.4) this necessary condition is actually equivalent to saying that the G-action on M is cocompact. Indeed, we do find a converse to Theorem 5 in general under the cocompactness assumption:

Theorem 6.   *Let $\rho \in \mathcal{R}(G, \mathbb{G})$ be such that $\rho(G)$ acts properly discontinuously and cocompactly on M. If the kernel of ρ is of type F_n then $\rho \in S\Sigma^n(G)$.*

The assumption in Theorem 6 that $\rho(G)$ be cocompact may be too strong. But we have doubts that the weaker condition $\rho \in S\Sigma^0(G)$ would be sufficient. We do, however, have necessary and sufficient conditions in terms of the subspace $h(X^n) \subseteq M$.

Proposition 7.   *(a) If $ker(\rho)$ is of type F_n then X and $h : X^n \to M$ can be chosen so that $h^{-1}(h(\sigma))$ is $(n-1)$-connected for each cell σ of X of dimension $\leq n - 1$. (b) If the choice referred to under (a) is made then $\rho \in S\Sigma^n(G)$ if and only if $h(X^n) \subseteq M$ is $(n-1)$-connected in every direction $e \in E_*$.*

Here, the definition of connectivity of $h(X^n)$ in the direction e has to be translated from the definition for X^n in §3.1 by using the filtration $(h(X^n_{(e,t)}))_{t \geq 0}$ for $e \in E_*$.

5. Horo-connectivity

5.1. horo-p-connectedness

We continue to discuss the set up of §3.1. A map $f : Z \to X^n$ is *proper in the direction* e if for each $t \geq 0$ there is a compact subset $C \subset Z$ such that $f(Z - C) \subset X^n_{(e,t)}$. We say X is *horo-p-connected rel* e if for every $0 \leq q \leq p$ and every map $f : \mathbb{R}^q \to X^q$ which is proper in the direction e there is a map $\tilde{f} : \mathbb{R}^q \times [0, \infty) \to X^{q+1}$ extending f which is also proper in the direction e. For $p < n$ this is well-defined; but we shall also need it for $p = n$, and until now $X^{n+1}_{(e,t)}$ has not been defined. To get around this, simply extend the map h to the (possibly not locally finite) complex X^{n+1} in the manner described in §3.1. This will not cause problems in view of Theorem 8 below.

5.2. Alternative description of $S\Sigma^n(G)$

In the flat case, when $M = \mathbb{E}^1$ and $\mathbb{G} = $ translations, it was shown (see [Bi], [BS$_2$]) that *X is horo-n-connected rel e if and only if X^n is $(n-1)$-connected in the direction e*, and this established an alternative description of the "flat" invariant $\Sigma^n(G)$. But in the present generality we are unable to prove either implication in that theorem. However, we can prove the desired result if we look at all directions e simultaneously:

Theorem 8. *X is horo-n-connected rel e for every $e \in E_*$ if and only if X^n is $(n-1)$-connected in all directions $e \in E_*$. Hence*

$$S\Sigma^n(G) = \{\rho \in \mathcal{R}(G, \mathbb{G}) \mid X \text{ is horo-}n\text{-connected in every direction } e\}.$$

5.3. How to check for horo-p-connectedness

If we pick a proper base ray α in $\mathbb{R}^q$ and a base ray $\bar{e}$ in X^1 which is proper in the direction e, the proper homotopy classes of proper maps in the direction e from $(\mathbb{R}^q, \alpha)$ to $(X^{n+1}, \bar{e})$ form *horo-homotopy groups* $\pi_q^e(X, \bar{e})$ for $q \geq 1$ (pointed sets for $q = 1$). The details are so precisely analogous to the discussion of the groups $\pi_q^S(X, e)$ in [BT] that we refer the reader to §2 of that paper for proofs of all assertions in this subsection. Clearly, X is horo-p-connected rel e if and only if X is horo-0-connected rel e, and $\pi_q^e(X, \bar{e})$ is trivial for $1 \leq q \leq p \leq n$. As in [BT; 2.9-11], there are short exact sequences of groups for $2 \leq q \leq n$, and pointed sets for $q = 1$:

$$\{1\} \to \varprojlim_i{}^1 \{\pi_q(X_{(e,i)}^{n+1}, \bar{e})\} \to \pi_q^e(X, \bar{e}) \to \varprojlim_i \{\pi_{q-1}(X_{(e,i)}^{n+1}, \bar{e})\} \to \{1\}.$$

Here, i ranges over $\mathbb{Z}^+$, and the inverse sequences of ordinary homotopy groups have base points along $\bar{e}$, and bonds induced by $\bar{e}$ in the obvious way. An inverse sequence of groups $G_1 \leftarrow G_2 \leftarrow \cdots$ is *semistable* or *Mittag-Leffler* if for any k the images in G_k of the groups G_j, $j \geq k$, are almost all the same. If those images are almost all trivial, the inverse sequence is *pro-trivial*. Thus, (see [MS; II §6]) we have, for $p \geq 1$:

Proposition 9. *X is horo-p-connected rel e if and only if there exists exactly one proper homotopy class of proper rays in the direction e, and, using any such ray $\bar{e}$ as base ray, the groups $\{\pi_q(X_{(e,i)}^{n+1}, \bar{e})\}_i$ are pro-trivial for $2 \leq q \leq p - 1$ and semistable for $q = p$.*

6. Homological companion results

All results in the previous sections have homological analogues which we collect here.

6.1. The homological invariants

We continue to discuss the set up of Section 3.1.

Fixing a commutative ring R with $1 \neq 0$, we say that X^n is *p-acyclic* (over R) *in the direction* e if there is a constant $\lambda \geq 0$ such that for each $t \geq 0$ the inclusion $X^n_{(e,t+\lambda)} \rightarrowtail X^n_{(e,t)}$ induces the zero map on R-homology in dimensions $\leq p$. Analogous to Proposition 1 we have that *for $p \leq n-1$ p-acyclicity (over R) in the direction e depends only on R, G, its action on M and the ray e.* Given a group $\mathbb{G}$ of isometries of M we can now consider the subset $\Sigma^p(G,R) \subseteq \mathcal{R}(G,\mathbb{G}) \times E_*$ consisting of all pairs (ρ,e) with the property that X^n is $(p-1)$-acyclic in the direction e.

The *symmetrized invariant* is the subset $S\Sigma^p(G;R) \subseteq \mathcal{R}(G,\mathbb{G})$ defined by

$$S\Sigma^p(G;R) = \{\rho \mid (\rho,e) \in \Sigma^p(G;R), \text{ for all } e \in E_*\},$$

6.2. The main results

Theorem 4'. $S\Sigma^p(G;R)$ is an open subset of $\mathcal{R}(G,\mathbb{G})$.

Theorem 5'. Let $\rho \in \mathcal{R}(G,\mathbb{G})$ be such that $\rho(G)$ acts properly discontinuously on M. If $\rho \in S\Sigma^n(G;R)$ then the kernel of ρ is of type FP_n over R.

Theorem 6'. Let $\rho \in \mathcal{R}(G,\mathbb{G})$ be such that $\rho(G)$ acts properly discontinuously and cocompactly on M. If the kernel of ρ is of type FP_n then $\rho \in S\Sigma^n(G;R)$.

6.3. Horo-acyclicity

We orient the cells of X^{n+1} so that $C^\infty_*(X^{n+1})$, the R-chain complex of possibly infinite locally finite cellular chains of X^{n+1} is defined. Here, in dimension $(n+1)$ it is understood that we only include chains whose boundary is well-defined. Let $C^e_*(X^{n+1})$ denote the subcomplex of $C^\infty_*(X^{n+1})$ consisting of those chains which are locally finite with respect to the filtration $(X^{n+1}_{(e,t)})_{t\geq 0}$, i.e., for each t $X^{n+1}_{(e,t)}$ contains all but a compact subset of the support of the chain. The q-th homology of this chain complex, $H^e_q(X)$, is the qth *horo-homology group* of X rel e (with R-coefficients). We say that X is *horo-p-acyclic rel e* if $H^e_q(X) = 0$ for all $q \leq p$, (reduced homology for $q=0$).

Theorem 8′. *X is horo-n-acyclic rel e for every direction $e \in E_*$ if and only if X is $(n-1)$-acyclic in all directions $e \in E$. Hence*

$$S\Sigma^n(G;R) = \{\rho \in \mathcal{R}(G,\mathbb{G}) \mid X \text{ is horo-}n\text{-acyclic in every direction } e\}.$$

6.4. How to check for horo-p-acyclicity

Everything in §5.3 has a homological analogue, changing homotopy groups to homology groups. The relevant exact sequence can be found in [Ma].

6.5. Entrance of group homology

The action $\rho : G \to \mathbb{G}$ together with an end $e \in E_*$ defines a filtration of G by "e-horoballs": for each $t \geq 0$ we put

$$G_{(e,t)} = \{g \in G \mid g* \in HB_{(e,t)}\}.$$

Let $\overline{RG}$ denote the set of all formal (possibly infinite) R-linear combinations of elements of G, and $RG^e \subseteq \overline{RG}$ the subset of those whose support is *locally finite in the direction e*, in the sense that, for each t, all but finitely many of its points lie in $G_{(e,t)}$. This is a right RG-module which we call the *Novikov module*, (compare [N], [S]). It is clear that there is a natural homomorphism $\phi_p : H_p(G; RG^e) \to H_p^e(X)$ which is an isomorphism for $p < n$ and an epimorphism for $p = n$. Hence, the vanishing of $H_p(G; RG^e)$ for $p \leq n$ implies that X is horo-n-acyclic in all directions e. In the flat case, the converse is true (see [BS$_2$]) but in the present non-positively curved case we only have

Theorem 10. *X is horo-n-acyclic in all directions $e \in E_*$ if and only if $H_p(G; RG^e) = 0$ for all $p \leq n$ and all $e \in E_*$. Hence $S\Sigma^n(G;R)$ is the set*

$$\{\rho \in \mathcal{R}(G,\mathbb{G}) \mid H_p(G; RG^e) = 0 \text{ for all } p \leq n \text{ and all } e \in E_*\}.$$

References

[Be]   A. F. Beardon, *The geometry of discrete groups,* Springer-Verlag New York (1983).

[Bi]   R. Bieri, *The geometric invariants of a group: a survey with emphasis on the homotopical approach,* in: Geometric Group Theory vol. 1 LMS Lecture Note Series, **181** Cambridge University Press Cambridge (1993), pp. 24–36 .

[Bi]   R. Bieri and R. Strebel, *Valuations and finitely presented metabelian groups,* Proc. Lond. Math. Soc. (3) **41** (1980), pp. 439–464.

[BS$_2$]   R. Bieri and R. Strebel, *Geometric invariants for discrete groups,* preprint of monograph, Frankfurt and Freiburg (1992).

[BNS]   R. Bieri, W. D. Neumann and R. Strebel, *A geometric invariant of discrete groups,* Invent. Math. **90** (1987), pp. 451–477.

[BR]   R. Bieri and B. Renz, *Valuations on free resolutions and higher geometric invariants of groups,* Comment. Math. Helvetici **63** (1988), pp. 464–497.

[Bo]   B. H. Bowditch, *Minkowskian subspaces of non-positively curved metric spaces,* Bull. London Math. Soc. **27** (1995), pp. 575–584.

[Br]   M. Bridson, *Geodesics and curvature in metric simplicial complexes,* in: "Group theory from a geometric viewpoint Proceedings of ICTP, Trieste World Scientific, Singapore and Teaneck (1991), pp. 373–463.

[BT]   M. G. Brin and T. L. Thickstun, *On the proper Steenrod homotopy groups, and proper embeddings of planes into 3-manifolds,*   Trans. Amer. Math. Soc. **289** (1983), pp. 737–755.

[H]   P. K. Hotchkiss, *The boundary of a Busemann space,* Proc. Amer. Math. Soc. **125** (1997). pp. 1903–1912.

[MS]   S. Mardešić and J. Segal, *Shape Theory,* North-Holland, Amsterdam (1982).

[Ma]   W. S. Massey, *Homology and cohomology theory,* Marcel Dekker, New York and Basel (1978).

[N]   S. P. Novikov, *Multivalued functions and functionals - an analogue of the Morse theory,* Soviet Math. Dokl. **24** (1981), pp. 222–225.

[R$_1$]   B. Renz, *Geometric invariants and HNN-extensions,* in: Group Theory (Singapore 1987) de Gruyter, Berlin (1989), pp. 465–484.

[R$_2$]   B. Renz, *Geometrische Invarienten und Endlichkeitseigenschaften von Gruppen,* Dissertation, Frankfurt (1988).

[S]   J-C. Sikorav, *Homologie de Novikov associé à une classe de cohomologie réelle de degré un,* preprint, Toulouse (1989).

[W]   A. Weil, *Remarks on the cohomology of groups,* AnnMath **80** (1964), pp. 149–157.

Cyclic Groups Acting on Free Lie Algebras

R. M. Bryant

Department of Mathematics, UMIST, P. O. Box 88, Manchester, M60 1QD.

1. Introduction

The purpose of this note is to study a conjecture of M. W. Short [5] which
is concerned with some special cases of the problem of determining the fixed
points of a finite group acting on a free Lie algebra. Short formulated his
conjecture on the basis of evidence obtained by hand and computer calculations. The main results of this note explain the theoretical significance of
the numbers occurring in Short's conjecture and show that this conjecture is
equivalent to some simple and natural conjectures concerning free Lie rings.

For any positive integer m and any commutative ring R with identity let
$L(R, m)$ denote the free Lie algebra over R on m free generators $x_1, \ldots, x_m$.
Let $G(m)$ be a cyclic group of order m generated by an element g and let
$G(m)$ act (on the right) as a group of Lie algebra automorphisms of $L(R, m)$ in
such a way that $x_i g = x_{i+1}$ $(1 \leqslant i \leqslant m - 1)$ and $x_m g = x_1$. For each positive
integer n let $L_n(R, m)$ denote the R-submodule of $L(R, m)$ spanned by all
monomials in $x_1, \ldots, x_m$ of degree n: thus $L(R, m) = \bigoplus_{n \geqslant 1} L_n(R, m)$ and
each $L_n(R, m)$ is $G(m)$-invariant. This note is concerned with the problem
of obtaining information about $L_n(R, m)^{G(m)}$, the set of elements of $L_n(R, m)$
which are fixed by all elements of $G(m)$ (or, equivalently, fixed by g). We
shall regard $L(R, m)$ and $L_n(R, m)$ as right $RG(m)$-modules in the obvious
way.

Let $L(m) = L(\mathbb{Z}, m)$ where $\mathbb{Z}$ is the ring of integers. Furthermore, if p is a
prime number, write $\mathbb{F}_p$ for the field of p elements and $M(p) = L(\mathbb{F}_p, p)$. For
positive integers m and n let

$$f(m, n) = \frac{1}{mn} \sum_{\substack{d|n \\ (d,m)=1}} \mu(d) m^{n/d}.$$

Here μ denotes the Möbius function and (d, m) denotes the greatest common
divisor of the positive integers d and m.

Conjecture A, (M. W. Short [5]). *For any prime number p and any positive integer n, $M_n(p)^{G(p)}$ has dimension $f(p,n)$ over $\mathbb{F}_p$.*

It is easy to see that the conjecture is true in those cases where p does not divide n (see [5]) and Short gave further evidence obtained partly by computer. He has verified that the conjecture is true in the following cases: $p = 2$, $n \leqslant 20$; $p = 3$, $n \leqslant 9$; and $p = 5$, $n \leqslant 5$. The evidence is therefore quite strong in the case where $p = 2$. This was in fact the motivating case because the problem of determining $\dim(M_n(2)^{G(2)})$ was posed some time ago by L. G. Kovács (see Problem 11.47 of [4]).

For any finite group G let ν_G be the (norm) element $\sum_{h \in G} h$ of the group ring $\mathbb{Z}G$. Clearly, for any (right) $\mathbb{Z}G$-module U, $U\nu_G \subseteq U^G$, where U^G is the set of elements of U fixed by G. Also, for all $u \in U^G$, $u\nu_G = |G|u$, so $U^G/U\nu_G$ has exponent dividing $|G|$. In particular, when G is cyclic of order p, $U^G/U\nu_G$ may be regarded as a vector space over $\mathbb{F}_p$.

It will be shown in this note that Conjecture A is equivalent to the following conjecture.

Conjecture B. *For any prime number p, $L(p)^{G(p)} = L(p)\nu_{G(p)}$.*

To be more precise, it will be shown that Conjecture A holds for given values of p and n if and only if $L_n(p)^{G(p)} = L_n(p)\nu_{G(p)}$. Since $L(p)^{G(p)}/L(p)\nu_{G(p)}$ is isomorphic to $\hat{H}^0(G(p), L(p))$, the zero-dimensional Tate cohomology group of $G(p)$ with coefficients in $L(p)$ (see Chapter XII of [2]), Conjecture B asserts that $\hat{H}^0(G(p), L(p)) = 0$. More general versions of Conjecture B might also be considered. For example, it may be true that $L(m)^{G(m)} = L(m)\nu_{G(m)}$ for every positive integer m.

It will also be shown that Conjecture B is equivalent to the following conjecture.

Conjecture C. *For any prime number p, the $\mathbb{Z}G(p)$-module $L(p)$ has no non-zero module direct summand on which $G(p)$ acts trivially.*

In the context of Conjecture C the case $p = 2$ is particularly interesting. The integral representation theory of $G(2)$ shows that, for each n, $L_n(2)$ (like any $\mathbb{Z}G(2)$-module which is free of finite rank as $\mathbb{Z}$-module) is the direct sum of $\mathbb{Z}G(2)$-modules of $\mathbb{Z}$-rank one (with one or other of the two possible $G(2)$-actions) and regular $\mathbb{Z}G(2)$-modules (see Section 74 of [3]). Thus $L(2)$ has a basis Ω with the property that $\Omega \cup -\Omega$ is $G(2)$-invariant. But can one actually find a basis with this degree of symmetry?

The author and R. Stöhr have now shown that Conjecture A is true in the case $p = 2$ (*Fixed points of automorphisms of free Lie algebras*, Arch. Math. **67** (1996), pp. 281–289). Thus Conjectures B and C are also true for $p = 2$.

Problem D. Find a basis Ω of $L(2)$ such that $\Omega \cup -\Omega$ is $G(2)$-invariant.

An effective solution of this problem would yield a resolution of Conjecture C (and hence Conjecture A) for $p = 2$ because Conjecture C holds in this case if and only if no element of Ω is fixed by $G(2)$.

The connection between Conjectures A, B and C is accomplished by means of some results which will be stated now and proved in Section 2.

Proposition 1. *Let G be a cyclic group generated by an element g of prime order p and let U be a right $\mathbb{Z}G$-module which is free of finite rank as $\mathbb{Z}$-module. Let ξ be a complex primitive p-th root of unity and let $d_\xi(U)$ denote the dimension over $\mathbb{Q}[\xi]$ of the ξ-eigenspace of g acting on $\mathbb{Q}[\xi] \otimes_{\mathbb{Z}} U$ (with $\mathbb{Q}[\xi] \otimes_{\mathbb{Z}} U$ regarded as a $\mathbb{Q}[\xi]G$-module). Then*

$$\dim_{\mathbb{F}_p}(U/pU)^G = d_\xi(U) + \dim_{\mathbb{F}_p}(U^G/U\nu_G).$$

Furthermore, $\dim(U^G/U\nu_G)$ is equal to the largest number which is the $\mathbb{Z}$-rank of a module direct summand of U on which G acts trivially.

Proposition 2. *Let m be a positive integer, $m \geqslant 2$, and let ξ be a complex primitive m-th root of unity. Let $G(m)$ be the cyclic group of order m generated by an element g acting on $L(\mathbb{Q}[\xi], m)$ as previously described. Then, for every positive integer n, the ξ-eigenspace of g acting on $L_n(\mathbb{Q}[\xi], m)$ has dimension $f(m, n)$.*

Now let $G = G(p)$ and $U = L_n(p)$, so that $\mathbb{Q}[\xi] \otimes_{\mathbb{Z}} U \cong L_n(\mathbb{Q}[\xi], p)$. By Proposition 2, $d_\xi(U) = f(p, n)$. Also, in this case, $U/pU \cong M_n(p)$, so Proposition 1 gives the following result.

Corollary 3. *Let p be a prime number, n a positive integer, and $G = G(p)$. Then*

$$\dim(M_n(p)^G) = f(p, n) + \dim(L_n(p)^G/L_n(p)\nu_G).$$

Thus $\dim(M_n(p)^G) = f(p, n)$ if and only if $L_n(p)^G = L_n(p)\nu_G$. This establishes the equivalence of Conjectures A and B. The equivalence of Conjectures B and C follows from the last statement of Proposition 1.

I am very grateful to Dr L. G. Kovács for introducing me to Short's conjecture and for discussing it with me on many occasions during my visit to the Australian National University in 1993. I also gratefully acknowledge the financial support for the visit which I received from the ANU. My further thanks are due to Dr Kovács for his comments on a preliminary draft of this note which enabled me to simplify the presentation and proofs very considerably. Finally I thank Dr Short for telling me of his own work on the conjecture and for sending me a copy of [5].

2. Proofs

Proof of Proposition 1 We write

$$\lambda(U) = d_\xi(U) + \dim(U^G/U\nu_G) - \dim(U/pU)^G$$

and aim to prove that $\lambda(U) = 0$. We shall make use of the integral representation theory of G as described in Section 74 of [3].

Let $R = \mathbb{Z}[\xi]$ and $K = \mathbb{Q}[\xi]$. We regard K as a $\mathbb{Q}G$-module in which g acts on K by multiplication by ξ. Hence K is also a $\mathbb{Z}G$-module. By a *fractional ideal* of K we mean a non-zero finitely generated R-submodule of K. Each fractional ideal A of K is a $\mathbb{Z}G$-submodule of K which is free of rank $p - 1$ as $\mathbb{Z}$-module.

We regard the group ring $\mathbb{Z}G$ as the right regular $\mathbb{Z}G$-module and write T for a trivial $\mathbb{Z}G$-module of rank 1, that is, a free $\mathbb{Z}$-module of rank 1 with the trivial action of G. Note that R regarded as a $\mathbb{Z}G$-module can be identified with the augmentation ideal of $\mathbb{Z}G$.

It is easy to verify that $\lambda(R) = 0$, $\lambda(\mathbb{Z}G) = 0$ and $\lambda(T) = 0$.

Let A be a fractional ideal of K regarded as $\mathbb{Z}G$-module. Since A has $\mathbb{Z}$-rank $p - 1$, it is easily verified that $\mathbb{Q} \otimes_{\mathbb{Z}} A$ is isomorphic to K as $\mathbb{Q}G$-module. But K is isomorphic to the augmentation ideal of $\mathbb{Q}G$. Hence $d_\xi(A) = 1$. Clearly $A^G = \{0\}$, so $\dim(A^G/A\nu_G) = 0$. It is easy to verify that $A/(1-\xi)A$, $(1-\xi)A/(1-\xi)^2A, \ldots, (1-\xi)^{p-2}A/(1-\xi)^{p-1}A$ are isomorphic to each other as G-modules and

$$(A/(1-\xi)^{p-1}A)^G = (1-\xi)^{p-2}A/(1-\xi)^{p-1}A.$$

Now $(1-\xi)^{p-1} = pu$ where u is a unit of R (see (21.11) of [3]). Hence $(1-\xi)^{p-1}A = pA$. But A/pA has order p^{p-1}. Thus each factor $(1-\xi)^iA/(1-\xi)^{i+1}A$ has order p. Hence $\dim(A/pA)^G = 1$. (In fact, A/pA is isomorphic as $\mathbb{F}_pG$-module to the augmentation ideal of $\mathbb{F}_pG$.) Therefore $\lambda(A) = 0$.

Now let U be an arbitrary $\mathbb{Z}G$-module which is free of finite rank as $\mathbb{Z}$-module. Then, by (74.9) of [3], the $\mathbb{Z}G$-module $U \oplus R$ has a direct sum decomposition

$$U \oplus R = U_1 \oplus U_2 \oplus \cdots \oplus U_s$$

where, for each i, either $U_i \cong \mathbb{Z}G$, $U_i \cong R$, $U_i \cong T$ or $U_i \cong A$ for some fractional ideal A of K. (By considering $U \oplus R$ instead of U we have ensured that $r \neq n$ in the notation of [3].) Thus

$$\lambda(U) + \lambda(R) = \lambda(U \oplus R) = \lambda(U_1) + \cdots + \lambda(U_s) = 0.$$

Hence $\lambda(U) = 0$ as required.

It remains to prove the last statement of Proposition 1. It is easily verified that $\dim(R^G/R\nu_G) = 0$, $\dim(\mathbb{Z}G^G/\mathbb{Z}G\nu_G) = 0$, and $\dim(A^G/A\nu_G) = 0$ for each fractional ideal A. On the other hand $\dim(T^G/T\nu_G) = 1$. Thus, with

$$U \oplus R = U_1 \oplus U_2 \oplus \cdots \oplus U_s,$$

as before, $\dim(U^G/U\nu_G)$ is the number of values of i such that $U_i \cong T$. As shown in the proof of (74.3) of [3], the number of trivial modules in a direct sum decomposition into indecomposables of any integral $\mathbb{Z}G$-module is an invariant. The result follows.

Proof of Proposition 2 Let χ be the character of the representation of $G(m)$ on $L_n(\mathbb{Q}[\xi], m)$. By Brandt's character formula [1],

$$\chi(g^i) = \frac{1}{n}\sum_{d|n}\mu(d)(\mathrm{tr}(g^{id}))^{n/d}, \quad i = 0, 1, \ldots, m-1,$$

where $\mathrm{tr}(g^{id})$ denotes the trace of g^{id} in the representation of $G(m)$ on the space $L_1(\mathbb{Q}[\xi], m)$ spanned by $x_1, \ldots, x_m$. Note that $\mathrm{tr}(g^{id}) = m$ if $m|id$ and $\mathrm{tr}(g^{id}) = 0$ otherwise.

By the inner product formula for characters, the ξ-eigenspace of g acting on $L_n(\mathbb{Q}[\xi], m)$ has dimension

$$\frac{1}{mn}\sum_{i=0}^{m-1}\sum_{d|n}\mu(d)(\mathrm{tr}(g^{id}))^{n/d}\xi^i.$$

Suppose d is a divisor of n satisfying $(d, m) = k > 1$ and write $m = km'$. Then

$$\sum_{i=0}^{m-1}(\mathrm{tr}(g^{id}))^{n/d}\xi^i = \sum_{j=0}^{k-1}m^{n/d}\xi^{jm'} = 0.$$

If d is a divisor of n satisfying $(d, m) = 1$, then $\mathrm{tr}(g^{id}) = 0$ for $i = 1, \ldots, m-1$. Thus

$$\sum_{i=0}^{m-1}(\mathrm{tr}(g^{id}))^{n/d}\xi^i = m^{n/d}.$$

Therefore the required dimension is

$$\frac{1}{mn}\sum_{\substack{d|n \\ (d,m)=1}}\mu(d)m^{n/d},$$

which is $f(m, n)$.

References

[1] A. Brandt, *The free Lie ring and Lie representations of the full linear group*, Trans. Amer. Math. Soc. **56** (1944), pp. 528–536.

[2] H. Cartan and S. Eilenberg, *Homological Algebra*, Princeton University Press, Princeton, (1956).

[3] C. W. Curtis and I. Reiner, *Representation Theory of Finite Groups and Associative Algebras*, Wiley (Interscience), New York, London, Sydney, (1962).

[4] *Unsolved Problems in Group Theory, The Kourovka Notebook, 12th revised and augmented edition*, (V. D. Mazurov and E. I. Khukhro, eds.), Institute of Mathematics, Siberian Branch of the Russian Academy of Sciences, Novosibirsk, (1992).

[5] M. W. Short, *A conjecture about free Lie algebras*, Communications in Algebra **23** (1995), pp. 3051–3057.

Cohomology, Representations and Quotient Categories of Modules

Jon F. Carlson

Department of Mathematics, University of Georgia, Athens, Georgia 30602.
email:jfc@sloth.math.uga.edu

0. Introduction

The purpose of this article is to report on some recent developments in the area of quotient categories of modules filtered by complexity. The quotient category construction is certainly not new but its application to the representation theory of finite groups was only recently begun in joint work with Peter Donovan and Wayne Wheeler. In [CDW] we considered the filtrations on the category of modules given by the complexity invariant. It was shown that the sets of morphisms in the quotients are localizations of cohomology. As a direct consequence, the endomorphism ring of the trivial module decomposes as a direct product of rings with the factors corresponding to the components of the maximal ideal spectrum of the cohomology ring of the group. Specifically it was shown that the endomorphism ring has a set of orthogonal central indempotents which correspond to the components of the variety.

At the same time, the trivial module is indecomposable in the quotient category. All of this indicates that the quotient categories have no Krull-Schmidt theorem, no uniqueness of decompositions of modules. This has been verified in general for the trivial module and shown to hold for other modules as well [C,CW].

The failure of Krull-Schmidt in quotients set the stage for several unusual developments in a different direction. It seems that the uniqueness of decompositions can be recovered if we move to the category of infinitely generated modules. Simply stated, the problems lies in the fact that the quotients are triangulated but not abelian. The usual method of splitting indempotents

Partly supported by a grant from NSF

in an abelian category involves looking at the kernel and cokernels of the idempotents. However these constructions are not available in a triangulated category, and the substitute construction of homotopy limits requires the taking of infinite direct sums. On the other hand many of the main techniques from the module theory of modular group algebras are based on finiteness conditions such as finite generation of modules and finite generation of cohomology. Indeed, the definition of the complexity of a module, which defines the filtration on the module category, depends on the modules having finite dimension over the base field.

A major focus of this survey is the recent results aimed at the development of a theory of varieties and complexity for infinitely generated modules. We will stick to the case in which G is a finite group and k is an algebraically closed field of characteristic $p > 0$. The requirements on the coefficient ring k are not entirely necessary, but occasional adjustments must be made in hypotheses of theorems if the algebraic closure condition is relaxed. Many of the older results on complexity and varieties of modules have been extended to certain classes of infinite groups such as compact Lie Groups or groups of finite virtual cohomological dimension. The same may be possible here. In fact, our hope is that a module theory for infinitely generated modules may aid in the development of a theory of modules for infinite groups, based on their finite subgroups.

The paper is organized roughly as follows. Section 1 is mainly a survey of background material on cohomology rings, varieties and complexity. We consider only finitely generated modules except that we give some examples to illustrate a few of the differences which should be expected when we pass to the category of infinitely generated modules. Section 2 begins with a review of the basic quotient category construction. An outline of the results of [CDW] is presented with an emphasis on the structure of endomorphism rings and the failure of the Krull-Schmidt theorem in the complexity quotient categories. In Section 3 we introduce homotopy limits. It and other limit techniques play a vital role in the analysis to come. We offer several equivalent definition of the extended notion of complexity. Each has its own peculiar advantages, and the proof of their equivalence is straight forward. More complicated is the definition of the variety of an infinitely generated module. The initial attempt at such a definition in [BCR1] has been refined and improved in later work.

In Section 4 we outline some of the latest work on indempotent modules and decompositions of the categories [Ric], [B2], [BCR2]. The main idea is that the category of all kG-modules has idempotent modules, that is, nonprojective modules M such that $M \otimes M \cong M \oplus P$ where P is projective. For those of us accustomed to working only with finitely generated modules the existence of indempotent modules was a surprise. A finitely generated module M is idempotent if and only if $M \cong k \oplus P$ with P projective.

1. Finiteness Conditions

Suppose that G is a finite group and that k is an algebraically closed field of characteristic $p > 0$. In this section we consider the category mod-kG of all finitely generated kG-modules. The cohomology ring $H^*(G, k)$ plays a large role in the general study of kG-modules. One of its most important features is a theorem of Evens which asserts that the cohomology ring is finitely generated as a k-algebra. In particular, it is a noetherian ring. Moreover, if M and N are in mod-kG then $\mathrm{Ext}^*_{kG}(M, N)$ is a finitely generated module over $H^*(G, k)$. To get a feeling for the structure we consider a few examples. First suppose that $G = \langle x \rangle$ is a cyclic group of order p. Then the trivial kG-module has a minimal projective resolution of the form

$$\cdots \longrightarrow P_3 \xrightarrow{\partial_3} P_2 \xrightarrow{\partial_2} P_1 \xrightarrow{\partial_1} P_0 \xrightarrow{\epsilon} k \longrightarrow 0$$

where $P_i = kG$ for all i. Here

$$\partial_i(\alpha) = \begin{cases} (x - 1)\alpha & \text{if } i \text{ is odd,} \\ (x - 1)^{p-1}\alpha & \text{if } i \text{ is even.} \end{cases}$$

Now $\mathrm{Hom}_{kG}(P_i, k) \cong k$ and hence $H^i(G, k) = H^i(\mathrm{Hom}_{kG}(P_*, k))$ has dimension 1 in every degree. Notice further that the resolution repeats every two places and there is a surjective chain map of degree 2 (or degree 1 if $p = 2$). The chain map represents an element ζ in $H^2(G, k)$ (in $H^1(G, k)$ if $p = 2$). So we have that

$$H^*(G, k) = \begin{cases} k[\zeta, \eta]/(\eta^2) & \text{if } p > 2, \\ k[\zeta] & \text{if } p = 2. \end{cases}$$

It is a technicality to check that the element $\eta \in H^1(G, k)$ for $p > 2$ squares to zero.

Next suppose that we have an elementary abelian p-group $G = \langle x_1, \ldots, x_n \rangle \cong (\mathbb{Z}/p)^n$. It can be seen that $kG \cong k\langle x_1 \rangle \otimes k\langle x_2 \rangle \otimes \ldots \otimes k\langle x_n \rangle$. As a result, the trivial kG-module, k, has a projective resolution of the form

$$P_* = P_{1,*} \otimes P_{2,*} \otimes \cdots \otimes P_{n,*} \xrightarrow{\epsilon_1 \otimes \cdots \otimes \epsilon_n} k$$

Here $P_{i,*} \xrightarrow{\epsilon_i} k$ is a $k\langle x_i \rangle$-projective resolution as in the previous paragraph. The tensor product is over the coefficient ring k and the formula for the tensor product of the complexes is the usual Künneth formula. Thus

$$\mathrm{Hom}_{kG}(P_i, k) = \sum_{i_1 + \cdots + i_n = i} \mathrm{Hom}_{kG}(k\langle x_i \rangle, k)$$

$$= \sum_{i_1 + \cdots + i_n = i} k. \qquad\qquad 1$$

and

$$H^*(G, k) = H^*((x_1), k) \otimes \cdots \otimes H^*((x_n), k).$$

The cup products satisfy the anticommutativity rule $\theta\sigma = (-1)^{\deg(\theta)\deg(\sigma)}\sigma\theta$.
So we have that

$$H^*(G, k) = \begin{cases} k[\zeta_1, \ldots, \zeta_n] \otimes \Lambda(\eta_1, \ldots, \eta_n) & \text{if } p \text{ ¿ } 2 ,\\ k[\zeta_1, \ldots, \zeta_n] & \text{if } p = 2. \end{cases}$$

Here Λ is the exterior algebra generated by the degree 1 elements $\eta_1, \ldots, \eta_n$.
Notice that the ideal generated by $\eta_1, \ldots, \eta_n$ is a nilpotent ideal and hence is
in the Jacobson radical of $H^*(G, k)$ and is contained in every maximal ideal.
Therefore we have that

$$H^*(G, k)/\mathrm{Rad}H^*(G, k) \cong k[\zeta_1, \ldots, \zeta_n], \qquad\qquad 1.1$$

a polynomial ring. If $p = 2$ the generators are in degree 1, while if $p > 2$,
they are in degree 2.

If G is not elementary abelian then the structure of the cohomology ring is
more complex. For example suppose that $G = \langle x, y \mid x^2 = y^2 = (xy)^4 = 1 \rangle$ is
a dihedral group of order 8, $p = 2$. Then the cohomology ring has the form

$$H^*(G, k) = k[\eta_1, \eta_2, \zeta]/(\eta_1\eta_2), \qquad\qquad 1.2$$

with η_1, η_2 in degree one and ζ in degree 2. This is not a polynomial ring.
It has two minimal prime ideals generated by η_1 and η_2. The quotient
$H^*(G, k)/(\eta_1) = k[\eta_2, \zeta]$ is a polynomial ring, though the generators are not
both in the same degree. The group G has two maximal elementary abelian
2-subgroups: $H_1 = \langle x, (xy)^2 \rangle$ and $H_2 = \langle y, (xy)^2 \rangle$. In fact, with proper choice
of generators, the prime ideals (η_1) and (η_2) can be assumed to be the kernels
of the restriction homomorphisms to the subgroups H_1 and H_2, respectively.
The example illustrates the general situation. It is always the case that the
minimal prime ideals of $H^*(G, k)$ are the radicals of the kernels of the restrictions to the maximal elementary abelian p-subgroups of G. All of this was
contained in the fundamental work of Quillen.

(1.3) Quillen's Theorem [Q]. *(see also [QV]). Let $\mathcal{A}_G$ be the collection of
all elementary abelian p-subgroups of G. Then the kernel of the map*

$$\prod_{E \in \mathcal{A}_G} \mathrm{res}_{G,E} : H^*(G, k) \longrightarrow \prod_{E \in \mathcal{A}_G} H^*(E, k)$$

is a nilpotent ideal. So $\mathrm{kernel}(\prod_E \mathrm{res}_{G,E})$ is in the Jacobson radical of $H^*(G, k)$
and, most importantly, it is in every maximal ideal of $H^*(G, k)$.

Let $V_G(k)$ denote the maximal ideal spectrum of $H^*(G, k)$. Because $H^*(G, k)$
is finitely generated as a k-algebra and almost commutative (if $p > 2$ then

the elements of odd degree, which anticommute, square to zero and hence are in every prime ideal of $H^*(G,k)$), $V_G(k)$ is an affine variety. Also it is a homogeneous variety since $H^*(G,k)$ is a graded ring. Indeed, it is often more convenient to consider the projective variety $\overline{V}_G(k)$ of lines in $V_G(k)$, since almost all of the ideal which we consider are graded ideals.

Notice that if $G \cong (\mathbb{Z}/p)^n$ is an elementary abelian group, then $V_G(k) = k^n$. This is because k is algebraically closed and $H^*(G,k)/\mathrm{Rad}H^*(G,k)$ is a polynomial ring. Thus every maximal ideal $\mathbf{m} \in V_G(k)$ is the kernel of a homorphism $H^*(G,k) \to k$ given by a point evaluation, $f \to f(\alpha)$, for f in the polynomial ring and $\alpha \in k^n$ a point corresponding to $\mathbf{m}$. Then the result of Quillen may be interpreted as saying that

$$V_G(k) = \cup \; \mathrm{res}_{G,E}^*(V_E(k)) \qquad\qquad 1.4$$

is a union of "folded" copies of k^t where $V_E(k) = k^t$ for t the p-rank of E. From Evens' Theorem on finite generation it can be deduced that $\mathrm{res}_{G,E}^*$: $V_E(k) \to V_G(k)$ is always finite-to-one, hence the "folded copy of k^t". But more importantly we get that

$$\dim V_G(k) = r = p\text{-rank}(G). \qquad\qquad 1.5$$

The p-rank of G is the largest integer r such that G has an elementary abelian subgroup $E \cong (Z/p)^r$.

All of the above has an extension to finitely generated modules. Suppose that M is a finitely generated kG-module. Then $H^*(G,k)$ acts on

$$\mathrm{Ext}_{kG}^*(M,M) \cong H^*(G, \mathrm{Hom}_k(M,M)).$$

So let $J(M)$ be the annihilator in $H^*(G,k)$ of $\mathrm{Ext}_{kG}^*(M,M)$. Notice that for any kG-module N, $\mathrm{Ext}_{kG}^*(M,N)$ is a right $\mathrm{Ext}_{kG}^*(M,M)$-module and $\mathrm{Ext}_{kG}^*(N,M)$ is a left $\mathrm{Ext}_{kG}^*(M,M)$-module. An element in $J(M)$ annihilates the identity element of $\mathrm{Ext}_{kG}^*(M,M)$ and hence by associativity it annihilates $\mathrm{Ext}_{kG}^*(M,N)$ and $\mathrm{Ext}_{kG}^*(N,M)$. That is, $J(M)$ is the annihilator of all cohomology of M in either variable. So we let

$$V_G(M) = V_G(J(M)) = \mathrm{MaxSpec}(H^*(G,k)/J(M)) \subset V_G(k) \qquad 1.6$$

be the closed set of all maximal ideals in $H^*(G,k)$ which contain $J(M)$. $V_G(M)$ is an important invariant of the module M. When we write $V_G(k)$ we mean the variety of the trivial module k, which is the same as the maximal ideal spectrum of $H^*(G,k) \cong \mathrm{Ext}_{kG}^*(k,k)$. A few of the properties of the variety, $V_G(-)$, are listed below.

(1.7.1) $V_G(M) = \{0\}$ if and only if M is projective.

(1.7.2) $V_G(M)$ has dimension 1 if and only if M is periodic. By periodic we mean that M has a projective resolution

$$\cdots \longrightarrow P_2 \xrightarrow{\partial_2} P_1 \xrightarrow{\partial_1} P_0 \longrightarrow M \longrightarrow 0$$

which repeats itself, i.e., for some n; $P_i \cong P_{i+n}$ and $\partial_i = \partial_{i+n}$ for all $i \geq 0$. $V_G(M)$ having dimension one means that $V_G(M)$ is a union of a finite number of lines.

(1.7.3) In general, $\dim V_G(M)$ is the complexity of M, as defined by Alperin (see Section 5.1 of [B1]). The complexity of M is the polynomial rate of growth of the terms of a minimal projective resolution of M. That is, if

$$\cdots \longrightarrow P_2 \xrightarrow{\partial_2} P_1 \xrightarrow{\partial_1} P_0 \longrightarrow M \longrightarrow 0$$

is a minimal projective resolution of M, then M has complexity $t \geq 0$ if and only if t is the least nonnegative integer such that there exists a polynomial f of degree $t - 1$ with $\mathrm{Dim}(P_n) < f(n)$ for n sufficiently large. Notice that the zero polynomial, $f(n) = 0$, has degree -1 and hence the complexity of a projective module is zero.

(1.7.4) $V_G(M) = V_G(\Omega^n(M)) = V_G(M^*)$. Here M^* is the k-dual module $M^* = \mathrm{Hom}_k(M, k)$ with G-action given by $(g\lambda)(m) = \lambda(g^{-1}m)$ for $\lambda \in M^*$, $m \in M$. The modules of $\Omega^n(M)$ are defined by the minimal projective resolution of M in (1.7) in that $\Omega^n(M) = \partial^n(P_n) = \mathrm{Kernel}(\partial_{n-1})$ for $n > 0$. For $n < 0$, we can give a minimal injective resolution

$$0 \longrightarrow M \longrightarrow Q_0 \xrightarrow{\partial_{-1}} Q_{-1} \xrightarrow{\partial_{-2}} Q_{-2} \longrightarrow \cdots$$

and the same definition applies (i.e. $\Omega^{-n}(M) = \partial_{-n}(Q_n)$). It should be noted that kG is a self-injective ring so that injective modules are projective and vice versa. Consequently, $\Omega^j(\Omega^i(M)) \cong \Omega^{i+j}(M)$ modulo projectives for all i, j. Here we must define $\Omega^0(M) = \Omega^{-1}(\Omega(M))$ to be the nonprojective part of M so that $M \cong \Omega^0(M) \oplus P$ where P is a projective module, possibly zero.

(1.7.5) If $0 \longrightarrow M_1 \longrightarrow M_2 \longrightarrow M_3 \longrightarrow 0$ is an exact sequence of finitely generated kG-modules then $V_G(M_i) \subset V_G(M_j) \cup V_G(M_k)$ whenever $\{i, j, k\} = \{1, 2, 3\}$. For any modules M_1 and M_2 we have that $V_G(M_1 \oplus M_2) = V_G(M_1) \cup V_G(M_2)$.

(1.7.6) $V_G(M \otimes N) = V_G(M) \cap V_G(N)$ for any finitely generated kG-modules M and N. Here $\otimes$ means $\otimes_k$ and the G-action on $m \otimes n$ is the diagonal one, $g(m \otimes n) = gm \otimes gn$.

(1.7.7) $V_G(M) = \cup_{E \in A(G)}(\mathrm{res}^*_{G,E}(V_E(M)))$. This is the natural analog of Quillen's Theorem, that was first proved by Alperin and Evens and, independently, by Avrunin.

Many of the properties of the variety $V_G(-)$ can be verified easily once the equivalence with the rank variety is established. The rank variety of a finitely generated module M is a closed set of k^n which can be computed directly from the structure of M in the case that $G \cong (\mathbb{Z}/p)^n$ is an elementary abelian p-group. Its definition is as follows.

Suppose that $G = \langle x_1, \ldots, x_n \rangle \cong (\mathbb{Z}/p)^n$ is an elementary abelian group of order p^n. For $\alpha \in k^n$, $\alpha = (\alpha_1, \ldots, \alpha_n)$, let $u_\alpha = 1 + \sum_{i=1}^n \alpha_i(x_1 - 1)$. Notice that, because k has characteristic p, $u_\alpha^p = 1$, and u_α is a unit of order p in kG. The important thing is that kG as a $k\langle u_\alpha \rangle$-module is projective. That is, $\langle u_\alpha \rangle$ as a group of units in kG looks like any cyclic subgroup $\langle x \rangle$ where $x \in G$. Indeed, the group algebra kG does not remember what the group elements are. If $u_1, \ldots, u_n$ are elements of the form u_α for different $\alpha(1), \ldots, \alpha(n) \in k^n$ with $\alpha(1), \ldots, \alpha(n)$ linearly independent in k^n, then the inclusion $u_i \mapsto kG$ induces an isomorphism $k\langle u_1, \ldots, u_n \rangle \longrightarrow kG$.

Now suppose that M is a finitely generated kG-module. It makes sense to think about the restriction, $M \downarrow_{\langle u_\alpha \rangle}$, of M to a $k\langle u_\alpha \rangle$-module. So we can define the rank variety of M to be

$$V_G^r(M) = \{\alpha \in k \mid M \downarrow_{\langle u_\alpha \rangle} \text{ is not free }\} \cup \{0\}. \qquad 1.8$$

The connection with the cohomological variety is expressed by the fact that when $G = \langle x_1, \ldots, x_n \rangle$ is an elementary abelian p-group and M is a finitely generated kG-module then $V_G(M) \cong V_G^r(M)$. This is not really an isomorphism but rather an isogency. The reason is that for p-odd, the map induced by restriction

$$H^*(G, k)/\mathrm{Rad}H^*(G, k) \longrightarrow H^*(\langle u_\alpha \rangle, k)/\mathrm{Rad}H^*(\langle u_\alpha \rangle, k)$$

takes a polynomial $f = f(\zeta_1, \ldots, \zeta_n)$ to the element $f(\alpha_1^p, \ldots, \alpha_n^p)\zeta^d$ (as in (1.1)) where $\alpha = (\alpha_1, \ldots, \alpha_n) \in k^n$, $\zeta \in H^2(\langle u_\alpha \rangle, k)$ is a carefully chosen generator, and d is the degree of the polynomial f. Hence the map on varieties also involves a twist by the Frobenius automorphism when $p > 2$.

We end this section with a discussion of a couple of examples of infinitely generated modules. The examples illustrate some of the constructions we have just discussed and also give a hint of the difficulties encountered in moving to the category of all kG-modules. In both examples we let $G = \langle x, y \rangle \cong (\mathbb{Z}/2)^2$ be a fours group, and k an algebraically closed field of characteristic 2. In the examples we represent the modules by diagrams in which the vertices represent elements of a k-basis for the module while the arrows represent multiplications by the elements of the group algebra as in

So the modules have the following diagrams.

Example A

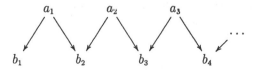

Example B

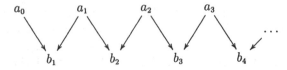

That is, example A has a k-basis $\{a_i, b_i \mid i > 0\}$ and the action of the group is given by

$$(x-1)a_i = b_i, \ (y-1)a_i = b_{i+1}, (x-1)b_i = (y-1)b_i = 0.$$

Example B has one extra basis element, a_0, with the same relations except that $(x-1)a_0 = 0$. Now if $U = \langle u \rangle$ is a group of order 2 then the free module kU has a basis $\{1, u-1\}$ such that (obviously) $(u-1) \cdot 1 = u-1$ and $(u-1)^2 = 0$. In general a kU-module N is free if and only if we have that $w \in (u-1)N$ for any $w \in N$ with $(u-1)w = 0$. With this in mind consider Example B. We claim that if $\alpha = (\alpha_1, \alpha_2) \in k^2$ and if $\alpha_2 \neq 0$, then $B \downarrow_{\langle u_\alpha \rangle}$ is a free module. The reason is that

$$b_1 = (u_\alpha - 1)\left(\frac{1}{\alpha_2}a_0\right), b_2 = (u_\alpha - 1)\left(\frac{1}{\alpha_2}a_1 + \frac{\alpha_1}{\alpha_2^2}a_0\right),$$

$$b_3 = (u_\alpha - 1)\left(\frac{1}{\alpha_2}a_2 + \frac{\alpha_1}{\alpha_2^2}a_1 + \frac{\alpha_1^2}{\alpha_2^2}a_2\right), \text{etc.}$$

and the set of U-fixed points of B is exactly the subspace spanned by the set $\{b_1, b_2, b_3, \ldots\}$. On the other hand if $\alpha_2 = 0$ then $(u_\alpha - 1)a_0 = 0$ and B is not a free $k\langle u_\alpha \rangle$-module. Hence if we compute a "rank variety" by the method of (1.8) we get that $V_G^r(B) = \{(0, \alpha_2) \mid \alpha_2 \in k\}$ which is a closed set, a line, in k^2.

Now suppose we consider the module A restricted to $\langle u_\alpha \rangle$, $\alpha = (\alpha_1, \alpha_2) \in k^2$. Notice that if $\alpha_2 = 0$, $\alpha_1 \neq 0$, then $A \downarrow_{\langle u_\alpha \rangle}$ is free since $b_i = (u_\alpha - 1)\left(\frac{1}{\alpha_1}a_i\right)$. However if $\alpha_2 \neq 0$ then it is impossible to write b_1 as a linear combination of a finite number of the elements $(u_\alpha - 1)a_i$. So $A \downarrow_{\langle u_\alpha \rangle}$ is not free if $\alpha_2 \neq 0$. Thus the "rank variety",

$$V_G^r(A) = k^2 \setminus \{(\alpha_1, 0) \mid \alpha_1 \in k\}$$

is an open set in the Zariski topology of k^2.

For both of these modules, $\Omega(M) = M$. That is, both are periodic of period 1. For example, for the module A, we have a projective cover $\phi : P \longrightarrow A$ with P being a free kG-module with kG-basis $\{c_i\}_{i>0}$ and $\phi(c_i) = a_i$. Then the kernel of ϕ has basis

$$a_i' = (y-1)c_i - (x-1)c_{i+1}, \quad b_i' = (x-1)(y-1)c_i, \quad i > 0.$$

It is an easy exercise to check that $\{a_i'\}$, $\{b_i'\}$ satisfy the same relations as the chosen basis for A.

In Section 3, we will see that the module A has complexity 2 while module B has complexity 1. This is still the dimension of the "rank variety", but it seems to be no longer tied to the "rate of growth" of the projective resolutions of the modules. The other thing which should be pointed out here is that both A and B are idempotent modules in the sense that, for A,

$$A \otimes_k A \cong A \oplus P$$

for some projective module P. We will show later that in the category of finitely generated modules there are no idempotents other than the trivial module.

2. Quotient Categories

In this section we introduce the quotient category construction on categories of kG-modules. We confine ouselves to the case of finitely generated modules, but our aim is to set the stage for the extension to infinitely generated modules. We begin with the ordinary module category mod-kG of all finite generated kG-modules and kG-module homomorphisms. We briefly survey the definition of a triangulated category and the construction of quotient categories. Details can be found in the books by Happel [H] or Weibel [W]. The applications of the quotient categories to the study of modular group representations can be found mostly in [CDW].

The stable category stmod-kG of finitely generated kG-modules modulo projectives has exactly the same objects as the category mod-kG. The morphisms in stmod-kG are given by the formula

$$\mathrm{Hom}_{\text{stmod-}kG}(M, N) = \mathrm{Hom}_{kG}(M, N)/\mathrm{PHom}_{kG}(M, N)$$

where $\mathrm{PHom}_{kG}(M, N)$ is the set of all homomorphisms $\alpha : M \to N$ such that α factors through a projective kG-module, i.e., $\alpha = \gamma \circ \beta$ where $\beta : M \to P, \gamma : P \to N$ for some kG-homomorphisms β, γ, and some projective module P. It might be important to recall that kG is a self-injective ring so that a kG-module is injective if and only if it is projective. Hence factoring through a projective is the same as factoring through an injective. There are two properties of morphisms in the stable categories which we would like to emphasize. The first is the fact that cohomology really takes place in the stable category.

(2.1) For any kG-modules M, N and any $n > 0$, we have that

$$\mathrm{Hom}_{\text{stmod-}kG}(\Omega^n(M), N) = \mathrm{Ext}^n_{kG}(M, N).$$

That is, suppose that (P_*, ϵ) is a minimal projective resolution of M and that $\zeta : \Omega^n(M) \to N$ is a homomorphism. Then we have a commutative diagram

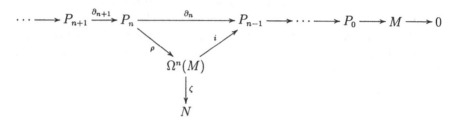

with ρ the surjection with kernel $\partial_{n+1}(P_{n+1})$ and i the inclusion. The fact that $(\zeta\rho)\partial_{n+1} = 0$ says that $\zeta\rho$ is a cocyle. It is an exercise to check that ζ factors through a projective if and only if $\zeta\rho$ is a coboundary, i.e., if and only if there exist $\phi : P_{n-1} \to N$ such that $\zeta = \phi i$. So the class of ζ in $\mathrm{Hom}_{\mathrm{stmod}\text{-}kG}(M, N)$ determines a unique element of $\mathrm{Ext}^n_{kG}(M, N)$.

(2.2) $\mathrm{Hom}_{\mathrm{stmod}\text{-}kG}(\Omega^n(M), \Omega^n(N)) \cong \mathrm{Hom}_{\mathrm{stmod}\text{-}kG}(M, N)$ for any $n > 0$.

That is, any homomorphism θ from M to N lifts to a chain map from a projective resolution of M to a projective resolution of N, thus inducing a homomorphism $\Omega^n(\theta)$ from $\Omega^n(M)$ to $\Omega^n(N)$. The class of $\Omega^n(\theta)$, modulo projectives, is uniquely determined by the class of θ. It is also true that the isomorphisms in (2.1) and (2.2) can be used to define the cup product. That is, if $\gamma \in \mathrm{Ext}^m(L, M)$, $\zeta \in \mathrm{Ext}^n(M, N)$ then the cup product $\zeta\gamma$ is the class of the composition $\zeta \circ \Omega^n(\gamma)$. Furthermore if we are willing to consider Tate cohomology and injective as well as projective resolutions, then (2.1) and (2.2) hold for all n.

For our purposes, one of the most important things about the stable category is that it is triangulated. Every morphism is a part of a triangle of modules and morphisms. We can illustrate the principle with the following exercise. Suppose that $\alpha : M \to N$ is any kG-homomorphism. Then we can find projective modules P and Q and homomorphisms $\sigma : P \longrightarrow N$, $\tau : M \longrightarrow Q$ such that $\alpha' = (\alpha, \sigma) : M \oplus P \longrightarrow N$ is onto and $\alpha'' = \binom{\alpha}{\tau} : M \longrightarrow N \oplus Q$ is one-to-one. For example, P might be a projective cover of N or Q an injective hull for M. So we have exact sequences

$$0 \longrightarrow U \xrightarrow{\gamma'} M \oplus P \xrightarrow{\alpha'} N \longrightarrow 0$$

and

$$0 \longrightarrow M \xrightarrow{\alpha''} N \oplus Q \xrightarrow{\beta'} V \longrightarrow 0$$

where U and V are the appropriate kernel and cokernel. It can be shown that $U \cong \Omega(V) \oplus (\mathrm{proj})$, that is, U is $\Omega(V)$ plus a projective direct summand. For convenience of language we actually think of this as $V \cong \Omega^{-1}(U) \oplus (\text{proj})$. So then we get a triangle in the stable category

$$U \xrightarrow{\gamma} M \xrightarrow{\alpha} N \xrightarrow{\beta} \Omega^{-1}(U)$$

with vertices U, M and N, and edges γ, α, β. Everytime we go around the triangle we must apply the translation functor Ω^{-1}. The principle is that exact sequences in the module category mod-kG correspond to triangles in the stable category exactly as in the illustration.

The definition of a triangulated category is given by several axioms (See [H] or [W]). Most of these express well known properties of modules and exact sequences and hence there is not much point in going into the details here. Rather we consider next the quotient category construction which is made possible by the triangulation.

Suppose that $\mathcal{M}$ is a triangulated subcategory of stmod-kG, meaning that if two of the objects in a triangle in stmod-kG are in $\mathcal{M}$ then so is the third. The quotient category $\mathcal{Q} = \text{stmod-}kG/\mathcal{M}$ has the same objects as stmod-kG, but the sets of morphism in $\mathcal{Q}$ are obtained by inverting any morphism in stmod-kG if the third object in the triangle of the morphisms is in the subcategory $\mathcal{M}$. Thus a morphism from M to N in $\mathcal{Q}$ might be represented by a diagram

$$M \xleftarrow{s} U \xrightarrow{f} N \qquad\qquad 2.3$$

where, s is invertible as above. We think of the morphism as being $f \circ s^{-1}$. It is actually an equivalence class since it is necessary to identity $f \circ s^{-1}$ with $(ft)(st)^{-1} = f \circ t \circ t^{-1} \circ s^{-1}$ whenever $t : W \longrightarrow U$ is a morphism in stmod-kG such that the third object in the triangle of t is in $\mathcal{M}$. The composition of two morphisms

$$L \xleftarrow{s} U \xrightarrow{f} M \quad \text{and} \quad M \xleftarrow{t} V \xrightarrow{g} N$$

is the splice $g \circ t^{-1} \circ f \circ s^{-1}$. In the cases we consider, it is not difficult to show using pullbacks that the product can be put in the form of (2.3) as above.

For any nonnegative integer c, let $\mathcal{M}_c$ be the full subcategory of stmod-kG of all finitely generated kG-modules with complexity at most c. By (1.5), $\mathcal{M}_r = \text{stmod-}kG$ if r is the p-rank of G. By (1.7.5) every $\mathcal{M}_c$ is a triangulated subcategory of stmod-kG. As a result we have that the quotients $\mathcal{M}_c/\mathcal{M}_d$ are defined.

Suppose that $\zeta \in H^n(G, k)$ is a nonzero cohomology element and that the dimension of $V_G(\zeta)$ is less than r. The element ζ is represented by a cocycle $\zeta' : \Omega^n(k) \to k$ (see (2.1)) which is obviously a surjection. Thus we have an exact sequence

$$0 \longrightarrow L_\zeta \longrightarrow \Omega^n(k) \xrightarrow{\zeta'} k \longrightarrow 0. \qquad\qquad 2.4$$

The important point is that:

(2.5) $V_G(L_\zeta) = V_G(\zeta)$, the set of all maximal ideals in $V_G(k)$ which contain ζ (see [B1], or [E]).

Hence the assumptions on ζ guarantee that L_ζ has complexity less than r and hence the morphism ζ is invertible in the quotient category $\mathcal{M}_r/\mathcal{M}_{r-1}$. This leads us to the following observation

(2.6) Every object of $\mathcal{M}_c/\mathcal{M}_{c-1}$ is periodic.

In the triangulated category $\mathcal{M}_c/\mathcal{M}_{c-1}$ the periodicity is measured by the translation functor Ω^{-1}, or by its inverse Ω. So suppose that M is a kG-module of complexity c. Then $V_G(M) \subset V_G(k)$ has dimension c. It is not hard to see that for some $n > 0$ there exists an element $\zeta \in H^n(G,k)$ such that $V_G(\zeta) \cap V_G(M)$ has dimension $c-1$. So we have that

$$V_G(L_\zeta \otimes M) = V_G(L_\zeta) \cap V_G(M) = V_G(\zeta) \cap V_G(M)$$

has dimension $c-1$ (by (1.7.6) and (2.5)) and hence $L_\zeta \otimes M \in \mathcal{M}_{c-1}$. So suppose we tensor the sequence in (2.4) with the module M. We get an exact sequence

$$0 \longrightarrow L_\zeta \otimes M \longrightarrow \Omega^n(k) \otimes M \xrightarrow{\zeta' \otimes Id_M} k \otimes M \longrightarrow 0.$$

Of course $k \otimes M \cong M$, but also $\Omega^n(k) \otimes M \cong \Omega^n(M) \oplus (\text{proj})$. This last is a consequence of the fact that tensoring with a projective module always yields a projective module. Hence we have that, in the quotient category, the class of $\zeta \otimes Id_M : \Omega^n(M) \to M$ is invertible and is an isomorphism. This proves (2.6).

The situation we encounter in the proof of (2.6) really does represent the general situation. For suppose that U and M are in $\mathcal{M}_c$ and that $s : U \to M$ is invertible modulo $\mathcal{M}_{c-1}$. Let $W \in \mathcal{M}_{c-1}$ be the third object in the triangle of the morphism s. It is always possible to find an element $\zeta \in H^n(G,k)$, some n, such that $V_G(\zeta) \cap V_G(M)$ has dimension $c-1$ and ζ annihilates the cohomology of W (implying that $V_G(W) \subset V_G(\zeta)$). Then it can be shown that

(2.7) There exists $t : \Omega^n(M) \to U$ such that $st : \Omega^n(M) \to M$ is equivalent to $\zeta \otimes Id_M$ in stmod-kG and is invertible modulo $\mathcal{M}_{c-1}$.

Thus we have that any morphism

$$M \xleftarrow{s} U \xrightarrow{f} N$$

is equal in $\mathcal{M}_c/\mathcal{M}_{c-1}$ to one of the form

$$M \xleftarrow{st} \Omega^n(M) \xrightarrow{ft} N$$

So by (2.1) we have that

(2.8) In $\mathcal{Q}_c$, $\mathrm{Hom}_{\mathcal{Q}_c}(M, N) = [\mathrm{Ext}^*_{kG}(M, N) \cdot S^{-1}]^0$ where $S = \{\zeta \in H^*(G, k) \mid \zeta$ homogeneous, $\dim V_G(\zeta) \cap V_G(M) < c\}$ and $[\ ]^0$ indicates the elements of degree zero.

Now consider the case $c = r$ and $M = N = k$, the trivial kG-module. We recall that $V_G(k) = \bigcup_{i=1}^{\ell} V_i$ where $V_i = \mathrm{res}^*_{G, E_i}(V_{E_i}(k))$, and the elements E_i run through a set of representatives of conjugacy classes of maximal elementary abelian p-subgroups of G (see (1.4)). The closed sets V_i are the components of the variety and hence no V_i is contained in the union of the others. Appealing to some results of commutative algebra we notice that there are elements $\zeta_1, \dots, \zeta_\ell$ in $H^n(G, k)$, for some n, such that

(1) $V_G(\zeta_i) \cap V_i < V_i$,

(2) $V_G(\zeta_i) \cap V_j = V_j$ for $j \neq i$,

(3) $\zeta_i \zeta_j = 0$ in $H^{2n}(G, k)$ if $i \neq j$.

Notice in (3) that $V_G(\zeta_i \zeta_j) = V_G(k)$ and hence we are assured that $\zeta_i \zeta_j$ is nilpotent. Now let

$$\zeta = \zeta_1 + \cdots + \zeta_n \in H^n(G, k).$$

Then the elements $e_i = \zeta_i \cdot \zeta^{-1}$ are orthogonal idempotents in $\mathrm{Hom}_{\mathcal{Q}_r}(k, k)$. It follows that

$$\mathrm{Hom}_{\mathcal{Q}_r}(k, k) = R_1 \oplus \cdots \oplus R_s \qquad\qquad 2.9$$

where each R_i is a local ring corresponding to a conjugacy class E_i of elementary abelian p-subgroups of maximal rank r in G.

Hence we have that the endomorphism ring of the trivial module k in $\mathcal{Q}_r$ decomposes. On the other hand it can be shown [CDW], using subadditive functions on the category, that the trivial module is indecomposable in $\mathcal{Q}_r$. This violates the usual intuition of people studying module categories over artinian rings. The indication is that there is no Krull-Schmidt Theorem in $\mathcal{Q}_r$, i.e., no uniqueness of decompositions of objects as direct sums of indecomposable objects. The validity of these indications is confirmed in the following example. This example is not the same as the one in [CDW]. That is, even if the ultimate decomposition is the same, the example of the triangle (exact sequence) is chosen to indicate a more general decomposition which is discussed further in Section 4.

Suppose that $G = \langle x, y \mid x^2 = y^2 = (xy)^4 = 1\rangle$ is a dihedral group of order 8. Of course, the coefficient field k has characteristic 2. Notice that

$$kG = k\langle A, B\rangle/(A^2, B^2, (AB)^2 - (BA)^2)$$

where $A = 1 + x, B = 1 + y$ and $k\langle A, B\rangle$ indicates the polynomial ring in noncommuting variables A and B. The diagrams which we draw for kG-

module are explained in detail in [Rig]. The free module kG has diagram

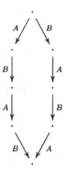

where again each vertex represents a k-basis element and an edge (arrow) represents multiplication by A or by B as indicated. Now G has two maximal elementary abelian 2-subgroups: $H_1 = \langle x, (xy)^2 \rangle$ and $H_2 = \langle y, (xy)^2 \rangle$. The induced modules from the two subgroups also have diagrams

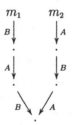

From the diagram for kG we see that $\Omega(k)$ has a diagram (obtained by removing the top vertex and its adjoining edges)

$$
\begin{array}{cc}
m_1 & m_2 \\
B\downarrow & \downarrow A \\
\cdot & \cdot \\
A\downarrow & \downarrow B \\
\cdot & \cdot \\
\end{array}
$$

That is $\Omega(k)$ is generated as a kG submodule of kG by $m_1 = x + 1$ and $m_2 = y + 1$. Then we have an exact sequence

$$0 \longrightarrow L \longrightarrow \Omega(k) \oplus \Omega(k) \xrightarrow{\theta} M_1 \oplus M_2 \longrightarrow 0$$

given in diagram form as

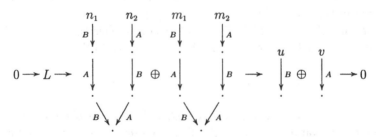

where $\theta(m_1) = u, \theta(m_2) = v, \theta(n_1) = Bu, \theta(n_2) = Av$. The kernel L is generated by $Bm_1 + n_1$ and $Am_2 + n_2$ and has diagram

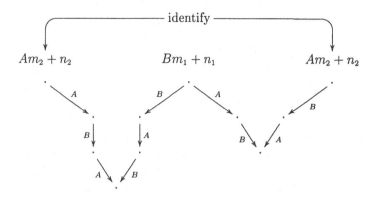

(the two ends of the diagram are the same.) It is not difficult to check that L is a periodic module. Moreover as in (2.6) we can show that $\Omega(k) \cong k$ in the quotient category $\mathcal{Q}_2 = \mathcal{M}_2/\mathcal{M}_1$. Because $L \in \mathcal{M}_1$ we have that in $\mathcal{Q}_2$,

$$k \oplus k \cong k^{\uparrow G}_{H_1} \oplus k^{\uparrow G}_{H_2}.$$

This is an honest example of the failure of Krull-Schmidt, since the three modules in the equation can all be shown to be indecomposable in $\mathcal{Q}_2$.

Finally we should mention that the morphisms in quotients of greater difference $\mathcal{M}_c/\mathcal{M}_d$, $c = d \geq 2$, can also be characterized [CW2], though the answer is far more complicated. The problem is to find objects to take the place of the modules $\Omega^n(M)$ in (2.7). The objects which we create are paramerterized by $(c-d)$-tuples $(\zeta_1, \ldots, \zeta_{c-d})$ of homogeneous elements of $H^*(G, k)$ such that $\cap V_G(\zeta_i)$ has dimension $r - (c-d)$. The homomorphism group, $\mathrm{Hom}_{kG}(U, k)$, for such an object U is given by a spectral sequence whose E_1-page is a truncated Koszul complex determined by the elements $\zeta_1, \ldots, \zeta_{c-d}$. It appears that the E_2-page is a sort of Čech cohomology of a certain sheaf. However, the elements of $\mathrm{Hom}_{kG}(U, k)$ only represent elements of $\mathrm{Hom}_{\mathcal{M}_c/\mathcal{M}_d}(k, k)$. To characterize $\mathrm{Hom}_{\mathcal{M}_c/\mathcal{M}_d}$ we must take a limit of the spectral sequences.

3. Extensions to infinitely generated modules

In the previous section we discussed the fact that the quotient categories $\mathcal{M}_c/\mathcal{M}_{c-1}$ do not in general have uniqueness of decompositions of objects. As one example we showed that the endomorphism ring of the trivial module in $\mathcal{Q}_r = \mathcal{M}_r/\mathcal{M}_{r-1}$ decomposed as a direct sum of rings with the summands corresponding to the components of maximal dimension r in the variety $V_G(k)$.

In particular, we stated that there were idempotents $e_i \in \text{Hom}_{\mathcal{Q}_r}(k, k)$ corresponding to the components of the variety. The usual proof of the Krull-Schmidt Theorem, which expresses the unique decomposition of modules into indecomposables, proceeds by finding an idempotent in the endomorphism ring of the module and showing that the module is the direct sum of the kernel and the image of that idempotent. That is, the idempotent splits the module. The problem with the quotient category $\mathcal{Q}_r$ is that it is not an abelian category. So even if we have idempotents in the endomorphism ring, there are no kernels or images. The category $\mathcal{Q}_r$ is triangulated but not abelian.

In this section we discuss some of the constructions which can be made in a triangulated category to imitate the properties of an abelian category. Most of the new results of this section are contained in the manuscript [BCR1]. The first is the technique of using idempotents to split modules. The method is called a homotopy limit and is well known to the experts in stable homotopy theory [BN]. The only problem with the construction is that it requires the taking of infinite direct sums of modules and hence a passage to the category of all kG-modules, even the infinitely generated ones. In order to put the method to use we must extended our concepts of complexity and varieties to the category of all kG-modules.

Suppose that we have an idempotent $e = f/s$ in $\text{Hom}_{\mathcal{M}_c/\mathcal{M}_d}(M, M)$ for some kG-module M. Let $1 - e = (s - f)/s = g/s$. Then $f + g = s$ and $fg = gf = 0$. For convenience of notation we assume here that $s = 1$. We will state later the adjustments which must be made if $s \neq 1$. The question here is how do we split the idempotent e in the triangulated category. The trick is to produce a triangle whose third object is the "image" or the "kernel" of e. This is the homotopy limit and we get it in the following fashion. Let $M' = \oplus_{i=1}^{\infty} M$ be a direct sum of copies of M and consider the map ψ_1 given diagrammatically by

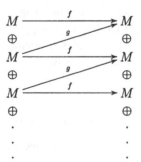

That is,

$$\psi_1(m_1, m_2, \ldots) = (f(m_1) + g(m_2), f(m_2) + g(m_3), \cdots).$$

The reader may wish to notice here, that if this were an ordinary category of modules with $f + g = 1$, then ψ_1 would be a surjection. But in the triangulated category (assuming we can allow the infinite direct sum) we have a triangle of the morphism ψ_1,

$$M_1 \longrightarrow M' \xrightarrow{\psi_1} M' \longrightarrow \Omega^{-1}(M_1),$$

which defines a new object M_1.

Now notice that, if $f + g = s \neq 1$, but s is invertible in the quotient category, then we need to let M_1 be the third object in the triangle of the morphism ψ_1' where ψ_1' is given by $\psi_1' = \psi_1/\psi_1''$

$$M' \xleftarrow{\psi_1''} M' \xrightarrow{\psi_1} M'$$

for ψ_1 as before and $\psi_1'' = (s, s, s, \ldots)$.

Next suppose that ψ_2 is the morphism constructed in the same way but with f and g interchanged. Then we have a triangle

$$M_2 \longrightarrow M' \xrightarrow{\psi_2} M' \longrightarrow \Omega^{-1}(M_2).$$

Now take the direct sum of the two maps ψ_1 and ψ_2. By rearrangment of the sums, we have a diagram

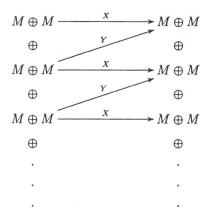

where $X = \begin{pmatrix} f & 0 \\ 0 & g \end{pmatrix}$ and $Y = \begin{pmatrix} g & 0 \\ 0 & f \end{pmatrix}$. The next step is to rearrange the module again by applying the map $A = \begin{pmatrix} f & g \\ -g & f \end{pmatrix}$ to the sum $M \oplus M$. Notice that $\begin{pmatrix} f & -g \\ g & f \end{pmatrix}$ is the inverse of A. It is a routine check to see that

$$\begin{pmatrix} f & g \\ -g & f \end{pmatrix} \begin{pmatrix} f & 0 \\ 0 & g \end{pmatrix} \begin{pmatrix} f & -g \\ g & f \end{pmatrix} = \begin{pmatrix} 1 & 0 \\ 0 & 0 \end{pmatrix}$$

and

$$\begin{pmatrix} f & g \\ -g & f \end{pmatrix} \begin{pmatrix} g & 0 \\ 0 & f \end{pmatrix} \begin{pmatrix} f & -g \\ g & f \end{pmatrix} = \begin{pmatrix} 0 & 0 \\ 0 & 1 \end{pmatrix}$$

We conclude that the third object in the triangle of the morphism

$$M' \oplus M' \xrightarrow{\psi_1 \oplus \psi_2} M' \oplus M'$$

is isomorphic to M. But then $M \cong M_1 \oplus M_2$ because of the uniqueness of the third object in the triangle of a specified morphism.

So we see, that in a triangulated category, it is possible to split idempotents provided we can take infinite direct sum. However the definitions of complexity and varieties for modules are based on the finiteness conditions of section 1. Hence we wish to try to extend these notions to the category of all kG-modules. For this purpose let StMod-kG denote the stable cateogry of all kG-modules. We need to define subcategories $\underline{\mathcal{M}}_c$, of StMod-$kG$, of all kG-modules of "complexity" c. Before attempting such a thing let us emphasize the properties which we wish $\underline{\mathcal{M}}_c$ to have. Specifically, we would hope to be able to get the following.

(3.2 i) $\underline{\mathcal{M}}_c$ should be a full triangulated subcateogry of StMod-kG.

(3.2.ii) The collection of finitely generated module in $\underline{\mathcal{M}}_c$ should be precisely the objects in $\mathcal{M}_c$. That is, $\underline{\mathcal{M}}_c \cap \mathrm{stmod}\text{-}kG = \mathcal{M}_c$.

(3.2.iii) $\underline{\mathcal{M}}_c$ should be closed under infinite direct sums and summand.

Finally, and most importantly, we would like to be assured that nothing is lost in the passage to the infinitely generated category. That is, we wish that

(3.2.iv) The natural functors $\mathcal{M}_c/\mathcal{M}_d \to \underline{\mathcal{M}}_c/\underline{\mathcal{M}}_d$ should be fully faithful for all $d < c$.

With these conditions in mind we offer several candidates for the subcategories defining complexity.

(3.3.a) Let $\mathcal{M}in_c$ denote the smallest triangulated subcategory of StMod-kG containing $\mathcal{M}_c$ and also closed under the taking of direct summands and countable direct sums. Notice that the intersection of two triangulated subcategories, which are closed under summands and countable direct sums, also has all of these properties. So there is a smallest subcategory with these properties. The categories $\mathcal{M}in_c$ clearly satisfy conditions (i) - (iii) of (3.2). But condition (iv) seems to be difficult to show directly. Rather we prove it as a consequence of the $\mathcal{M}in_c$'s being subcategories of the other candidates.

(3.3.b) Let $\mathcal{U}_c$ be the full subcategory of StMod-kG consisting of all modules M satisfying the condition

(*) if $\gamma : X \to M$ with $X \in \mathrm{mod}\text{-}kG$, then γ factors as the composition $X \xrightarrow{\eta} Y \xrightarrow{\zeta} M$, $\gamma = \zeta\eta$, with $Y \in \mathcal{M}_c$.

This is the condition which we need to achieve condition (iv) of (3.2). For suppose that we have a morphism

$$M \xleftarrow{s} U \xrightarrow{f} N$$

in $\mathcal{U}_c/\mathcal{U}_d$ with $M, N, \in \mathcal{M}_c$ and s invertible. Let W be the third object in the triangle of the morphism s. Then we have a triangle

$$W \longrightarrow U \xrightarrow{s} M \xrightarrow{\gamma} \Omega^{-1}(W),$$

where $\Omega^{-1}(W) \in \mathcal{U}_d$. Now by (*), γ factors as $M \xrightarrow{\eta} Y \xrightarrow{\zeta} \Omega^{-1}(W)$ for $Y \in \mathcal{M}_d$. So we get a diagram of triangles

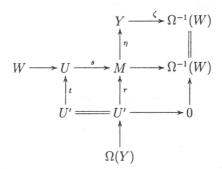

where U' is the third object in the triangle of η. Hence $U' \in \mathcal{M}_c$ and $r = st$ is invertible modulo $\mathcal{M}_d$. The existence of the map t is a conseqence of the axioms of a triangulated category ($\zeta \cdot \eta \cdot r = 0$). However it is probably just as easy to understand it by pulling the diagram back to a diagram of exact sequences in mod-kG.

The existence of t, which is invertible modulo $\mathcal{M}_d$ says that our original morphism f/s is equivalent to a morphism,

$$M \xrightarrow{st} U' \xrightarrow{ft} N,$$

which takes place entirely in $\mathcal{M}_c$. Hence the mapping

$$\operatorname{Hom}_{\mathcal{M}_c/\mathcal{M}_d}(M, N) \longrightarrow \operatorname{Hom}_{\mathcal{U}_c/\mathcal{U}_d}(M, N)$$

is surjective. A similar argument shows that it is also injective. Conditions (i) and (ii) of (3.2) can also be verified without much difficulty. Condition (iii) is a little more difficult.

(3.3.c) Let $\mathcal{E}_c$ be the full subcategory of stmod-kG consisting of all modules M with the property
(**) Suppose that $S \subset M$ is a finite set of elements. Then there exists a projective kG-module P and a submodule $Y \in M \oplus P$ such that $S \times \{0\} \in Y$ and $Y \subseteq \mathcal{M}_c$.

The condition defining $\mathcal{E}_c$ is the one which we usually think is most easily recognized. However we shall see later in the section that there are some problems with the recognition. It is easy to verify that the subcategories $\mathcal{E}_c$ satisfy conditions (i) - (iii) of (3.2).

The last candidate for a definition of complexity is a variation on (3.3.c). The limit construction is very useful in creating examples and plays an extremely important role in the development of the idempotent modules of the next section.

(3.3.d) Let $\mathcal{C}_c$ be the full subcategory of StMod-kG consisting of all modules which are filtered colimits of modules in $\mathcal{M}_c$.

To put all of this together we prove the following

(3.4) Theorem [BCR1]. *For* $0 < c \le r$,

$$Min_c \subset \mathcal{U}_c = \mathcal{C}_c = \mathcal{E}_c.$$

The proof is reasonably straight-forward. Once the equality of $\mathcal{U}_c, \mathcal{C}_c$ and $\mathcal{E}_c$ is verified then the inclusion of Min_c is obvious. The inclusion of $\mathcal{U}_c$ into $\mathcal{C}_c$ requires all of the technicalities of the definition of filtered colimits. An object M in $\mathcal{C}_c$ must be in $\mathcal{E}_c$ because any finite subset of M is in the image of one of the modules in the system giving the colimit. Finally an object M from $\mathcal{E}_c$ is in $\mathcal{U}_c$ because the image of any $\gamma : X \to M$ with $X \in$ mod-kG is generated by the finite set which is the image of the generators of X.

We should note that if we limit ourselves to countably generated modules then Min_c coincides with the other subcategories. In any case we have that $\mathcal{U}_r = $ StMod-kG, if r is the p-rank of G, by (1.5) and the definition. The complexity of a module $M \in$ StMod-kG is defined to be c if $M \in \mathcal{U}_c$ but $M \notin \mathcal{U}_{c-1}$. The Alperin-Evens Theorem extends to the catgegory of all kG-modules. That is, the complexity of a module is the maximum of the complexities of the restrictions to the elementary abelian p-subgroups.

Using the arguments from the first part of this section we can split idempotents in $\mathcal{U}_c/\mathcal{U}_{c-1}$. To get a Krull-Schmidt Theorem we need only that idempotents can be split and also that endomorphism rings be artinian rings. To this end, we define $\mathcal{M}_{c,c-1}$ to be the smallest subcategory of $\mathcal{U}_c/\mathcal{U}_{c-1}$ which contains the quotient $\mathcal{M}_c/\mathcal{M}_{c-1}$ and is closed under the taking of direct summands. The fact that endomorphims rings of objects in $\mathcal{M}_{c,c-1}$ are artinian was proved in [CW1].

(3.5) Theorem (Krull-Schmidt). *Up to order and isomorphism every object in* $M_{c,c-1}$ *is uniquely a direct sum of a finite number of indecomposble objects.*

The effort to find a definition for the variety of a module is still in progress. Clearly some modification must be made to the definition in section 1. For an infinitely generated module M the cohomology ring $\text{Ext}^*_{kG}(M, M)$ is not necessarily finitely generated as an $H^*(G, k)$-module. So it is possible to have a situation in which, for a particular element ζ in $H^*(G, k)$, each element of $\text{Ext}^*_{kG}(M, M)$ is annihilated by some power of ζ, but no power of ζ annihilates every element. This actually happens for the example (B) of section 1. In addition, a close look at the example (A) of section 1 indicates that variety of a module might be an open rather than a closed set. One attempt at solving this problem is proposed in [BCR1]. It defines the variety, $\mathcal{V}_G(M)$, of a module M to be the collection of all $V_G(J) \subset V_G(k)$ where J runs through the set of all annihilators of all finitely generated $H^*(G, k)$-submodules of $\text{Ext}^*_{kG}(X, M)$ for X a finitely generated kG-modules. Thus $\mathcal{V}_G(M)$ is a collection of closed subsets of $V_G(k)$. This definition works well relative to the problem encountered in example (B). In particular, the maximum of the dimensions of the closed sets in $\mathcal{V}_G(M)$ is the complexity of M, as just defined. In the case of (B) that complexity is 1. In addition if M is finitely generated then $\mathcal{V}_G(M)$ has a unique maximal element which is $V_G(M)$ and which contains every other closed set in the collection $\mathcal{V}_G(M)$. However, the example does not work so well for example (A). In this case $V_G(k)$ is one of the closed sets in the collection. Among the other things, it can be seen from this that the new variety $\mathcal{V}_G()$ does not satisfy the tensor product theorem (1.6.6). Other variations on this theme are explored in [BCR2]. Some of the problems, particularly the lack of a tensor product theorem are resolved with appropriate modificiations in the definition.

It can be shown that example (A) has complexity 2 in the new definition. For one thing the map $\phi : k \longrightarrow A$, given by $\phi(1) = b_1$, does not factor through any finitely generated module of complexity one. So $A \notin \mathcal{U}_1$. Similarly we can see that $A \notin \mathcal{E}_1$, because the set $\{b_1\}$ is not in any finitely generated submodule of A or of $A \oplus (\text{proj})$ such that the submodule has complexity 1.

There are several other interesting examples in the appendix of [BCR1]. Some of them show that the condition in (**) of (3.3.c), which allows the addition of a projective module P, can not be eliminated. That is, in general, there exist infinitely generated modules of complexity c which have no finitely generated submodules of complexity c. The point is that the maps which present a module as a direct limit or filtered colimit of modules of complexity c, need not be injective.

Another example from [BCR1] illustrates the decomposition of the trivial module which we obtain by extending the category to include all modules. At the end of Section 2 we showed how the direct sum of two copies of the trivial module is isomorphic to the direct sum of the two modules induced from the trivial modules on the maximal elementary abelian 2-subgroups. In this case the p-rank of G is $r = 2$, and $k \in \mathcal{M}_2$. All $H^1(G, k)$ has two elements η_1, η_2 with $\eta_1 \cdot \eta_2 = 0$. Here η_i is represented by a cocycle $\eta_i' : \Omega(k) \to k$ with

$\eta_i'(m_j) = \delta_{ij}$ in the notation of (2.10). It is not difficult to show that $\eta_1 + \eta_2$ is an isomorphism in $\mathcal{Q}_2 = \mathcal{M}_2/\mathcal{M}_1$. That is, the kernel of $\eta_1' + \eta_2'$ has complexity 1. So let $\eta = \eta_1 + \eta_2$. Then in $\mathcal{Q}_2$, $e_1 = \eta_1/\eta$, $e_2 = \eta_2/\eta$ are orthogonal idempotents. It remains to put these elements into the homotopy limit scheme of the first part of the section, and to write the trivial module k as the direct sum of two infinitely generated modules. The answer is that $k \cong U_1 \oplus U_2$, and this decomposition is represented in diagrams as

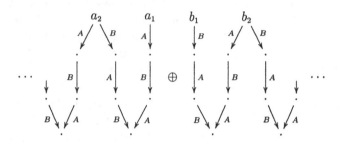

where U_1 is generated by $a_1, a_2, \ldots$ and U_2 is generated by $b_1, b_2, \ldots$. Then we have homomorphism $\theta : U_1 \oplus U_2 \to k$ given by

$$\theta(a_1) = 1 = \theta(b_1) \quad \text{and} \quad \theta(a_i) = 0 = \theta(b_i) \text{ for } i > 1.$$

The kernel of θ is obtained diagrammatically by indentifying a_1 and b_1. To verify that θ is an isomorphism in $\mathcal{Q}_2$ it is only necessary to check that its kernel is in $\mathcal{U}_1$ and hence has complexity 1.

4. Decomposition, Idempotents and rank varieties

We saw in the last section, that in a triangulated category, such as $\mathcal{M}_c/\mathcal{M}_{c-1}$, a decomposition which is only indicated by the endomorphism ring of an object is actually realized when the category is extended to permit infinite direct sums. Results obtained in the last few months have demonstrated the principle several times over. The most striking fact to be discovered is that the stable category StMod-kG of all kG-modules has idempotent objects, that is non-projective modules M such that $M \otimes M \cong M \oplus P$ where P is projective. In fact, the modules A and B at the end of Section 1 are idempotent modules. The idempotent modules, separate the category into orthogonal subcategories and tensoring with an idempotent module is a localizing functor in the sense of homotopy theory.

To set some background, we begin with another decomposition of the quotient category $\mathcal{M}_r/\mathcal{M}_{r-1}$. For notation let $E_1, \ldots, E_t$ be a complete set of representatives of the conjugacy classes of maximal elementary abelian p-subgroups of G. Let $V_i = \operatorname{res}_{G,E_i}^*(V_G(E_i)) \subset V_G(k)$, and recall that $V_1, \ldots, V_t$ are the components of the variety $V_G(k)$ (see (1.4)). For each i, let $D_i = D_G(E_i)$ be the "diagonalizer" of E_i. That is, the normalizer $N_G(E_i)$ acts by $\mathbb{F}_p$-linear transformations on E_i, which is an $\mathbb{F}_p$-vector space, and the diagonalizer is the set of elements of $N_G(E_i)$ which act on E_i by scalar matrices. Thus an element $x \in N_G(E_i)$ is in D_i if and only if there exists $a \in \mathbb{Z}/p = \mathbb{F}_p$ such that $xyx^{-1} = y^a$ for all $y \in E_i$.

The following theorem is stated in full detail. However this discussion will require only the more easily understood implications for the quotient categories. These will be explained later.

(4.1) Theorem [C]. *Suppose that $m = \operatorname{LCM}_i\{|N_G(E_i) : D_i|\}$, and for each $i = 1, \ldots, t$, let $m_i = m/|N_G(E_i) : D_i|$. Then for some $n > 0$ and all $\ell > 0$, there is a projective module P (depending on n, ℓ) and an exact sequence*

$$0 \longrightarrow L \longrightarrow (\Omega^{n\ell}(k))^m \oplus P \overset{\psi}{\longrightarrow} \sum_{i=1}^{t}(k_{D_i}^{\uparrow G})^{m_i} \longrightarrow 0$$

such that $V_G(L) \cap V_i \neq V_i$ for all $i = 1, \ldots, t$.

Notice that the Theorem takes place entirely within the category mod-kG of finitely generated kG-modules. The statement that $V_G(L) \cap V_i \neq V_i$ implies, in particular, that $\dim V_G(L) < r = \dim V_G(k)$ where r is the p-rank of G. Hence $L \in \mathcal{M}_{r-1}$. By (2.6), k is periodic and for some value of ℓ we have that $\Omega^{n\ell}(k) \cong k$ in the quotient category $\mathcal{Q}_r = \mathcal{M}_r/\mathcal{M}_{r-1}$. Composing this with the isomorphism ψ proves the following.

(4.2) Corollary. *In $\mathcal{Q}_r$, $k^m \cong \sum_{i=1}^{t}((k_{D_i}^{\uparrow G})^{m_i}$ where m, m_i are given as in the Theorem.*

Frobenius reciprocity for modules over group algebras says that if $H \subset G$ is a subgroup then $M \otimes k_H^{\uparrow G} \cong (M \otimes k_H)^{\uparrow G} = M_H^{\uparrow G}$ where M_H means the restriction of M to a kH-module. So in $\mathcal{Q}_r$, for any module M,

$$M^m \cong \sum_{i=1}^{t}(M_{D_i}^{\uparrow G})^{m_i}.$$

Hence one way of expressing the corollary is to say that, except for a "complexity factor" (embodied in L) the module theory of kG is controlled at the level of the diagonalizers of the maximal elementary abelian p-subgroups. Even if the "complexity factor" is not well understood, it is not difficult to derive several consequences of the idea. For example it explains why the cohomology ring $\operatorname{Ext}^*_{kG}(k_{D_i}^{\uparrow G}, k_{D_i}^{\uparrow G})$ must have irreducible modules of dimension $m/m_i = |N_G(E_i) : D_i|$ (see [C]). In a more complicated piece, it was used to prove restrictions on the varieties of modules M in the principal kG-block with $H^*(G, M) = \{0\}$ (see [CR]).

Several months ago, Dave Benson [B2] observed that the corollary implies a decomposition of the quotient category itself. Remember that $\mathcal{M}_{r,r-1}$ is the closure under taking direct summands of $\mathcal{M}_r/\mathcal{M}_{r-1}$ in the quotient $\mathcal{U}_r/\mathcal{U}_{r-1}$. In $\mathcal{M}_{r,r-1}$ we can split idempotents and the trivial module has a decomposition into a direct sum of indecomposable modules corresponding to the idempotents in $\mathrm{Hom}_{\mathcal{M}_{r,r-1}}(k,k)$. In turn the idempotents correspond to the components V_i of maximal dimension in $V_G(k)$. For notation here let $E_1, \ldots, E_s$ be the representatives of the conjugacy classes of maximal elementary abilian p-subgroups of rank r. So if $s < i \leq t$ then the rank of E_i, is less than r, V_i has dimension less than r and $k_{D_i}^{\uparrow G}$ has complexity less than r.

The identity element in $\mathrm{Hom}_{\mathcal{M}_{r,r-1}}(k,k)$ is a sum of s orthogonal primitive idempotents corresponding to the components $V_1, \ldots, V_t$. Then k decomposes in $\mathcal{M}_{r,r-1} \subset \mathcal{U}_r/\mathcal{U}_{r-1}$ as

$$k \cong \pi_1 \oplus \ldots \oplus \pi_s \qquad (4.3)$$

where each π_i is indecomposable and $V_G(\pi_i) = V_i$. The fact that $k \otimes k \cong k$ implies, without much difficulty, that $\pi_i \otimes \pi_i \cong \pi_i$ modulo $\mathcal{U}_{r-1}$. The key to the argument is that for $i \neq j$, $\pi_i \otimes \pi_j \in \mathcal{U}_{r-1}$. Then, given the two decompositions, it can be shown that $k_{D_i}^{\uparrow G} \cong (\pi_i)^{m/m_i}$ modulo $\mathcal{U}_{r-1}$. More than that, the category decomposes

$$\mathcal{U}_r/\mathcal{U}_{r-1} = \mathcal{V}_1 \cup \cdots \cup \mathcal{V}_s$$

as a direction sum of full triangulated subcategories $\mathcal{V}_i = (\mathcal{U}_r/\mathcal{U}_{r-1}) \otimes \pi_i$. The category $\mathcal{V}_i$ is equivalent to the corresponding category for the normalizer of E_i. For finitely generated quotients, this fact could only be proved with heavy restrictions [CW2].

Before proceeding further we might notice that the stable category stmod-kG of finitely generated kG-modules has no idempotent modules other than the identity. Hence we are really seeing an unusual new phenomenon in the infinitely generated quotients.

(4.4) Proposition. *Suppose that M is a finitely generated nonprojective kG-module and that $M \otimes M \cong M \oplus P$ for some projective module P. Then $M \cong k \oplus Q$ for some projective module Q.*

The proof is based on a lemma which can be found in [AC]. For the convenience of the reader we reproduce both the Lemma and a sketch of its proof here. We write $U|V$ to mean that U is a direct summand of V.

(4.5) Lemma. *Suppose that M is a finitely generated indecomposable kG-module. If $p \nmid \mathrm{Dim}_k(M)$ then $k|M \otimes M^*$. If $p|\mathrm{Dim}_k M$ then $M \oplus M|M \otimes M^* \otimes M$.*

Proof of Lemma Let $\{m_i\}$ be a basis for M and let $\{\lambda_i\} \subset M^*$ be a dual basis. So $\lambda_i(m_j) = \delta_{ij}$. Let $\theta : M \otimes M^* \to k$ and $\gamma : k \longrightarrow M \otimes M^*$ be given by $\theta(m \otimes \lambda) = \lambda(m)$ and $\gamma(a) = a \cdot \sum m_i \otimes \lambda_i$. The reader should check that θ and γ are homomorphisms and that $\theta \circ \gamma = (\text{Dim}_k M) \cdot Id_k$. If $\text{Dim}_k M$ is not zero in k, then $(1/\text{Dim}_k M)\theta$ is a one-sided inverse for γ. So suppose that $p | \text{Dim}_k M$. Let $\theta_1 = 1 \otimes \theta$, $\theta_2 = \theta \otimes 1 : M \otimes M^* \otimes M \to M$ and let $\gamma_1 = 1 \otimes \gamma$, $\gamma_2 = \gamma \otimes 1 : M \longrightarrow M \otimes M^* \otimes M$. Of course, the above notation requires that we make the switch of $M \otimes M^*$ with $M^* \otimes M$ when needed. We need to check that $\theta_1 \gamma_1 = 0 = \theta_2 \gamma_2$ while $\theta_2 \gamma_1 = Id_M = \theta_1 \gamma_2$; then the composition

$$M \oplus M \xrightarrow{(\gamma_1, \gamma_2)} M \otimes M^* \otimes M \xrightarrow{\binom{\theta_2}{\theta_1}} M \oplus M$$

is the identity on $M \oplus M$. This proves the lemma.

Proof of Proposition The first thing to notice is that if M is an idempotent module, then so is M^*, as $M^* \otimes M^* \cong (M \otimes M)^*$, and so also is $M \otimes M^*$. We have two cases.

Assume first that $p \nmid \text{Dim} M$. Then $k | M \otimes M^*$ and $M \cong (M \otimes k) | (M \otimes M \otimes M^*) \cong M \otimes M^* \oplus (\text{proj})$. Hence if $M \not\cong k \oplus (\text{proj})$ then $k \oplus M | M \otimes M \otimes M^*$ and $M^* \oplus k | M \otimes M \otimes M^* \otimes M^* \cong M \otimes M^* \oplus (\text{proj})$. Clearly this leads to a contradiction.

Simlarly, if $p | \text{Dim} M$, then $M \oplus M | M \otimes M \otimes M^*$ and $(M \otimes M^* \oplus M \otimes M^*) | (M \otimes M \otimes M^* \otimes M^*) \cong M \otimes M^* \oplus (\text{proj})$. This also is impossible.

Most recently Jeremy Rickard showed us that the stable category StMod-kG of all kG-module has idempotent objects corresponding to closed subsets of the variety $V_G(k)$, [R]. The construction is again by a limit process which looks something like the one we use to split idempotents. This time we take a colimit of triangles in StMod-kG or of exact sequences in Mod-kG. We illustrate the technique for the special case that the closed set $V = V_G(\zeta)$ is the subvariety corresponding to the ideal generated by a single homogeneous element $\zeta \in H^*(G, k)$.

Suppose that $\zeta \in H^n(G, k)$. As we saw in Section 1, ζ is represented by a cocycle $\zeta : \Omega^n(k) \to k$. We may translate by the translation functor Ω^{-n} and add a projective P to k so that we have a surjection $\zeta : k \oplus P_1 \to \Omega^{-n}(k)$ which also represents ζ. So we have an exact sequence

$$0 \to \Omega^{-n}(L_\zeta) \to k \oplus P_1 \to \Omega^{-n}(k) \to 0 \qquad 4.6$$

From the viewpoint of the stable category, it is the translation by Ω^{-n} of the triangle

$$L_\zeta \longrightarrow \Omega^n(k) \longrightarrow k \longrightarrow \Omega^{-1}(L_\zeta)$$

(see (2.4)) which was considered earlier. Of course, there also exists a sequence such as (4.6) for any power of ζ, and with careful choice of representatives we get a commutative diagram of the form

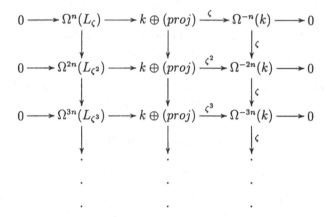

where the maps in the middle column are all congruent to the identity on k modulo homomorphisms which factor projectives. Actually the maps in the middle column are only isomorphisms modulo projective modules. But that is sufficient. Now take the colimit of the diagram of the corresponding system of triangles. The colimit is an exact sequence whose corresponding triangle is

$$\longrightarrow L(V) \xrightarrow{\mu} k \xrightarrow{\nu} F(V) \longrightarrow \qquad\qquad 4.8$$

in StMod-kG.

Suppose that M is a finitely generated kG-module such that $V_G(M) \subset V = V_G(\zeta)$. Then for all m sufficiently large ζ^m annihilates $\operatorname{Ext}_{kG}^*(M, M)$. If we take the tensor product of M with everything in the diagram (4.7), then any sufficient long composition of map in the right hand column factors through a projective and hence is zero in the stable category. That is, for any t

$$\zeta^m \otimes 1 : \Omega^{-tn}(k) \otimes M \longrightarrow \Omega^{(-t-m)n}(k) \otimes M$$

is a translate of the cup product $\zeta^m \cup Id_M \in \operatorname{Ext}^{nm}(M, M)$ and is zero. The conclusion is that the colimit of the system in the right hand of column of (4.7) tensored with M is a projective module. Hence $L(V) \otimes M \cong k \otimes M \oplus (\text{proj}) \cong M \oplus (\text{proj})$.

The next observation is that for any m, the module L_{ζ^m} satisfies the property that $V_G(L_{\zeta^m}) \subset V$. So from the standpoint of the stable category the system in the left hand column of diagram (4.7) is unchanged when tensored with $L(V)$. Thus it follows that

$$L(V) \otimes L(V) \cong L(V) \oplus (\text{proj}).$$

It is not difficult to see also that the "rank variety" of $L(V)$, if computed by the method of (1.8), is exactly V. As a result the "rank variety" of $F(V)$

is precisely $V_G(k) \setminus V$, which is an open set. In fact, the two examples at the end of Section 1 have this property. That is, $A \cong F(V_1)$ and $B \cong L(V_2)$ where $V_1 = \{(\alpha, 0) | \alpha \in k\}$ and $V_2 = \{(0, \beta) | \beta \in k\}$.

Several other observations are immediate. First notice that the remarks of the last paragraph can be interpreted as saying that the map

$$L(V) \otimes L(V) \xrightarrow{1 \otimes \mu} L(V) \otimes k$$

for μ as in (4.8) is a stable isomorphism. Thus $L(V) \otimes F(V)$ must be a projective module. Tensoring the triangle (4.8) with $F(V)$, we get that

$$F(V) \otimes k \xrightarrow{1 \otimes \nu} F(V) \otimes F(V)$$

is also a stable isomorphism and so $F(V)$ is an idempotent module. Similar constructions can be made for any closed set. If $V = V(\zeta_1, \ldots, \zeta_t)$ is the variety of the ideal generated by $\zeta_1, \ldots, \zeta_t$ then $L(V) \cong L(V(\zeta_1)) \otimes \cdots \otimes L(V(\zeta_t))$ and $F(V)$ is the third object in the triangle of the map $L(V) \to k$. Also $F(V)$ can be created by taking a limit of the modules $U(\zeta_1, \ldots, \zeta_n)$ defined in [CW2].

One of the surprises is that some of the ideas can be extended to infinite collections of closed sets. For example, suppose that $V_c = \{V_i\}_{i \in I}$ is the collection of all closed subvarieties $V_i \in V_G(k)$ of dimension c, indexed by some set I. With some care it is possible to define a tensor product $F(V_c) = \underset{i \in I}{\otimes} F(V_i)$ and a triangle

$$L(V_c) \xrightarrow{\mu} k \xrightarrow{\nu} F(V_c)$$

as before. Again $L(V_c)$ is an idempotent module and tensoring $L(V_c)$ is the identity functor on $\underline{\mathcal{U}}_c$. On the other hand, tensoring with $F(V_c)$ is a localizing functor and gives us an embedding of the quotient category $\underline{\mathcal{U}}_r / \underline{\mathcal{U}}_c$ as $F(V_c) \otimes$ StMod-kG in StMod-$kG = \underline{\mathcal{U}}_r$.

In the case that $c = 1$, $F(V_c)$ is an infinitely generated nonprojective module whose rank variety, as defined in (1.8), contains only the zero point. This is one of many indications that the definition of the rank variety needs some alteration in order to be useful in the category of infinitely generated modules. So we define a new rank variety as follows (see [BCR2]).

(4.9) Let K be the algebraic closure of a large transcendental extension of k (transcendence degree n^2 is certainly enough). Let $G = \langle x_1, \ldots, x_n \rangle$ be an elementary abelian p-group. Then the rank variety of a kG-module M is the set

$$V_G^r(M) = \{0\} \cup \{\alpha \in K^n \mid (K \otimes M)_{\langle u_\alpha \rangle} \text{ is not free }\}$$

where, for $\alpha = (\alpha_1, \ldots, \alpha_n) \in K^n$, $u_\alpha = 1 + \sum \alpha_i (x_i - 1)$ as before.

In other words, the "new" rank variety of M is the same as the old rank variety of the extension $K \otimes_k M$ of M to KG-module. The point of the extension is that every point of K^n is a generic point for a subvariety of k^n. If two points $\alpha, \beta \in K^n$ are generic for the same subvariety of k^n then for any kG-module M, either both α and β are in $\mathcal{V}_G^r(M)$ or neither of them is in $\mathcal{V}_G^r(M)$. Consequently the rank variety of a module M can be thought of as a subset of the prime ideal spectrum of $H^*(G, k)$. Of course, $\mathcal{V}_G^r(M)$ is not really a variety but just a subset of K^n. It is not clear, at this point, what subsets of K^n can occur as rank varieties of indecomposable kG-modules.

When actually doing the calculation of $\mathcal{V}_G^r(M)$ for a specific module M, it is easiest to think in terms of homogeneous varieties. We know that $\alpha \in k^n$, $\alpha \neq 0$, is in $\mathcal{V}_G^r(M)$ if and only if every point on the line through α is also in $\mathcal{V}_G^r(M)$. So consider the projective variety $\overline{\mathcal{V}}_G^r(M) \subset KP^n$. Now notice that any line $\overline{\alpha} \in KP^n$ contains at least one point which is generic for a uniquely defined, homogeneous subvariety of k^n. So $\overline{\alpha}$ is generic for some uniquely defined subvariety of the projective space KP^n. Hence we can see that the variety of $L(V_c)$ consists of precisely those points α such that $\overline{\alpha}$ is generic for a subvariety of dimension at most $c - 1$ in kP^{n-1}. The rank variety of $F(V_c)$ is $K^n \setminus \mathcal{V}_G^r(L(V_c))$. In particular, the rank variety of $F(V_c)$ is not zero so long as $c < r$.

(4.10) Theorem [BCR2]. *Suppose that G is an elementary abelian p-group.*

(a) (Dade's Lemma) Suppose that M is a kG-module and that $\mathcal{V}_G^r(M) = 0$. Then M is a projective module.

(b) (Tensor Product Theorem) Suppose that M and N are kG-modules. Then

$$\mathcal{V}_G^r(M \otimes N) = \mathcal{V}_G^r(M) \cap \mathcal{V}_G^r(N).$$

It is also possible to define cohomological varieties as in the case of finitely generated module. One such definition is as follows. Let $\mathcal{V}_G(M)$ be the collection of all closed homogeneous subsets V of $V_G(k)$ such that the complexity of $M \otimes E(V)$ is the same as the dimension of V. For elementary abelian groups we have that $\mathcal{V}_G(M)$ is equivalent to $\mathcal{V}_G^r(M)$ in the sense that a closed subvariety $V \in k^n$ is in $\mathcal{V}_G(M)$ if and only any corresponding generic point for V is in $\mathcal{V}_G^r(M)$.

References

[AC]   M. Auslander and J F Carlson, *Almost split sequences and group algebras,* J. Algebra **103** (1986), pp. 122–140.

[B1]   D. J. Benson, *Representations and Cohomology II: Cohomology of Groups and Modules,* Cambridge University Press, Cambridge (1991).

[B2]   D. J. Benson, *Decomposing the complexity quotient category,* Math. Proc. Camb. Philos. Soc. **120** (1996), pp. 589–595.

[BCR1] D. J. Benson, J. F. Carlson and J. Rickard, *Complexity and varieties for infinitely generated modules,* Math. Proc. Camb. Philos. Soc. **118** (1995), pp. 223–243.

[BCR2] D. J. Benson, J. F. Carlson and J. Rickard, *Complexity and varieties for infinitely generated modules, II,* Math. Proc. Camb. Philos. Soc. **120** (1996), pp. 597–615.

[BN]   M. Bökstedt and A. Neeman, *Homotopy limits in triangulated categories,* Composition Mathematica **86** (1993), pp. 209–234.

[C]   J. F. Carlson, *The decomposition of the trivial module in the complexity quotient category,* J. Pure Appl. Algebra **106** (1996), pp. 23–44.

[CDW]   J. F. Carlson, P. W. Donovan and W. W. Wheeler , *Complexity and quotient categories for group algebras,* J. Pure Appl. Algebra **93** (1994), pp. 147–167.

[CR]   J. F. Carlson and G. R. Robinson, *Varieties and modules with vanishing cohomology,* Math. Proc. Camb. Philos. Soc. **116** (1995), pp. 245–251.

[CW1]   J. F. Carlson, and W. W. Wheeler, *Varieties and localizations of module categories,* J. Pure Appl. Algebra **102** (1995), pp. 137–153.

[CW2]   J. F. Carlson, and W. W. Wheeler, *Homomorphisms in higher complexity quotient categories,* Proc. Symp. Pure Math. (to appear)

[E]   L. Evens, *The Cohomology of Groups,* Oxford University Press, New York (1991).

[H]   D. Happel, *Triangulated Categories in the Representation Theory of Finite-Dimensional Algebras,* Cambridge University Press, Cambridge (1991).

[Q]   D. Quillen, *The spectrum of an equivariant cohomology ring I,II,* Ann. of Math. (2) **94** (1971), pp. 549–602.

[QV]   D. Quillen and B. B. Venkov, *Cohomology of finite groups and elementary abelian subgroups,* Topology **11** (1972), pp. 552–556.

[Ric]   J. Rickard, *Idempotent modules in the stable category,* J. London Math. Soc. **178** (1997), pp. 149–170.

[Rig]   C. M. Ringel , *The indecomposable representations of the dihedral 2-group,* Math. Ann. **214** (1975), pp. 19–34.

[W]   C. A. Weibel, *An Introduction to Homological Algebra,* Cambridge University Press, Cambridge (1994).

Protrees and Λ-trees

I. M. Chiswell

School of Mathematical Sciences, Queen Mary and Westfield College, University of London, Mile End Road, London E1 4NS.

1. Introduction

The idea of a protree is due to M.J. Dunwoody, and they first appear in [7], although the name protree was not used until later, in [8]. Two major advances in combinatorial group theory in the late 1960's were the Bass-Serre theory (see [10]), and Stallings' work on ends of groups (there is an account of this work in [11], and from a different perspective in [5]; a more recent and more general account can be found in [6]). Together, these imply that a finitely generated group with more than one end acts on a tree with finite edge stabilisers. This raised the problem of giving a direct construction of the tree, and it was in solving this problem that Dunwoody introduced protrees. Under certain circumstances (the finite interval condition, which will be considered in §3 below), a protree gives rise to an ordinary simplicial tree.

We show here that any protree arises in a simple way from a Λ-tree, for some suitable ordered abelian group Λ. For information on Λ-trees, see [1]. This is part of a programme, started in [4], to demonstrate that any notion of generalised tree which occurs in the literature is a manifestation of some suitable Λ-tree.

A protree is, by definition, a partially ordered set $(E, \leq)$ together with an order-reversing involution $E \to E$, $e \mapsto e^*$, such that, for all $e, f \in E$, exactly one of $e > f$, $e > f^*$, $e^* > f$, $e^* > f^*$, $e = f$, $e = f^*$ holds. Note that this implies $e^* \neq e$ for all $e \in E$.

For example, if E is the set of oriented edges of a tree, and e^* is the oppositely oriented edge to e, then we can obtain a protree by defining $e > f$ to mean that e points towards f (i.e. there is a reduced path in the tree starting with e and ending with f), as illustrated in the following picture.

In view of this example, we shall occasionally refer to elements of a protree as edges. It is well-known that a tree determines a $\mathbb{Z}$-tree (see, for example [4; Lemma 4]), whose points are the vertices of the tree, and whose metric is the path metric of the tree. The oriented edges of the tree are in one-to-one correspondence with the ordered pairs of vertices at distance 1 apart (associate to each edge e the ordered pair consisting of the initial and terminal vertices of e).

This can be generalised as follows. Let Λ be an ordered abelian group with a least positive element, denoted by 1, and let (X, d) be a Λ-tree. We define $E(X) = \{(u, v) \in X \times X \mid d(u, v) = 1\}$. We put $(u, v)^* = (v, u)$, and if $e = (u, v)$, $f = (x, y)$, then we define $e > f$ to mean $[u, y] = [u, v, x, y]$ (this notation is explained in (2.10) of [1], and will be used extensively). This defines a protree $(E(X), \leq)$, a fact we shall prove shortly. (We are, of course, using the usual conventions for partial orders, so $e \leq f$ means $f > e$ or $f = e$).

We shall often suppress the partial ordering and write E instead of $(E, \leq)$, and use $\leq$ to denote any partial orderings under consideration. A mapping $\phi : E \to E'$ of protrees will be called an embedding if it is one-to-one, $e \leq f$ implies $\phi(e) \leq \phi(f)$ and $\phi(e^*) = \phi(e)^*$, for all e, $f \in E$, and an isomorphism is a bijective embedding. A *subprotree* of a protree E' is a subset E of E' such that $e \in E$ implies $e^* \in E$; it then follows that E is a protree by restriction, and the inclusion map $E \hookrightarrow E'$ is an embedding. The notions of automorphism and of group action on a protree are defined in the obvious way. If E is a G-protree (i.e. G is a group acting as automorphisms of E), then E is said to be G-finite if there are only finitely many G-orbits of edges. We shall show that any protree can be embedded in a protree $(E(X), \leq)$ for some Λ-tree (X, d) and ordered abelian group Λ having a least positive element. However, before doing so, we shall consider the vertices, in the generalised sense defined in [8], for a protree of the form $(E(X), \leq)$. The result is what one would expect intuitively, that vertices arise from points of X, open ends of X and open cuts in X (these terms are defined in [1]).

Before proving that $(E(X), \leq)$ is a protree, we need a remark, which will also be useful later.

Remark. If $(s, t) \in E(X)$ and $r \in X$, then either $t \in [s, r]$, so $[s, r] = [s, t, r]$, or $s \in [t, r]$, so $[t, r] = [t, s, r]$. For if $t \notin [s, r]$, then $[s, r] \cap [s, t] = \{s\}$ (since $[s, t] = \{s, t\}$), so $[t, r] = [t, s, r]$, using one of the axioms for a Λ-tree.

Lemma 1. *Let (X, d) be a Λ-tree, where Λ has a least positive element. Then $(E(X), \leq)$, with the involution $e \mapsto e^*$ defined above, is a protree.*

Proof Let $e = (u, v)$, $f = (x, y)$ and $g = (s, t)$ be elements of $E(X)$, and suppose $e > f > g$. Then $[u, y] = [u, v, x, y]$ and $[x, t] = [x, y, s, t]$. Since $x \neq y$, it follows from the Piecewise Geodesic Proposition ([1; 2.14(c)]) that

$[u, t] = [u, v, x, y, s, t] = [u, v, s, t]$, hence $e > g$.

If $e > e$ then $[u, v] = [u, v, u, v]$, which implies $d(u, u) = 2d(u, v) > 0$, a contradiction. It follows that $\leq$ is a partial order.

Clearly $e^{**} = e$, and $e > f$ implies $f^* > e^*$. It remains to show that exactly one of $e > f$, $e > f^*$, $e^* > f$, $e^* > f^*$, $e = f$, $e = f^*$ holds. We first show that at least one of these possibilities occurs. By the remark preceding the lemma,

$$[u, y] = [u, v, y] \quad \text{or} \quad [v, y] = [v, u, y]$$

$$\text{and} \quad [u, y] = [u, x, y] \quad \text{or} \quad [u, x] = [u, y, x]$$

giving four cases, which we consider in turn.

Case 1. $[u, y] = [u, v, y]$ and $[u, y] = [u, x, y]$. Then either $[u, y] = [u, x, v, y]$ and $v \neq x$, or $[u, y] = [u, v, x, y]$. In the first case, $e = f$ (since $d(u, v) = d(x, y) = 1$), and in the second, $e > f$.

Case 2. $[u, y] = [u, v, y]$ and $[u, x] = [u, y, x]$. Then $[u, x] = [u, v, y, x]$, so $e > f^*$.

Case 3. $[v, y] = [v, u, y]$ and $[u, y] = [u, x, y]$. Then $[v, y] = [v, u, x, y]$, that is, $e^* > f$.

Case 4. $[v, y] = [v, u, y]$ and $[u, x] = [u, y, x]$. By the remark preceding the lemma, either $[v, x] = [v, y, x]$, in which case $[v, x] = [v, u, y, x]$, that is, $e^* > f^*$, or $[v, y] = [v, x, y]$. Then either $[v, y] = [v, u, x, y]$, which implies $x = y$, a contradiction, or $[v, y] = [v, x, u, y]$ with $u \neq x$, which implies $u = y$ (since $[u, x] = [u, y, x]$) and $v = x$ (since $d(u, v) = 1$). Thus $e = f^*$.

We have to show that exactly one of the six possibilities occurs. We have already noted that $e > e$ is never true, and clearly $e^* \neq e$. Also, $e > e^*$ is never true. For otherwise, $[u, u] = [u, v, v, u]$, which implies $v = u$, a contradiction. Thus it suffices to show that, if $e > f$, then none of $e > f^*$, $e^* > f$, $e^* > f^*$ holds, for we can replace e by e^* or f by f^*, or both in the argument. Assume, then, that

$$[u, y] = [u, v, x, y] \qquad (*).$$

If $e^* > f$ then $[v, y] = [v, u, x, y]$, so $d(v, y) = d(v, u) + d(u, y)$, and from $(*)$, $d(u, y) = d(v, y) + d(v, u)$. This gives $d(u, v) = 0$, a contradiction. Similarly if $e^* > f^*$, so $[v, x] = [v, u, y, x]$, we obtain from $(*)$ the contradiction $d(u, v) = 0$. If $e > f^*$, then we obtain from $(*)$ the contradiction $d(x, y) = 0$. This completes the proof of the lemma.

Note that, if G is a group acting as isometries on a Λ-tree (X, d), where Λ has a least positive element, then G preserves segments, so G acts as automorphisms of the protree $(E(X), \leq)$.

2

Let $(E, \leq)$ be a protree. An orientation of E is a subset O of E such that O contains exactly one element of the pair $\{e, e^*\}$, for every $e \in E$. The set $\overline{V}E$ of generalised vertices of E is the set of all orientations O of E satisfying

$$e \in O \text{ and } f > e \text{ implies } f \in O.$$

These definitions are given in [8], although we are using the notation of [9]. In the case that E is the set of edges of a tree, each vertex gives rise to an element of $\overline{V}E$, namely, the orientation consisting of all edges which point towards the vertex. Similarly, every end of the tree gives rise to an element of $\overline{V}E$. As noted in [9], these are the only elements of $\overline{V}E$ in this case. We shall show that there is a similar classification of elements of $\overline{V}E$ in the case that $E = E(X)$, where (X, d) is a Λ-tree and Λ has a least positive element, denoted by 1. In general, there are three kinds of elements in $\overline{V}E(X)$, the third kind corresponding to open cuts of X, which do not occur in the case $\Lambda = \mathbb{Z}$. We begin by describing the three types of elements of $\overline{V}E$ in detail.

(1) Let $w \in X$. Define a subset O_w of $E(X)$ by: $e \in O_w \Longleftrightarrow [u, w] = [u, v, w]$, where $e = (u, v)$. This is an orientation by the remark preceding Lemma 1. Suppose $e = (u, v)$, $e \in O_w$, $f = (x, y) \in E(X)$ and $f > e$. Then $[x, v] = [x, y, u, v]$ and $[u, w] = [u, v, w]$. Since $u \neq v$, it follows from [1; 2.14(c)] that $[x, w] = [x, y, u, v, w] = [x, y, w]$, hence $f \in O_w$. Thus $O_w \in \overline{V}E$.

(2) Let ε be an end of X (see [1; 2.23]). Define O_ε by $e \in O_\varepsilon \Longleftrightarrow v \in [u, \varepsilon)$, where $e = (u, v)$ (the notation $[u, \varepsilon)$ is explained in (2.24)(a) and the paragraph preceding it in [1]). It follows easily from (2.22)(c) and the remarks in (2.21) of [1] that $[u, \varepsilon) \cap [v, \varepsilon) = [w, \varepsilon)$ for some $w \in X$, and $[u, w] \cap [v, w] = \{w\}$. By the axioms for a Λ-tree, $[u, v] = [u, w, v]$, and since $d(u, v) = 1$, either $w = u$ or $w = v$. Hence O_ε is an orientation of $E(X)$.
Suppose $e = (u, v)$, $e \in O_\varepsilon$, $f = (x, y) \in E(X)$ and $f > e$. Then $[x, v] = [x, y, u, v]$ and $v \in [u, \varepsilon)$. Since $u \neq v$, it follows easily that $[x, v] \subseteq [x, \varepsilon)$, so $y \in [x, \varepsilon)$, hence $f \in O_\varepsilon$ and we have shown that $O_\varepsilon \in \overline{V}E$.

(3) Let (X_1, X_2) be an open cut of X; this means that X_1, X_2 are disjoint subtrees of X whose union is X, and neither of them is a closed subtree of X (see (2.25) and (2.27) in [1]). Let ε_1, ε_2 be the open ends of X_1, X_2 respectively which meet at the cut. By (2), O_{ε_i} is an orientation of X_i for $i = 1, 2$. We claim that $O_{X_1, X_2} = O_{\varepsilon_1} \cup O_{\varepsilon_2}$ is in $\overline{V}E$. To see this, we make two remarks.

Remark 1. There is no edge $e = (u, v)$ with $u \in X_1$, $v \in X_2$.
For suppose e is such an edge. Then, as noted in (2.27) of [1], it follows from [1; 2.26] that $[u, v] = [u, \varepsilon_1) \cup [v, \varepsilon_2)$, a disjoint union. Since ε_1 is an open

end of X_1, there exists $w \in [u, \varepsilon_1)$ with $w \neq u$. then $w \neq v$ since $v \in X_2$, and $[u, v] = [u, w, v]$, which implies $d(u, v) > 1$, a contradiction.

It follows from Remark 1 that O_{X_1, X_2} is an orientation of $E(X)$.

Remark 2. If $e = (u, v) \in O_{X_1, X_2}$, $f = (x, y) \in E(X)$, $f \geq e$ and $e \in E(X_i)$, then $f \in E(X_i)$ (for $i = 1, 2$).

For without loss of generality assume $i = 1$. Suppose $f \in E(X_2)$, so $f > e$. Then again by [1; 2.27], $[x, v] = [x, y, u, v] = [v, \varepsilon_1) \cup [x, \varepsilon_2)$. Since $u \in X_1$, $u \in [v, \varepsilon_1)$, which means $e^* \in O_{X_1, X_2}$, contradicting the fact that O_{X_1, X_2} is an orientation of $E(X)$.

It follows from Remark 2 that $O_{X_1, X_2} \in \overline{V}E$.

Note that, if ε is an end of X which is not open, then there exists $w \in X$ such that, for $u \in X$, $[u, \varepsilon) = [u, w]$, so $O_\varepsilon = O_w$. (In the case that X is a $\mathbb{Z}$-tree arising from an ordinary tree T, it is the open ends of X which correspond to the ends of T in the usual sense, and the closed ends of X correspond to terminal vertices (leaves) of T). We shall show that, if we confine attention to open ends in (2) above, then we have defined three different kinds of elements of $\overline{V}E$. In order to do so, we first recall the definition of directions in an arbitrary protree $(E, \leq)$ ([8]). For $O \in \overline{V}E$ and $e \in O$, put $\langle O, e] = \{ f \in O \mid f \leq e \}$. For $e, f \in O$, define $e \sim f$ to mean that there exists $g \in O$ such that $g \leq e$ and $g \leq f$ (equivalently, $\langle O, e] \cap \langle O, f] \neq \emptyset$). It is shown in [8] that this is an equivalence relation on O, and elements of $O / \sim$ are called directions at O.

Lemma 2. *Let (X, d) be a Λ-tree, where Λ has a least positive element denoted by 1. Then*

(1) *If ε is an open end of X, there is exactly one direction at O_ε in the protree $E(X)$.*

(2) *If (X_1, X_2) is an open cut of X, then there are exactly two directions at O_{X_1, X_2} in $E(X)$.*

Proof (1) Let $e = (u, v)$, $f = (x, y)$ be elements of O_ε. Then as we have noted previously, $[v, \varepsilon) \cap [y, \varepsilon) = [w, \varepsilon)$ for some $w \in X$. Since ε is an open end, there exists $w' \in [w, \varepsilon)$ such that $d(w, w') = 1$. It follows easily that $[u, w'] = [u, v, w, w']$ and $[x, w'] = [x, y, w, w']$, hence $e \geq (w, w')$ and $f \geq (w, w')$. Also, $(w, w') \in O_\varepsilon$, hence $e \sim f$.

(2) Let ε_1, ε_2 be the ends of X_1, X_2 respectively which meet at the cut. By (1), O_{ε_i} defines a unique direction in X_i, for $i = 1, 2$. Also, if $e \sim f$ with respect to O_{ε_i}, then $e \sim f$ with respect to O_{X_1, X_2}. Thus, it suffices to show that if $e \in O_{\varepsilon_1}$ and $f \in O_{\varepsilon_2}$, then $e \not\sim f$. Suppose on the contrary that there exists $g \in O_{X_1, X_2}$ such that $g \leq e$ and $g \leq f$. If $g \in O_{\varepsilon_1}$, then $f \in E(X_1)$ by Remark (2) above, a contradiction, and if $g \in O_{\varepsilon_2}$ then $e \in E(X_2)$ by Remark

(2), which is again a contradiction. These are the only possibilities for g, so this proves the lemma.

It is also easy to show that, if $w \in X$, then the number of directions at O_w in $E(X)$ is equal to $\mathrm{ind}_X(w)$, as defined in §1 of [2], but we shall not need this. In particular, (1) of Lemma 2 is true for any end of X, open or not.

Lemma 3. *Let $E = E(X)$, where (X,d) is a Λ-tree and Λ has a least positive element, denoted by 1. An element of $\overline{V}E$ cannot be of more than one of the following forms.*
(1) O_w for some $w \in X$
(2) O_ε for some open end ε of X
(3) O_{X_1,X_2} for some open cut (X_1, X_2) of X.

Proof If ε is an open end of X, then $O_\varepsilon \neq O_w$ for any $w \in X$. For suppose $O_\varepsilon = O_w$. Since ε is an open end, there exists $w' \in [w, \varepsilon)$ with $d(w, w') = 1$. Then if $e = (w, w')$, $e \in O_\varepsilon$ but $e^* \in O_w$, a contradiction.

If (X_1, X_2) is an open cut, then $O_{X_1,X_2} \neq O_w$ for any $w \in X$. For suppose $O_{X_1,X_2} = O_w$. We assume without loss of generality that $w \in X_1$. Then $O_w \cap E(X_1) = O_{\varepsilon_1}$, and $O_w \cap E(X_1)$ is the element of $\overline{V}E(X_1)$ determined by w, so this contradicts what has just been proved (with X_1 in place of X). Finally, $O_\varepsilon \neq O_{X_1,X_2}$ by Lemma 2.

We shall show that any element of $\overline{V}E(X)$ is of one of the three forms in Lemma 3. Before doing so, we pause to show that the endpoints of edges in $E(X)$ are as expected, given the example of a simplicial tree. Recall that ([8]), if E is any protree and $e \in E$, the endpoint ιe is the element of $\overline{V}E(X)$ defined by $f \in \iota e \iff f > e$ or $f \geq e^*$. (The endpoint τe is defined by $\tau e = \iota e^*$.)

Lemma 4. *Let $E = E(X)$, where (X,d) is a Λ-tree and Λ has a least positive element, denoted by 1. Let $e = (u, v) \in E$. Then $\iota e = O_u$.*

Proof Let $f = (x, y) \in E(X)$. Then by definition,
$$f \in \iota e \iff ([x, v] = [x, y, u, v] \text{ or } [x, u] = [x, y, v, u] \text{ or } (x = v \text{ and } y = u))$$
and
$$f \notin \iota e \iff f \leq e \text{ or } f < e^*$$
$$\iff ([u, y] = [u, v, x, y] \text{ or } [v, y] = [v, u, x, y] \text{ or } (y = v \text{ and } x = u)).$$

It follows that $f \in \iota e$ if and only if $y \in [x, u]$, that is, $f \in O_u$, as required.

Given an element $O \in \overline{V}E(X)$, the next lemma constructs a linear subtree $L_{O,e}$ of X, where $e \in O$, which will play a role in showing that O is of one of the forms described in Lemma 3. In the case that $O = O_\varepsilon$ for some open end ε, the subtree $L_{O,e}$ will be a ray in X representing ε. The terminology in the statement of the lemma is explained in (2.18) and (2.21) of [1].

Lemma 5. *Let $E = E(X)$, where (X, d) is a Λ-tree and Λ has a least positive element, denoted by 1. Let $O \in \overline{V}E$, and let $e = (x, y) \in O$. Put*

$$L_{O,e} = \{u \in X \mid \exists v \in X \text{ such that } (v, u) \in \langle O, e]\}.$$

Then $L_{O,e}$ is a linear subtree of X with y as an endpoint. Further, if $(u, v) \in E$, $v \in L_{O,e}$ and $(u, v) \le e$, then $(u, v) \in O$.

Proof We abbreviate $L_{O,e}$ to L. Take u, $v \in L$, and suppose $w \in [u, v]$. We show that $w \in L$. We can assume that $u \ne w \ne v$, so $u \ne v$. Choose u', $v' \in X$ such that $(u', u) \in \langle O, e]$ and $(v', v) \in \langle O, e]$. Since $\langle O, e]$ is linearly ordered (see [8]) and $u \ne v$, we can assume without loss of generality that $(u', u) < (v', v)$, that is, $[v', u] = [v', v, u', u]$. Since $d(u, u') = 1$ and $w \ne u$, $[v', u] = [v', v, w, u', u]$, and since $w \ne v$, there exists $w' \in X$ such that $d(w, w') = 1$ and $[v', u] = [v', v, w', w, u', u]$. Then $(w', w) < (v', v)$, hence $(w', w) \le e$, and $(u', u) < (w', w)$, so $(w', w) \in O$. Thus $(w', w) \in \langle O, e]$, and $w \in L$. We have shown that L is a subtree of X. (It is non-empty since $y \in L$).

Next, we show that L is linear. Let u, v, $w \in L$, pairwise distinct. By [1; 2.20], we need to show they are collinear. Choose u', v', $w' \in X$ such that $(u', u) \in \langle O, e]$, $(v', v) \in \langle O, e]$ and $(w', w) \in \langle O, e]$. again since $\langle O, e]$ is linearly ordered, we can assume without loss of generality that $(u', u) < (v', v) < (w', w)$. Thus $[w', v] = [w', w, v', v]$ and $[v', u] = [v', v, u', u]$. Since $v' \ne v$, it follows from [1; 2.14] that $[w', u] = [w', w, v', v, u', u]$. Hence $v \in [w, u]$, and u, v, w are collinear, as required.

Now we need to show that y is an endpoint of L. Suppose $[u, v] = [u, y, v]$, where u, $v \in L$. Take u', $v' \in X$ such that $(u', u) \in \langle O, e]$ and $(v', v) \in \langle O, e]$. Then $(u', u) \le (x, y)$, so either $u = y$ or $[x, u] = [x, y, u', u]$. Also, $(v', v) \le (x, y)$, so either $v = y$ or $[x, v] = [x, y, v', v]$. Thus, if $u \ne y$ and $v \ne y$, $[u, v] = [u, y, v] = [u, u', y, v', v]$, which implies $(u, u') > (v', v)$. By the axioms for a protree, this contradicts the fact that $\langle O, e]$ is linearly ordered. Hence either $u = y$ or $v = y$, and y is an endpoint of L.

Finally, suppose $(u, v) \in E$, $v \in L$ and $(u, v) \le (x, y)$. Take $w \in X$ such that $(w, v) \in \langle O, e]$. If $(u, v) = (x, y)$ then $(u, v) \in O$ as required, so assume $(u, v) < (x, y)$, that is, $[x, v] = [x, y, u, v]$. Then $v \ne y$ since $u \ne v$, so $(w, v) < (x, y)$, i.e. $[x, v] = [x, y, w, v]$. Since there is a unique point in $[x, v]$ at distance 1 from v, it follows that $u = w$, hence $(u, v) = (w, v) \in O$.

Before proceeding, the following simple observation is worth recording.

Remark. If, in Lemma 5, $w \in L_{O,e}$, then $[w, x] = [w, y, x]$. In particular, $x \notin L_{O,e}$.

Lemma 6. *Under the hypotheses of Lemma 5, suppose $L_{O,e}$ is an X-ray from y, so $L_{O,e} = [y, \varepsilon)$ for some end ε of X. Then $O = O_\varepsilon$.*

Proof Let $f = (u, v) \in E(X)$. We have to show that $v \in [u, \varepsilon)$ if and only if $f \in O$. Since both O and O_ε are orientations, it suffices to show that $v \in [u, \varepsilon) \Longrightarrow f \in O$. We again write L for $L_{O,e}$.

Now, as has been noted previously, $[u, \varepsilon) \cap L = [u, \varepsilon) \cap [y, \varepsilon) = [w, \varepsilon)$ for some $w \in X$. Also, it is easy to see that $[u, \varepsilon) = [u, w] \cup [w, \varepsilon)$, $[y, \varepsilon) = [y, w] \cup [w, \varepsilon)$ and $[u, w] \cap L = \{w\} = [u, w] \cap [y, w]$ (see [1; 2.21]). Assume $v \in [u, \varepsilon)$.

Case (1). Suppose $u \neq w$. Then $d(u, w) \geq 1$, so $[u, w] = [u, v, w]$. Also, $L \neq [y, w]$, otherwise $L \subsetneqq [y, u] = [y, w] \cup [w, u]$ (using one of the axioms for a Λ-tree), contradicting the assumption that L is an X-ray. Since L is linear, there exists $w' \in L$ such that $[y, w'] = [y, w, w']$ and $d(w, w') = 1$. By the remark preceding the lemma, $[x, w'] = [x, y, w'] = [x, y, w, w']$. Hence $(x, y) > (w, w')$, so $(w, w') \in O$ by Lemma 5.

If $w' \in [u, w]$ then $w' \in [u, w] \cap L = \{w\}$, i.e. $w = w'$, a contradiction. By the remark preceding Lemma 1, $w \in [u, w']$. Hence $[u, w'] = [u, w, w'] = [u, v, w, w']$, which implies $(u, v) > (w, w')$, so $(u, v) \in O$ since $O \in \overline{V}E$.

Case (2). Suppose $u = w$. Then $u \in L$, and $[u, \varepsilon) \subseteq L$, so $v \in L$. By the remark preceding the lemma, $[x, v] = [x, y, v]$. Also, $[y, \varepsilon) = [y, u] \cup [u, \varepsilon)$, so $[y, v] = [y, u, v]$ (because $[y, \varepsilon)$ is linear, has y as an endpoint and $v \in [u, \varepsilon)$). Thus $[x, v] = [x, y, u, v]$, hence $(x, y) > (u, v)$, and $(u, v) \in O$ by Lemma 5. This completes the proof.

The case where $L_{O,e}$ is not an X-ray is dealt with in the next two lemmas.

Lemma 7. *Under the hypotheses of Lemma 5, choose $z \in X \setminus L_{O,e}$. Define*

$$X_1 = \{u \in X \mid [u, z] \cap L_{O,e} \neq \emptyset\} \text{ and } X_2 = \{u \in X \mid [u, z] \cap L_{O,e} = \emptyset\}.$$

Then (X_1, X_2) is a cut in X.

Proof Clearly $X_1 \cap X_2 = \emptyset$ and $X_1 \cup X_2 = X$, so we have to show that X_1, X_2 are subtrees of X. As usual we write L for $L_{O,e}$.

Suppose $u, v \in X_1$. Then we can find $w, w' \in L$ such that $[v, z] = [v, w, z]$ and $[u, z] = [u, w', z]$. By [1; 2.14(a)], $[v, u] \subseteq [v, w] \cup [w, w'] \cup [w', u]$. If $p \in [v, w]$, then $[v, z] = [v, p, w, z]$, so $w \in [p, z]$, hence $p \in X_1$. Thus $[v, w] \subseteq X_1$, and similarly $[w', u] \subseteq X_1$. Also, $[w, w'] \subseteq L \subseteq X_1$ since L is a subtree of X, hence $[v, u] \subseteq X_1$, showing X_1 is a subtree of X (it is non-empty since $L \subseteq X_1$).

Suppose $u, v \in X_2$. Let $w \in [u, v]$ and suppose that $w \notin X_2$, so $[w, z] = [w, l, z]$ for some $l \in L$. By [1; 2.14(a)], either $w \in [u, z]$ or $w \in [v, z]$. If $w \in [u, z]$, then $[u, z] = [u, w, z] = [u, w, l, z]$, which implies $u \notin X_2$, a contradiction. If $w \in [v, z]$, then $[v, z] = [v, w, z] = [v, w, l, z]$, so $v \notin X_2$, which is also a contradiction. This shows that X_2 is a subtree of X (it is non-empty since $z \in X_2$).

Before stating the next lemma, we need another remark.

Remark. Suppose (X, d) is a Λ-tree, where Λ has a least positive element denoted by 1. Let (X_1, X_2) be a cut in X. Then either X_1, X_2 are both closed subtrees of X, or neither is closed, i.e. the cut is an open cut.

For suppose X_1 is closed, and choose $x \in X_1$, $y \in X_2$. Then $[x, y] \cap X_1 = [x, z]$ for some $z \in X_1$. Since $z \neq y$, there exists $w \in X$ such that $d(w, z) = 1$ and $[x, y] = [x, z, w, y]$. Then $[w, y] \cap X_1 = \emptyset$ and $[x, y] = [x, z] \cup [w, y]$, hence $X_2 \cap [x, y] = [w, y]$, and it follows easily that X_2 is a closed subtree of X.

Lemma 8. *Under the hypotheses of Lemma 5, suppose $L_{O,e}$ is not an X-ray from y, and let L' be a linear subtree from y such that $L_{O,e} \subsetneqq L'$. Choose $z \in L' \setminus L_{O,e}$. Let (X_1, X_2) be the cut defined in Lemma 7. Then*

(1) *If X_1 (so X_2) is a closed subtree of X, then $O = O_v$ for some $v \in X$*
(2) *If (X_1, X_2) is an open cut, then $O = O_{X_1,X_2}$.*

Proof It follows easily from the hypotheses that $L \subseteq [y, z]$, where L means $L_{O,e}$. Also, $[y, z] \cap X_1 = L \cap X_1 = L$. For clearly $L \subseteq X_1$, and if $u \in [y, z] \cap X_1$, then $[u, z] = [u, l, z]$ for some $l \in L$, hence $[y, z] = [y, u, l, z]$, so $u \in [y, l] \subseteq L$.

Case (1). Suppose X_1, X_2 are both closed subtrees of X. Then $L = [y, z] \cap X_1 = [y, v]$ for some $v \in X$. Also, $X_2 \cap [y, z] = [u, z]$ for some $u \in X$. It follows that $d(u, v) = 1$ (see (2.26) in [1]).

We claim that, if $d(v, w) = 1$, where $w \in X$, then $(w, v) \in O$. Suppose first that $w \notin L$. Then since $[y, v] = L$, it follows from the remark preceding Lemma 1 that $[y, w] = [y, v, w]$. Again by the remark preceding Lemma 1, either $[x, w] = [x, y, v, w]$, or $[y, w] = [y, x, v, w]$, or $x = w$. If $[x, w] = [x, y, v, w]$, then $(x, y) > (v, w)$, and it follows that $(v, w) \notin O$ (otherwise $w \in L$ by definition of L), hence $(w, v) \in O$. If $[y, w] = [y, x, v, w]$ then $x \in [y, v] \subseteq L$, contradicting the remark preceding Lemma 6. If $x = w$, then $[v, x] = [v, y, x]$ by the remark preceding Lemma 6, and $d(v, x) = 1$, so $v = y$. Therefore $(w, v) = (x, y) \in O$.

Now suppose $w \in L = [y, v]$. By the remark preceding Lemma 6, $[x, v] = [x, y, v] = [x, y, w, v]$, which implies $e = (x, y) > (w, v)$, hence $(w, v) \in O$ by Lemma 5. This establishes the claim.

We finish the proof of Case (1) by showing that $O = O_v$. As in the proof of Lemma 6, it suffices to show that $(p, q) \in E(X)$ and $q \in [p, v]$ implies $(p, q) \in O$. If $q = v$ this follows from the claim we have just proved. If $q \neq v$, then we can find $w \in X$ such that $d(v, w) = 1$ and $[p, v] = [p, q, w, v]$. By the claim, $(w, v) \in O$, and since $(p, q) > (w, v)$, we obtain $(p, q) \in O$ since $O \in \overline{V}E(X)$.

Case (2). Suppose the cut is an open cut, and let ε_1, ε_2 be the open ends of X_1 and X_2 respectively which meet at the cut. By the observations at the beginning of the proof and (2.26), (2.27) in [1], $L = [y, z] \cap X_1 = [y, \varepsilon_1)$, and

$[y, z] = [y, \varepsilon_1) \cup [z, \varepsilon_2)$. Note that, since $y \in L$, $y \in X_1$, hence $x \in X_1$ by Remark 1 preceding Lemma 2.

We show that $O \cap E(X_2) = O_{\varepsilon_2}$. As in the proof of Lemma 6, it suffices to show that $(u, v) \in E(X_2)$ and $v \in [u, \varepsilon_2)$ implies $(u, v) \in O$. Suppose $(u, v) \in E(X_2)$ and $v \in [u, \varepsilon_2)$. Again by [1; 2.27], $[u, y] = [y, \varepsilon_1) \cup [u, \varepsilon_2)$. It follows that $v \in [u, y]$, and that $x \notin [u, y]$. (For $x \notin [y, \varepsilon_1) = L$ by the remark preceding Lemma 6, and $x \notin [u, \varepsilon_2)$ since $[u, \varepsilon_2) \subseteq X_2$ and $x \in X_1$). By the remark preceding Lemma 1, $[u, x] = [u, y, x] = [u, v, y, x]$. It follows that $(x, y) > (v, u)$, hence $(v, u) \notin O$, otherwise $u \in L$, which is impossible. Therefore $(u, v) \in O$, as required.

It remains to show that $O \cap E(X_1) = O_{\varepsilon_1}$. Since $e \in X_1$ and L is the X-ray $[y, \varepsilon_1)$ of X_1, it suffices by Lemma 6 to show that $L = L_1$, where

$$L_1 = \{ u \in X_1 \mid \exists v \in X_1 \text{ such that } (v, u) \in \langle O \cap E(X_1), e] \}.$$

Clearly $L_1 \subseteq L$. Suppose $u \in L$, so $u \in X_1$ and there exists $v \in X$ such that $(v, u) \in \langle O, e]$. It follows from Remark 1 preceding Lemma 2 that $v \in X_1$. Thus $(v, u) \in \langle O \cap E(X_1), e]$, so $u \in L_1$. This finishes the proof.

We can now state a theorem classifying the elements of $\overline{V}E(X)$.

Theorem 1. *Let $E = E(X)$, where (X, d) is a Λ-tree and Λ has a least positive element. Then if $O \in \overline{V}E$, exactly one of the following is true.*
(1) *$O = O_w$ for some $w \in X$*
(2) *$O = O_\varepsilon$ for some open end ε of X*
(3) *$O = O_{X_1, X_2}$ for some open cut (X_1, X_2) of X.*

Proof This follows from Lemma 3, Lemma 6 and Lemma 8, except in the case that X consists of a single point, say $X = \{w\}$, in which case $E(X) = \emptyset$. But then $O_w = \emptyset$, which is the unique element of $\overline{V}E$ in this case.

3

A protree $(E, \leq)$ (indeed, any partially ordered set) is said to satisfy the finite interval condition if, for all e, $f \in E$, the set $\{g \in E \mid e \leq g \leq f\}$ is finite. If $(E, \leq)$ is a G-protree, where G is a group, we say $(E, \leq)$ is a nice G-protree if every G-finite subprotree has the finite interval condition. With these definitions, we can state the main result, which is slightly more general than suggested in the introduction, in that a nice group action on a protree can be extended to an action by isometries on a Λ-tree in which the protree embeds. The method of proof uses ultraproducts, and is similar to that used in [4].

Theorem 2. *Suppose, G is a group and $(E, \leq)$ is a nice G-protree. Then there exist an ordered abelian group Λ with a least positive element, a Λ-tree (X, d) on which G acts as isometries and a G-equivariant embedding of protrees $\phi : E \to E(X)$.*

Proof Let S be the set of all G-finite subprotrees of $(E, \leq)$. For $F \in S$, there is by [7; Theorem 2.1] (or [6; Ch. 2, Cor. 1.10]) a G-tree T_F realising $(F, \leq)$, in that F is the set of oriented edges of T_F, and $(F, \leq)$ is obtained from T_F in the manner described in §1. Further, as noted in §1, T_F determines a $\mathbb{Z}$-tree, say (X_F, d_F), and G acts as isometries on X_F. Thus there is an isomorphism of protrees $F \to E(X_F)$, $e \mapsto (s_F(e), t_F(e))$, where $s_F(e)$ is the initial vertex, and $t_F(e)$ the terminal vertex of e in T_F. We choose, for each $F \in S$, an element $v_F \in X_F$.

For $F \in S$, define $a_F = \{F' \in S \mid F \subseteq F'\}$. Then if $F, F' \in S$, $a_F \cap a_{F'} \supseteq a_{F \cup F'}$, and $F \cup F' \in S$. Also, $a_F \neq \emptyset$ since $F \in a_F$. It follows from Cor. 3.5, Ch. 1 in [3] that there is an ultrafilter $\mathcal{D}$ in $\mathcal{P}(S)$, the Boolean algebra of all subsets of S, such that $a_F \in \mathcal{D}$ for all $F \in S$. Form the ultraproducts

$$X = \textstyle\prod_{F \in S} X_F / \mathcal{D}, \quad {}^*G = \textstyle\prod_{F \in S} G / \mathcal{D} = G^S / \mathcal{D}, \quad \Lambda = \textstyle\prod_{F \in S} \mathbb{Z} / \mathcal{D} = \mathbb{Z}^S / \mathcal{D}$$

and let $d = \prod_{F \in S} d_F / \mathcal{D}$. If $(x_F)_{F \in S}$ is an element of the cartesian product $\prod_{F \in S} X_F$, its equivalence class in $\prod_{F \in S} X_F / \mathcal{D}$ will be denoted by $\langle x_F \rangle_{F \in S}$, and similar notation will be used for the elements of the other ultraproducts. Thus d is given by: $d(\langle x_F \rangle_{F \in S}, \langle y_F \rangle_{F \in S}) = \langle d_F(x_F, y_F) \rangle_{F \in S}$.

By Lemma 5 of [4], (X, d) is a Λ-tree, with ${}^*G$ acting as isometries, hence G acts as isometries via the canonical embedding of groups $G \hookrightarrow {}^*G$, $g \mapsto \langle g_F \rangle_{F \in S}$, where $g_F = g$ for all $F \in S$. Explicitly, $g \langle x_F \rangle_{F \in S} = \langle g x_F \rangle_{F \in S}$ for $g \in G$ and $x_F \in X_F$. Further, it is well-known (and easy to prove) that the canonical embedding of $\mathbb{Z}$ into Λ embeds $\mathbb{Z}$ as a minimal non-trivial convex subgroup of Λ, so the integer 1 is the least positive element of Λ. Thus we can form the protree $(E(X), \leq)$, and $E(X)$ is a ${}^*G$-protree, hence a G-protree. For $e \in E$, define $s(e) = \langle u_F(e) \rangle_{F \in S}$, where

$$u_F(e) = \begin{cases} s_F(e) & \text{if } e \in F \\ v_F & \text{otherwise} \end{cases}$$

and define $t(e) = \langle w_F(e) \rangle_{F \in S}$, where

$$w_F(e) = \begin{cases} t_F(e) & \text{if } e \in F \\ v_F & \text{otherwise.} \end{cases}$$

Thus $s(e), t(e) \in X$, and in fact $d(s(e), t(e)) = 1$. For let $F_0 = Ge \cup Ge^*$, so $F_0 \in S$ and for all $F \in a_{F_0}$, $e \in F$, hence $d_F(u_F(e), w_F(e)) = d_F(s_F(e), t_F(e)) = 1$, since d_F is the path metric on T_F. Since $a_{F_0} \in \mathcal{D}$, it follows that $d(s(e), t(e)) = 1$. Therefore there is a mapping $\phi : E \to E(X)$ given by $\phi(e) = (s(e), t(e))$.

Now for $e \in E$, we have just used the fact that, for all $F \in a_{F_0}$, $u_F(e) = s_F(e)$ and $w_F(e) = t_F(e)$. Since $a_{F_0} \in \mathcal{D}$ and $\mathcal{D}$ is closed under finite intersections, it follows that, if $x_F \in X_F$ for $F \in S$, then $\langle x_F \rangle_{F \in S} = s(e)$ if and only if $x_F = s_F(e)$ for almost all F, and $\langle x_F \rangle_{F \in S} = t(e)$ if and only if $x_F = t_F(e)$ for almost all F. (That is, the sets $\{F \in S \mid x_F = s_F(e)\}$ and $\{F \in S \mid x_F = t_F(e)\}$ belong to $\mathcal{D}$).

It follows easily that $s(e)$, $t(e)$ are independent of the choice of the elements v_F. It also follows that ϕ is one-to-one. For if e, $f \in E$, $s(e) = s(f)$ and $t(e) = t(f)$, then $s_F(e) = s_F(f)$ and $t_F(e) = t_F(f)$ for almost all F, in particular for at least one choice of $F \in S$. But this implies $e = f$ since T_F has no circuits of length 2. We have to show ϕ is an embedding of protrees. Suppose e, $f \in E$ and $e < f$. Let $F_1 = Ge \cup Ge^* \cup Gf \cup Gf^*$, so $F_1 \in S$. For all $F \in a_{F_1}$, we have $[s_F(f), t_F(e)] = [s_F(f), t_F(f), s_F(e), t_F(e)]$ and since $a_{F_1} \in \mathcal{D}$, it follows that

$$d(s(f), t(e)) = d(s(f), t(f)) + d(t(f), s(e)) + d(s(e), t(e))$$

hence, by (2.14)(b) in [1], $[s(f), t(e)] = [s(f), t(f), s(e), t(e)]$. Thus $\phi(e) < \phi(f)$, and ϕ is order preserving.

If $e \in E$, let $F_0 = Ge \cup Ge^*$, so $F_0 \in S$ and for all $F \in a_{F_0}$, $s_F(e^*) = t_F(e)$. Since $a_{F_0} \in \mathcal{D}$, it follows that $s(e^*) = t(e)$ for all $e \in E$, hence $s(e) = t(e^*)$ and $\phi(e^*) = \phi(e)^*$, as required.

Finally, we have to show that ϕ is G-equivariant. Let $g \in G$ and $e \in E$. Again let $F_0 = Ge \cup Ge^*$. For $F \in a_{F_0}$, T_F is a G-tree, so

$$(u_F(ge), v_F(ge)) = (s_F(ge), t_F(ge)) = (gs_F(e), gt_F(e)) = (gu_F(e), gv_F(e)).$$

By definition, $gs(e) = \langle gu_F(e) \rangle_{F \in S}$, and this is equal to $\langle u_F(ge) \rangle_{F \in S}$, i.e. to $s(ge)$, since $a_{F_0} \in \mathcal{D}$. Similarly $gt(e) = t(ge)$, so $g\phi(e) = \phi(ge)$, and the proof is complete.

We end by noting that Theorem 2 cannot be strengthened to the statement that any protree E is isomorphic to $E(X)$ for some suitable Λ-tree X. For example, let $E_0 = E(X)$, where X is $\mathbb{Z}$, viewed as a $\mathbb{Z}$-tree. Thus $E_0 = \{(m, n) \mid m, n \in \mathbb{Z}$ and $|m - n| = 1\}$. Let $O = \{(n, n + 1) \mid n \in \mathbb{Z}\}$, so $O \in \overline{V}E_0$ (indeed $O = O_\varepsilon$ where ε is one of the ends of $\mathbb{Z}$). Now take 3 copies of E_0, i.e. let $E = E_0 \times \{1, 2, 3\}$. For e, $e' \in E_0$ and k, $k' \in \{1, 2, 3\}$, define $(e, k) < (e', k')$ to mean that either $k = k'$ and $e < e'$ in E_0, or $k \neq k'$, $e' \in O$ and $e \notin O$. Also, define $(e, k)^*$ to be (e^*, k). It is easily checked that this makes E into a protree.

Let $O' = O \times \{1, 2, 3\}$, so O' is an orientation of E, and in fact $O' \in \overline{V}E$. For if $(e, k) \in O'$ and $(e, k) < (e', k')$, then $e \in O$, so $k = k'$ and $e < e'$, hence $e' \in O$ and $(e', k') \in O'$. We claim that there are three directions at O' in E, namely the equivalence classes of $e_i = ((0, 1), i)$ for $i = 1$, 2 and 3. For suppose $e \in O$ and $k \in \{1,2,3\}$. Then either $(e,k) \leq e_k$ or $e_k \leq (e,k)$, so $(e,k) \sim e_k$. If $e_i \sim e_j$, then there exists $g \in O$ and $k \in \{1,2,3\}$ such that $e_i \geq (g,k)$ and $e_j \geq (g,k)$. Since $g \in O$, this implies $i = k = j$, establishing the claim.

Suppose there is an isomorphism of protrees $\phi : E \to E(X)$, where (X,d) is a Λ-tree and Λ is an ordered abelian group with a least positive element 1. By Lemma 2 and Theorem 1, $\phi(O') = O_w$ for some $w \in X$. Now X has more than one point, since E is non-empty, so there is a point of X, say u, at distance 1 from w. Then $(u,w) \in O_w$, so $(u,w) = \phi(e)$ for some $e \in O'$. We can write $e = ((n, n+1), k)$ for some $n \in \mathbb{Z}$ and $k \in \{1,2,3\}$. Let $f = ((n+1, n+2), k)$, so $f \in O'$, and $\phi(f) = (u', w')$ for some $u', w' \in X$. Since $e > f$, $\phi(e) > \phi(f)$, that is, $[u, w'] = [u, w, u', w']$. It is easy to see that there is no edge $g \in E$ such that $e > g > f$, and it follows that $w = u'$, so $\phi(f) = (w, w')$. But then $\phi(f^*) \in O_w$, hence $f^* \in O'$, a contradiction, so no such isomorphism ϕ exists.

References

[1]   R.C. Alperin and H. Bass, *Length functions of group actions on Λ-trees,* in: Combinatorial Group Theory and Topology (S.M.Gersten and J.R.Stallings, eds.), Annals of Mathematics Studies, **111** Princeton University Press (1987), pp. 265–378.

[2]   H. Bass, *Group actions on non-archimedean trees,* in: Arboreal Group Theory (R.C. Alperin, ed.), MSRI Publications **19** Springer–Verlag, New York (1991) pp. 69–131.

[3]   J.L. Bell and A.B. Slomson, *Models and ultraproducts,* North-Holland, Amsterdam (1971).

[4]   I.M. Chiswell, *Generalised trees and Λ-trees,* in: Combinatorial and Geometric Group Theory, Edinburgh 1993 (A.J. Duncan, N.D. Gilbert and J. Howie, eds.), London Mathematical Society Lecture Notes, **204** Cambridge University Press (1994), pp. 43–55.

[5]   D.E. Cohen, *Groups of Cohomological Dimension One,* Lecture Notes in Mathematics, **245** Springer, Berlin, Heidelberg, New York (1972).

[6]   W. Dicks and M.J. Dunwoody, *Groups Acting on Graphs,* Cambridge University Press (1989).

[7]   M.J. Dunwoody, *Accessibility and groups of cohomological dimension 1,* Proc. London Math. Soc. **38** (1979), pp. 193–215.

[8]   M.J. Dunwoody, *Inaccessible groups and protrees,* J. Pure Appl. Algebra **88** (1993), pp. 63-78.

[9]   M.J. Dunwoody, *Groups acting on protrees,* J. London Math. Soc. **56** (1997), pp. 125–136.

[10] J.-P. Serre, *Trees*, Springer, New York (1980).

[11] J.R. Stallings, *Group Theory and Three-dimensional manifolds*, Yale University Press, New Haven and London (1971).

Homological Techniques for Strongly Graded Rings: A Survey

Jonathan Cornick

Department of Mathematics, The Ohio State University, 231 W18th Avenue, Columbus, Ohio 43210, USA.

email:cornick@math.ohio-state.edu

1. Introduction

Let G be a group and k be a commutative ring. The k-algebra R is said to be graded by G if it has a direct sum decomposition as k-modules

$$R = \bigoplus_{g \in G} R_g$$

such that $R_g R_h \subseteq R_{gh}$ for all $g, h \in G$. We call R_1 the base ring. For example, if one wants to study the group algebra $R = k\Gamma$ where $\Gamma/N \cong G$ then it is sometimes useful to consider R as being G-graded with $R_{Nx} = kNx$ and base ring kN. In this case $R = kN * G$ is a crossed product of G over kN and thus a *strongly* G-graded k-algebra since $kNxkNy = kNxy$ for all $Nx, Ny \in G$. We shall elaborate on this in Section 2. Most of what we describe is true for arbitrary k but for simplicity we shall assume that $k = \mathbb{Z}$ and from now on R will denote a strongly graded ring (i.e. $\mathbb{Z}$-algebra) unless otherwise stated. The purpose of this paper is to describe some techniques for studying certain homological properties of R and of the category of (right) R-modules. Much of what we shall say is already known but it seems worthwhile to collect this material for group theorists since some of the known (and hopefully future) applications have a homological group theory flavour. From a group theoretical point of view the main problem with (but also the main interest in) studying strongly graded rings is that although G is not in general isomorphic to a subgroup of $U(R)$, the units of R, its group structure has a great influence on the ring structure of R (which is not the case for arbitrary graded rings.) The unifying theme of the paper is to use homological algebra and group

The author was supported by a CRM Postdoctoral Fellowship

actions to translate properties of simpler objects , i.e. group rings and skew group rings where $G \subseteq U(R)$, into the more general setting.

In Section 3 we describe *Cohen-Montgomery duality* for rings which are graded by finite groups [12]. Let F be a finite group, let R be an F-graded ring, let $M_F(R)$ be the matrix ring over R with rows and columns indexed by F, and let $R\sharp F^*$ be the smash product which we shall define later. We shall sketch the proof of the following duality theorem [12, Theorem 2.12 and Theorem 3.2] .

Theorem 1.1. *(Cohen-Montgomery) The matrix ring $M_F(R)$ is isomorphic to the skew group ring $(R\sharp F^*)F$. Furthermore, if R is strongly graded then $R\sharp F^*$ is Morita equivalent to R_1.*

Therefore, since R is Morita equivalent to $M_F(R)$, we can study R-modules by studying modules over a skew group ring whose base ring is Morita equivalent to that of R. As an illustration we prove a version of Maschke's theorem for strongly graded rings.

If the group is infinite then the duality construction can be extended in various ways. Unfortunately, we then find ourselves faced with infinitely indexed matrix rings and thus lose Morita equivalence. The question is then: Which type of matrix ring to use in a given situation?

Quinn discovered one such construction which has proved useful for answering ring theoretic questions concerning (graded) ideals [31]. The construction we shall use differs slightly from Quinn's in that the matrix algebra is nonunital. However it is a direct limit of ordinary matrix rings and this is often enough to solve problems of a homological (though not cohomological) nature. We then show how the Hochschild homology and, if G is torsion free, the cyclic homology groups of strongly graded algebras can be computed by using duality together with the corresponding results for skew group rings in [27] and [19].

If S is a ring then there are Connes' shift operators $\sigma : HC_{2n}(S) \longrightarrow HC_{2n-2}(S)$ and Chern characters $ch_n : K_0(S) \longrightarrow HC_{2n}(S)$ for every integer n, with $ch_{n-1} = \sigma \cdot ch_n$. In particular the Hattori-Stallings rank function ch_0 factors through the higher cyclic homology groups. Eckmann has used this fact to prove the strong Bass conjecture for rational group algebras of some classes of groups [18]. We show how Eckmann's result can be slightly extended by computing the cyclic homology of certain crossed products.

In Section 4 we review the *tensor identity* for strongly graded rings (see [15]) and describe some applications. If R is G graded , M is a right R-module and V is a right $\mathbb{Z}G$-module then $M \otimes V$ has a right R-module structure and in particular $M \otimes \mathbb{Z}[H\backslash G]$ is an R-module for every subgroup H of G. Let R_H be the subring of R supported on the homogeneous components corresponding to elements of H. The tensor identity states that

Lemma 1.2. *If R is a strongly graded ring then $M \otimes \mathbb{Z}[H\backslash G] \cong M \otimes_{R_H} R$ as R-modules*

The main applications of the tensor identity so far have been in proving that an R-module M has finite projective dimension if it has finite projective dimension over R_H for every subgroup H of G belonging to a some class of groups $\mathfrak{x}$ [23] and [24]. The basic strategy is to find naturally occurring long exact sequences

$$0 \longrightarrow V_n \longrightarrow \cdots \longrightarrow V_1 \longrightarrow V_0 \longrightarrow \mathbb{Z} \longrightarrow 0$$

where each V_i is a direct sum of permutation modules of the form $\mathbb{Z}[H\backslash G]$ with $H \in \mathfrak{x}$, and then to tensor this sequence with M since $M \otimes V_i$ is a projective R-module by the tensor identity. This happens, for example, if $G \in \mathbf{H_1}\mathfrak{x}$.

Definition 1.3. $\mathbf{H_1}\mathfrak{x}$ is the class of groups which admit a cellular action on a finite dimensional contractible cell complex with isotropy subgroups in $\mathfrak{x}$. In particular $\mathbf{H_1}\mathfrak{F}$ is the class of groups which admit an action with finite isotropy groups.

The class $\mathbf{H_1}\mathfrak{F}$ (strictly) includes all groups of finite virtual cohomological dimension and seems to be the correct setting for studying such groups. For example Theorem 1.8 and Theorem 1.10 generalise known results about groups of finite virtual cohomological dimension to groups in $\mathbf{H_1}\mathfrak{F}$. Many of the results in this paper concerning groups in $\mathbf{H_1}\mathfrak{F}$ have also been proved for the much larger class $\mathbf{H}\mathfrak{F}$ if one restricts to modules of type FP_∞.

To illustrate how the tensor identity is used we prove the following version of Chouinard's theorem for strongly graded rings [2]:

Theorem 1.4. *(Aljadeff-Ginosar) Let F be a finite group, let R be an F-graded ring and let M be an R-module. Then M is a projective R-module if and only if M is projective as an R_E-module for all elementary abelian subgroups E of F.*

The idea in the proof is to use the tensor identity twice, first to reduce to Sylow p-subgroups of F and then to reduce from p-groups to elementary abelian p-subgroups by an induction argument.

Let k be a field of characteristic $p > 0$ and let M be a finitely generated kF-module. The theory of varieties of modules has been developed by several authors in recent years and has proved to be very powerful. See [7, Chapter 5] for an overview and references to other work. Basically the idea is to study the commutative ring

$$H^{\cdot}(F, k) = \begin{cases} \oplus_{n \geq 0} H^{2n}(F, k) & \text{if } p \neq 2, \\ H^{*}(F, k) & \text{if } p = 2 \end{cases}$$

using techniques from algebraic geometry. Let V_F denote the affine algebraic variety of the maximal ideal spectrum of $H^{\cdot}(F, k)$. One associates a subvariety $V_F(M)$ to the module and the dimension of this variety is equal to the complexity $c_F(M)$ of M. Let

$$\cdots \longrightarrow P_2 \longrightarrow P_1 \longrightarrow P_0 \longrightarrow M \longrightarrow 0$$

be a minimal projective resolution of M [6, p.29].

Definition 1.5. The complexity $c_F(M)$ of the module M is defined to be the least integer s such that there is a constant $\kappa > 0$ with

$$\dim_k P_n \leq \kappa.n^{s-1}.$$

Alperin and Evens have shown that $c_F(M)$ is equal to the maximal complexity $c_E(M)$ of the restriction of M to elementary subgroups E of F [1]. It would be interesting to develop a complexity theory for group algebras of arbitrary groups or even for more general types of ring, but this seems out of reach with present methods.

However, if $c_F(M) = 0$ then it follows that M is projective and this situation had already been covered by Chouinard [11]. Thus in some sense Theorem 1.4 can be considered as the 'complexity zero' case for strongly graded rings over finite groups.

As another application of the tensor identity we prove the following theorem [15].

Theorem 1.6. *(Cornick-Kropholler)*
Suppose $G \in \mathbf{H}_1 \mathfrak{F}$ and H is a subgroup of G of finite index. Let R be a G-graded ring and let M be an R-module. Then M is a projective as an R-module if and only if it is projective as an R_H-module and as an R_F-module for all finite subgroups F of G.

Using Theorems 1.4 and 1.6 we outline a proof of a theorem which was proved during the conference [3].

Conjecture 1.7. *(Moore) Let G be a group, let H be a subgroup of finite index and let S be a ring. Suppose that for every $g \in G - H$, either*
  1 g has finite order invertible in S, or
  2 there exists an integer n such that $g^n \in H - \{1\}$.
Then every SG-module M which is projective over SH is also projective over SG.

Notice that the conjecture makes sense for strongly graded rings where R takes the place of SG, R_H takes the place of SH and the condition that g has finite order invertible in S is replaced by the condition that g has finite order invertible in R_1. If these conditions are satisfied we say that the triple (R, G, H) satisfies the Moore condition.

Theorem 1.8.   *Let G be a group in $\mathbf{H}_1\mathfrak{F}$, let H be a subgroup of finite index and let R be a strongly G-graded ring. If (R, G, H) satisfies the Moore condition and M is an R-module which is projective over R_H then M is a projective R-module.*

Finally in Section 5 we turn our attention to the complete cohomology theory invented independently by Vogel (see Goichot's paper [22]) and Mislin [29]. Recall that for any ring S and S-modules M and N, there are complete cohomology groups $\widehat{\mathrm{Ext}}_S^n(M, N)$ defined for every integer n. There has been considerable progress in developing this theory. For example, $proj.dim_S(M) < \infty$ if and only if $\widehat{\mathrm{Ext}}_S^0(M, M)$ [23] and this fact has been used in several subsequent papers ([24], [25], [14], [15]). However it has proved difficult to actually compute the complete cohomology of specific examples.

When G is a group of finite virtual cohomological dimension and M is a $\mathbb{Z}G$-module then $\widehat{\mathrm{Ext}}_{\mathbb{Z}G}^*(\mathbb{Z}, M) \cong \widehat{H}^*(G, M)$ (Tate-Farrell cohomology.) The advantage with Tate-Farrell cohomology is that it is that it is defined using *complete resolutions* and so techniques such as Shapiro's Lemma and spectral sequences are available for computation [9, Chapter X]. Under what other conditions do complete resolutions exist, and when can they be used to compute complete cohomology groups?

The first observation to make is that complete cohomology disappears on projective modules. With this in mind we make the following definition.

Definition 1.9.   Let S be a ring and let M be an S-module. Then a complete resolution of M is an acyclic sequence of projective S-modules $P_\bullet = (P_*, \delta)$, indexed by the integers, such that

  $P_\bullet$ agrees with a projective resolution of M in sufficiently high dimensions.

  $\mathrm{Hom}_S(P_*, Q)$ is acyclic for every projective S-module Q.

We remark that only the first part of the definition is used in defining complete resolutions in Tate-Farrell cohomology, but it automatically follows that these cohomology groups vanish on projective (even induced) modules. Most importantly we first show that if $P_\bullet$ is a complete resolution of M then

$$\widehat{\mathrm{Ext}}_S^*(M, N) \cong H^*(\mathrm{Hom}_S(P_*, N))$$

for every S-module N. Returning to strongly graded rings we prove the following theorem which describes circumstances under which complete resolutions exist with an explicit construction.

Theorem 1.10.   *Let G be a group in the class $\mathbf{H}_1\mathfrak{F}$, let R be a strongly G-graded ring and let M be an R-module such that $proj.dim_{R_1} M < \infty$ Then M has a complete resolution.*

As a corollary to this theorem we prove a version of Shapiro's Lemma for the complete cohomology of strongly graded rings.

2. Strongly graded rings

In this section we define strongly graded rings and discuss some basic examples and properties which we will need later. The reader is referred to Dade's paper [17] or Passman's book [30] for further background information.

Definition 2.1. Let G be a group.

The ring R is strongly G-graded if it has a direct sum decomposition as abelian groups $R = \bigoplus_{g \in G} S_g$ such that $R_g R_h = R_{gh}$ for all $g, h \in G$.

The subring R_1 is called the base ring of R.

If X is subset of G then $R_X = \bigoplus_{g \in X} R_x$. In particular if H is a subgroup of G then R_H is subring of R.

As mentioned in the introduction strongly graded rings are a generalisation of other well known types of ring. We choose to work in this setting rather than with crossed products because the proofs seem to become more streamlined.

Definition 2.2. Let R be a strongly graded ring and let $U(R)$ denote the units of R.

If there exists $u_g \in U(R) \cap R_g$ for every $g \in G$ then R is a crossed product, and in this case $R = \bigoplus_{g \in G} R_1 u_g$.

If the set $\{u_g : g \in G\}$ forms a subgroup of $U(R)$ then $R = R_1 G$ is a skew group ring. In this case there is a group homomorphism $G \longrightarrow \mathrm{Aut}(R_1)$ defined by $g \mapsto (r \mapsto u_g^{-1} r u_g)$.

If the homomorphism is trivial then R is the group ring $R_1[G]$.

We have the following examples when S is a ring, $1 \longrightarrow N \longrightarrow \Gamma \longrightarrow G \longrightarrow 1$ is a group extension and $R = S\Gamma$ is the group ring.

Example 2.3. R is a strongly G-graded ring with base ring SN and Nx component SNx for $Nx \in \Gamma/N \cong G$. In fact

1 $R = SN * G$ is a crossed product of G over SN because $x \in SNx$ is a unit.

2 If the group extension is split then $R = (SN)G$ is a skew group ring since the coset representatives may be chosen to form a subgroup of Γ.

3 If $\Gamma \cong N \times G$ then $R = S[N \times G] \cong SN[G]$ is a group ring.

It has proved useful in some applications to use the following equivalent definition of G-graded rings [15].

Definition 2.4. A G-graded ring is a ring R together with a ring homomorphism $\gamma : R \longrightarrow R \otimes \mathbb{Z}G$ which makes R a $\mathbb{Z}G$-comodule, and in this case

$$R_g = \{r \in R : \gamma(r) = r \otimes g\}.$$

The advantage with this definition is that if M is an R-module and V is a $\mathbb{Z}G$-module then $M \otimes V$ has a right R-module structure via the *semi-diagonal action* $(m \otimes v) \cdot r = mr \otimes vg$ where $m \in M, v \in V$ and $r \in R_g$. We will use this fact to deduce the tensor identity for strongly graded rings in Section 4. Similarly if N is an R-module we can define a semi-diagonal action on $\mathrm{Hom}_{\mathbb{Z}}(V, N)$ by $\phi^r(v) = (\phi(vg^{-1}))r$. The functors $\_ \otimes V$ and $\mathrm{Hom}_{\mathbb{Z}}(V, \_)$ from R-modules to itself are adjoint [15, Lemma 3.2].

Lemma 2.5. $\mathrm{Hom}_R(M \otimes V, N) \cong \mathrm{Hom}_R(M, \mathrm{Hom}_{\mathbb{Z}}(V, N))$.

3. Cohen-Montgomery duality

Before we describe the duality construction it will be useful to recall some facts about Morita contexts [4, pp. 60-74].

3.1. Morita contexts

Let S and T be rings, let M be an S-T-bimodule and let N be a T-S-bimodule.

Definition 3.1. A Morita context is a 6-tuple $(S, T, M, N, \phi, \theta)$ where

$$\phi : M \otimes_T N \longrightarrow S \text{ and } \theta : N \otimes_S M \longrightarrow T$$

are bimodule maps satisfying the associativity conditions $\phi(m \otimes n) \cdot m = m \cdot \theta(n \otimes m)$ and $n \cdot \phi(m \otimes n) = \theta(n \otimes m) \cdot n$ for all $m \in M, n \in N$.

Theorem 3.2. *Let $(S, T, M, N, \phi, \theta)$ be a Morita context.*

1 If ϕ is a surjection then it is an isomorphism and M and N are finitely generated projective T-modules. Similarly for θ and S.

2 If ϕ and θ are both isomorphisms then S and T are Morita equivalent, and there is a category equivalence $\_ \otimes_S M : Mod\,S \longrightarrow Mod\,T$ with inverse $N \otimes_T \_$.

3 If in addition $S = T$ then P and Q are invertible bimodules. Thus

$$[P], [Q] \in Pic(S)$$

the Picard group of S, and $[P]^{-1} = [Q]$.

We illustrate this theorem with the standard example of matrix rings. Let T denote the matrix ring $M_n(S)$ over the unital ring S, let $\mathrm{Row}_n(S)$ be the S-T-bimodule of row matrices of length n and let $\mathrm{Col}_n(S)$ be the corresponding T-S-bimodule of column matrices. Then we have a Morita context

$$(S, T, \mathrm{Row}_n(S), \mathrm{Col}_n(S), \phi, \theta)$$

where ϕ and θ are given by matrix multiplication. Since S is unital it is easy to see that ϕ and θ are both surjections and so S and $M_n(S)$ are Morita equivalent.

3.2. Finite Groups and Maschke's Theorem

Definition 3.3. Let F be a finite group, let g and h be arbitrary elements of F and let R be an F-graded ring.

1 $M_F(R)$ is the matrix ring over R with rows and columns indexed by F. $\mathrm{Row}_F(R)$ and $\mathrm{Col}_F(R)$ are then defined in the obvious way.

2 $e_{g,h} \in M_F(R)$ is the element with 1 in the g,h-position and zeros elsewhere. Similarly for $\rho_g \in \mathrm{Row}_F(R)$ and $\kappa_g \in \mathrm{Col}_F(R)$.

3 $R\natural F^* = \sum_{g,h\in F} R_{gh^{-1}} e_{g,h}$ is a subring of $M_F(R)$ called the smash product.

Proof of Theorem 1.1 We identify R as a subring $R\natural F^*$ via the ring monomorphism μ where

$$\mu(r) = \sum_{h\in F} r e_{h,g^{-1}h}, r \in R_g.$$

Similarly we identify F as a subgroup of $U(M_F(R))$ via the group monomorphism ν where

$$\nu(g) = \sum_{h\in F} e_{h,hg},$$

and observe that F acts on $R\natural F^*$ by matrix conjugation. It is easy to check that $\sum_{g\in F} R\natural F^* g \subseteq M_F(R)$ is a direct sum and is in fact equal to $M_F(R)$. Thus the first part of the theorem is proved.

For the second part let $P = \bigoplus_{g\in F} R_{g^{-1}} \rho_g$ and $Q = \bigoplus_{g\in F} R_g \kappa_g$. Then we have a Morita context $(R_1, R\natural F^*, P, Q, \phi, \theta)$ where ϕ and θ are given by matrix multiplication. The map ϕ is clearly a surjection and θ is a surjection since R is strongly graded.

Although the calculations in the proof are easy to check, they are also tedious. Thus it is worthwhile to keep the next example in mind.

Example 3.4. *Let* $F = C_3 = \{1, g, h\}$, *let* $R = R_1 \oplus R_g \oplus R_h$ *be a strongly F-graded ring and let* $r = r_1 + r_g + r_h$ *be an element of R. Then*

$$R\natural F^* = \begin{bmatrix} R_1 & R_h & R_g \\ R_g & R_1 & R_h \\ R_h & R_g & R_1 \end{bmatrix} \text{ and } \mu(r_1 + r_g + r_h) = \begin{bmatrix} r_1 & r_h & r_g \\ r_g & r_1 & r_h \\ r_h & r_g & r_1 \end{bmatrix}.$$

The group F is embedded in $U(M_F(R))$ by the regular representation

$$\nu(1) = \begin{bmatrix} 1 & 0 & 0 \\ 0 & 1 & 0 \\ 0 & 0 & 1 \end{bmatrix}, \nu(g) = \begin{bmatrix} 0 & 1 & 0 \\ 0 & 0 & 1 \\ 1 & 0 & 0 \end{bmatrix}, \nu(h) = \begin{bmatrix} 0 & 0 & 1 \\ 1 & 0 & 0 \\ 0 & 1 & 0 \end{bmatrix},$$

and so

$$
M_F(R) \;=\;
\begin{bmatrix} R_1 & R_h & R_g \\ R_g & R_1 & R_h \\ R_h & R_g & R_1 \end{bmatrix}
\oplus
\begin{bmatrix} R_g & R_1 & R_h \\ R_h & R_g & R_1 \\ R_1 & R_h & R_g \end{bmatrix}
\oplus
\begin{bmatrix} R_h & R_g & R_1 \\ R_1 & R_h & R_g \\ R_g & R_1 & R_h \end{bmatrix}
$$

$$
\begin{array}{cccc}
\| & \| & \| & \| \\
(R\sharp F^*)F & R\sharp F^* & R\sharp F^* g & R\sharp F^* h
\end{array}
$$

As an illustration of the duality theorem we prove the following variation of Maschke's Theorem which we will need in Section 4 [30, Chapter 1].

Theorem 3.5. *Let F be a finite group and let R be a strongly F-graded ring such that $|F| \in U(R_1)$. Then every R-module M which is projective as an R_1-module is also projective as an R-module.*

Proof By Morita equivalence it is enough to show that $\mathrm{Row}_F(M)$ is a projective $M_F(R)$-module. Recall that $P = \bigoplus_{g \in F} R_{g^{-1}} \rho_g$ and that there is a category equivalence $\_ \otimes_{R_1} P : \mathrm{Mod}\, R_1 \longrightarrow \mathrm{Mod}\, R\sharp F^*$. It is easy to see that

$$
\mathrm{Row}_F(M)_{|R\sharp F^*} \cong M_{|R_1} \otimes_{R_1} P
$$

and so $\mathrm{Row}_F(M)_{|R\sharp F^*}$ is projective. Thus we may assume that $R = \bigoplus_{g \in F} R_1 u_g$ is a skew group ring.

Choose a free R-module V which maps onto M. Since M is projective over R_1 there is an R_1-homomorphism $\alpha : M \longrightarrow V$ which splits the surjection. But then, using the usual averaging trick, we can split the surjection with the R-homomorphism

$$
m \mapsto 1/|F| \sum_{g \in F} \alpha(m u_{g^{-1}}) u_g.
$$

3.3 Infinite groups: Hochschild and cyclic homology

Turning now to infinite groups we denote by $M_G(R)$ the ring of row and column finite matrices indexed by G with entries from R and by $M_G^*(R)$ the ideal of $M_G(R)$ whose elements have only finitely many non-zero entries. If $\mathcal{F}$ is the set of finite subsets of G then

$$
M_G^*(R) = \varinjlim_{X \in \mathcal{F}} M_X(R)
$$

is a direct limit of ordinary matrix rings over R. Of course, if G is finite then $M_G^*(R) = M_G(R)$ is the usual matrix ring.

The homomorphisms $\mu : R \longrightarrow M_G(R)$ and $\nu : G \longrightarrow U(M_G(R))$ still make sense if G is infinite and we put $R\sharp G^* = \sum_{g,h \in F} R_{gh^{-1}} e_{g,h} \cap M_G^*(R)$. There is an action of G on the non-unital ring $R\sharp G^*$ by matrix conjugation in $M_G(R)$ and one can deduce [13, Theorem 2.8].

Theorem 3.6. $M_G^*(R) = (R \sharp G^*)G$, *the skew group ring of G over $R \sharp G^*$.*

We use this theorem to to compute the Hochschild homology and, if G is torsion free, the cyclic homology groups of strongly graded rings. The reader unfamilar with these homology theories is referred to [26] for definitions and properties. We first reduce to the case of skew group rings [13, Theorem 2.11].

Corollary 3.7. $HH_*(R) \cong HH_*((R \sharp G^*)G)$ *and* $HC_*(R) \cong HC_*((R \sharp G^*)G)$.

Proof $HH_*(R)$ and $HC_*(R)$ are Morita invariants of the ring R (e.g. [27]) and commute with direct limits (since they are homology theories).

Let $T(G)$ denote the set of conjugacy classes in G with $[g] \in T(G)$ representing the conjugacy class of $g \in G$. Then the homology groups of R have a direct sum decomposition indexed by elements of $T(G)$:

$$HH_*(R) = \bigoplus_{[g] \in T(G)} HH_*(R)_{[g]} \text{ and } HC_*(R) = \bigoplus_{[g] \in T(G)} HC_*(R)_{[g]}.$$

Example 3.8. *If $S[G]$ is a group ring then $HH_0(S[G]) = HC_0(S[G]) = S^{T(G)}$.*

If $R = R_1 G = \bigoplus_{g \in G} R_1 u_g$ is a skew group ring and $g \in G$, then there is an action of $\mathbf{C}_G(g)$ on the Hochschild complex $C_*(R_1, R_1 u_g)$ by

$$(r \otimes r_1 \otimes \cdots \otimes r_n)^x = (u_x^{-1} r u_g u_x \otimes u_x^{-1} r_1 u_x \otimes \cdots \otimes u_x^{-1} r_n u_x), x \in \mathbf{C}_G(g).$$

Similarly there is an action on the acyclic Hochschild complex $C'(R_1)$ and thus an action on $\mathrm{Tot}\mathcal{C}(R_1)$. One can then prove the following [13].

Theorem 3.9. *Let R be a strongly G-graded ring. Then*

1 for all $g \in G$ there is a first quadrant spectral sequence

$$(E_{[g]}^2)_{p,q} = H_p(\mathbf{C}_G(g), HH_q(R_1, R_g)) \Longrightarrow HH_{p+q}(R)_{[g]}.$$

2 There is a first quadrant spectral sequence

$$(E_{[1]}^2)_{p,q} = H_p(G, HC_q(R_1)) \Longrightarrow HC_{p+q}(R)_{[1]}.$$

3 If $g \in G$ has infinite order then there is a first quadrant spectral sequence

$$(E_{[g]}^2)_{p,q} = H_p(\mathbf{C}_G(g)/\langle g \rangle, HH_q(R_1, R_g)) \Longrightarrow HC_{p+q}(R)_{[g]}.$$

Proof(sketch) This essentially follows from the corresponding results for skew group rings ([27, Proposition 2.6], [19, Theorem 4.1.1]) and Corollary 3.7. One must check the following:

    1 The isomorphisms in Corollary 3.7 respect $[g]$-components. This is an easy computation.

2 $HH_q(R_1, R_g) \cong HH_q(R\sharp G^*, R\sharp G^* g)$ as abelian groups. This follows from the fact that R_1 and $R\sharp G^*$ are Morita equivalent if G is finite, Morita invariance of Hochschild homology and a direct limit argument.

3 The existence of an action of $\mathbf{C}_G(g)$ on $HH_q(R_1, R_g)$ which makes the isomorphism into a $\mathbb{Z}\mathbf{C}_G(g)$-isomorphism. There is a Morita context

$$(R_1, R_1, R_g, R_{g^{-1}}, \phi, \theta)$$

where the maps are given by multiplication in R and are surjections since R is strongly graded. Thus R_g is an invertible bimodule by Theorem 3.2 and so there is a group homomorphism $G \longrightarrow Pic(R_1)$ defined by $g \mapsto [R_g]$. We have our action since finitely generated projective modules induce homorphisms on Hochschild homology [27].

Example 3.10. Let Γ be a countable group which is an extension of a locally finite group N by a torsion free group G, and let $T_0(\Gamma)$ (respectively $T_1(G)$) denote the set of conjugacy classes of elements of finite (respectively infinite) order. Using Burghelea's description of the cyclic homology of group algebras [10] we know that $HC_*(\mathbb{Q}\Gamma) = A \oplus B$ where

$$A = \bigoplus_{[\gamma] \in T_0(\Gamma)} H_*(\mathbf{C}_\Gamma(\gamma)/\langle\gamma\rangle, \mathbb{Q}) \otimes HC_*(\mathbb{Q})$$

and

$$B = \bigoplus_{[\gamma] \in T_1(\Gamma)} H_*(\mathbf{C}_\Gamma(\gamma)/\langle\gamma\rangle, \mathbb{Q}).$$

Alternatively we may regard $\mathbb{Q}\Gamma$ as the crossed product $R = \mathbb{Q}N * G$. In this case

$$HC_*(R) = HC_*(R)_{[1]} \oplus \bigoplus_{[g \neq 1] \in T(G)} HC_*(R)_{[g]}$$

and one can check that $HC_*(R)_{[1]} = A$ and $\bigoplus_{[g \neq 1] \in T(G)} HC_*(R)_{[g]} = B$. The advantage with this description is that we can use the properties of N and G to compute B.

To compute $HC_*(R)_{[g]}$ we put $g = Nx$ and observe that

$$HH_*(R_1, R_g) = H_*(N, \mathbb{Q}Nx)$$

where N acts on $\mathbb{Q}Nx$ by conjugation [28, p. 292]. Therefore the spectral sequence collapses , because N has homological dimension zero over $\mathbb{Q}$ and so

$$HC_*(R)_{[g]} = H_*(\mathbf{C}_G(g)/\langle g\rangle, H_0(N, \mathbb{Q}Nx)).$$

Now suppose further that G has finite homological dimension over $\mathbb{Q}$ and is either a linear group in characteristic 0 or a soluble group. Eckmann has shown that $\mathbf{C}_G(g)/\langle g\rangle$ has finite homological dimension over $\mathbb{Q}$ for all $g \in G$

[18] and deduced that the strong Bass conjecture holds for the rational group algebras of such groups. Thus in our situation $HC_n(R)_{[g]} = 0$ for $n \gg 0$ and hence the strong Bass conjecture holds for $\mathbb{Q}\Gamma$ (see [5] for the statement of this conjecture.)

4. The tensor identity

Let R be a strongly G-graded ring, let M be an R-module and let H be a subgroup of G. We first prove the tensor identity.

Proof of Lemma 1.2 There is an R-module homomorphism

$$M \otimes_{R_H} R \longrightarrow M \otimes \mathbb{Z}[H \backslash G]$$

defined by

$$m \otimes r \mapsto mr \otimes Hg$$

for $r \in R_g$. Since R is strongly graded we may choose finite collections $x_i \in R_{g^{-1}}$ and $y_i \in R_g$ such that $\sum_i x_i y_i = 1$. The inverse homomorphism

$$M \otimes \mathbb{Z}[H \backslash G] \longrightarrow M \otimes_{R_H} R$$

is defined by

$$m \otimes Hg \mapsto \sum_i m x_i \otimes y_i.$$

We have already seen in the previous section that R_g is projective as an R_1-module, and it follows that R is projective over R_1. As a first application of the tensor identity we show that more is true [15, Lemma 6.2].

Lemma 4.1. *R is projective as an R_H-module for all subgroups H of G.*

Proof Since R_H is a direct summand of R as a R_H-R_H-bimodule it suffices to show that $R \otimes_{R_H} R$ is a projective R-module. Using the tensor identity we see that

$$R \otimes_{R_H} R \cong R \otimes \mathbb{Z}[H \backslash G].$$

so it is enough to show that $\text{Hom}_R(R \otimes \mathbb{Z}[H \backslash G], \_)$ is exact. If N is any R-module then there is a natural isomorphism

$$\text{Hom}_R(R \otimes \mathbb{Z}[H \backslash G], N) \longrightarrow \text{Hom}_{\mathbb{Z}}(\mathbb{Z}[H \backslash G], N),$$

defined by

$$\theta \mapsto (Hg \mapsto \theta(1 \otimes Hg)),$$

and $\text{Hom}_{\mathbb{Z}}(\mathbb{Z}[H \backslash G], \_)$ is exact because $\mathbb{Z}[H \backslash G]$ is free as a $\mathbb{Z}$-module.

4.1. Chouinard's Theorem for strongly graded rings

Theorem 1.4 was originally proved by Aljadeff and Ginosar using trace maps and a variation on the tensor identity. We outline an alternative proof, due to Benson and Kropholler.

Reduction to p-groups:

Lemma 4.2. *Let F be a finite group, let $p_1, \ldots, p_n$ be the distinct primes dividing the order of F and let P_i be a Sylow p_i-subgroup. Then the trivial $\mathbb{Z}F$-module $\mathbb{Z}$ is a direct summand of the permutation module*

$$\bigoplus_{i=1}^{n} \mathbb{Z}[P_i \backslash F].$$

Proof Let $\iota : \mathbb{Z} \longrightarrow \bigoplus_{i=1}^{n} \mathbb{Z}[P_i \backslash F]$ be defined by $1 \longrightarrow (\xi_1, \ldots, \xi_n)$ where ξ_i denotes the sum of the distinct cosets of P_i in F. To split this map set $m_i := |F : P_i|$, choose integers l_i such that $\sum_{i=1}^{n} l_i m_i = 1$ and define

$$\varepsilon = (\varepsilon_1, \ldots, \varepsilon_n) : \bigoplus_{i=1}^{n} \mathbb{Z}[P_i \backslash F] \longrightarrow \mathbb{Z}$$

by

$$(\eta_1, \ldots, \eta_n) \mapsto l_1 \varepsilon_1(\eta_1) + \ldots + l_n \varepsilon_n(\eta_n)$$

where ε_i is the augmentation map. One checks that $\varepsilon\iota$ is the identity map on $\mathbb{Z}$ and the result follows.

Lemma 4.3. *Let R be a strongly F-graded ring and let M be an R-module which is projective over R_P for each Sylow p-subgroup P of F. Then M is projective over R.*

Proof Tensoring the map ι in the previous lemma with M and applying the tensor identity we see that M is a direct summand of

$$\bigoplus_{i=1}^{n} M \otimes_{R_{P_i}} R$$

and the result follows by the hypothesis on M.

Thus we may assume that $F = P$ is a finite p-group. Recall that (for any ring) elements of $\mathrm{Ext}^n_R(M, N)$ correspond to equivalence classes of sequences

$$0 \longrightarrow N \longrightarrow E_{n-1} \longrightarrow \cdots \longrightarrow E_1 \longrightarrow E_0 \longrightarrow M \longrightarrow 0$$

if $n > 0$ and multiplication in $\mathrm{Ext}^*_R(M, M)$ corresponds to concatenation of (equivalence classes of) sequences. We shall need the following fact about the correspondence:

Lemma 4.4. *If the exact sequence*

$$0 \longrightarrow N \longrightarrow E_{n-1} \longrightarrow \cdots \longrightarrow E_1 \longrightarrow E_0 \longrightarrow M \longrightarrow 0$$

represents zero in $\mathrm{Ext}_R^n(M, N)$, *and if in addition each* E_i *is a projective R-module, then* N *is a projective R-module.*

Proof Extend the sequence

$$E_{n-1} \longrightarrow \cdots \longrightarrow E_1 \longrightarrow E_0 \longrightarrow M \longrightarrow 0$$

to a projective resolution $E_\bullet$ of M. There is an induced map $\phi : E_n \longrightarrow N$ and by the correspondence ϕ is a coboundary. Therefore ϕ factors through E_{n-1} and so $N \longrightarrow E_{n-1}$ is split.

Let $\beta : H^1(P, \mathbb{Z}/p\mathbb{Z}) \longrightarrow H^2(P, \mathbb{Z})$ be the connecting homomorphism in the long exact cohomology sequence corresponding to the short exact sequence

$$0 \longrightarrow \mathbb{Z} \longrightarrow \mathbb{Z} \longrightarrow \mathbb{Z}/p\mathbb{Z} \longrightarrow 0.$$

A non-zero element x of $H^1(P, \mathbb{Z}/p\mathbb{Z})$ may be regarded as a homomorphism from P onto $\mathbb{Z}/p\mathbb{Z}$ with kernel H and the element $\beta(x)$ in $H^2(P, \mathbb{Z}) = \mathrm{Ext}_{\mathbb{Z}P}^2(\mathbb{Z}, \mathbb{Z})$ corresponds to a four term exact sequence

$$0 \longrightarrow \mathbb{Z} \longrightarrow \mathbb{Z}[H\backslash P] \longrightarrow \mathbb{Z}[H\backslash P] \longrightarrow \mathbb{Z} \longrightarrow 0.$$

Thus if $x_1, \ldots, x_n$ is a sequence of non-zero elements of $H^1(P, \mathbb{Z}/p\mathbb{Z})$ then

$$\beta(x_1) \cdot \ldots \cdot \beta(x_n) \in H^{2n}(P, \mathbb{Z})$$

corresponds to an exact sequence

$$0 \longrightarrow \mathbb{Z} \longrightarrow E_{2n-1} \longrightarrow \cdots \longrightarrow E_1 \longrightarrow E_0 \longrightarrow \mathbb{Z} \longrightarrow 0$$

where $E_{2i-1} = E_{2i-2} = \mathbb{Z}[H_i\backslash P]$ and H_i is the kernel of x_i.

Proof of Theorem 1.4 We have reduced to the case where P is a p-group. Let R be a strongly P-graded ring and let M be an R-module which is projective over R_E for all elementary abelian p-subgroups E of P. We need to show that M is projective over R. Clearly we may assume that P is not elementary abelian, and by induction that M is projective over R_H for all proper subgroups H of P.

By a theorem of Serre [7, Theorem 4.7.3], there exists a sequence $x_1, \ldots, x_n$ of non-zero elements in $H^1(P, \mathbb{Z}/p\mathbb{Z})$ such that

$$\beta(x_1) \cdot \ldots \cdot \beta(x_n) = 0 \in H^{2n}(P, \mathbb{Z}).$$

Thus there is a sequence

$$0 \longrightarrow \mathbb{Z} \longrightarrow E_{2n-1} \longrightarrow \cdots \longrightarrow E_1 \longrightarrow E_0 \longrightarrow \mathbb{Z} \longrightarrow 0$$

as above which represents the zero element in $H^{2n}(P, \mathbb{Z})$. Tensoring this sequence with M and letting R act semi-diagonally we have the exact sequence

$$0 \longrightarrow M \longrightarrow M \otimes E_{2n-1} \longrightarrow \cdots \longrightarrow M \otimes E_1 \longrightarrow M \otimes E_0 \longrightarrow M \longrightarrow 0$$

which represents the zero element in $\mathrm{Ext}_R^{2n}(M, M)$. By the tensor identity $M \otimes E_j$ is isomorphic to $M \otimes_{R_{H_i}} R$ (where $j = 2i - 1$ or $2i - 2$), and these modules are projective. Therefore M is projective by Lemma 4.4.

4.2. Moore's Conjecture

Let G be a group, let H be a subgroup of finite index and let R be a strongly G-graded ring. In this section we consider the question: If M is an R-module which is projective over R_H, then is it projective over R?

Of course this is not true in general. If F is a non-trivial finite group and S is a ring, then S is a projective $S[F]$-module if and only if the order of F is invertible in S by Maschke's Theorem.

Thus, it is necessary to impose additional restrictions on R, G and H. Recall from the introduction that (R, G, H) satisfies the *Moore condition* if for all $g \in G - H$ either

1 g has finite order invertible in R_1, or
2 there exists an integer n such that $g^n \in H - \{1\}$.

If G is finite and (R, G, H) satisfies the Moore condition then Theorem 1.4 provides a positive answer to the question. To deal with arbitrary groups one looks for situations where it is possible to reduce to the finite case. One such situation is described in Theorem 1.6.

Proof of Theorem 1.6 It is enough to show that M has finite projective dimension over R since $proj.dim_R M = proj.dim_{R_H} M$ if the former is finite [15, Lemma 6.6].

The group G acts on a finite dimensional contractible cell complex X with finite cell stabilisers. Let

$$0 \longrightarrow C_r \longrightarrow \cdots \longrightarrow C_1 \longrightarrow C_O \longrightarrow \mathbb{Z} \longrightarrow 0$$

be the cellular chain complex of X where each C_i is isomorphic to

$$\bigoplus_{\sigma \in \Sigma_i} \mathbb{Z}[G_\sigma \backslash G],$$

Σ_i is a set of G-orbit representatives of i-dimensional cells in X, and G_σ is the (finite) isotropy group of the cell $\sigma \in \Sigma_i$. Tensoring the sequence with M and letting R act semi-diagonally we have the exact sequence of R-modules

$$0 \longrightarrow M \otimes C_r \longrightarrow \cdots \longrightarrow M \otimes C_1 \longrightarrow M \otimes C_O \longrightarrow M \longrightarrow 0.$$

Applying the tensor identity we see that $M \otimes C_i$ is isomorphic to

$$\bigoplus_{\sigma \in \Sigma_i} M \otimes_{R_{G_\sigma}} R$$

which is projective by hypothesis and so $proj.dim_R(M)$ is finite.

Proof of Theorem 1.8 We need to show that M is projective over R_F for every finite subgroup F of G and then the result will follow from Theorem 1.6. Consequently, in view of Theorem 1.4, we need only show that M is projective as an R_E-module for every elementary abelian subgroup E of G.

First observe that M is projective over R_1 because R_H is a projective R_1-module. Let p be a prime and let E be an elementary abelian p-subgroup of G. If p is invertible in R_1 then M is projective over R_E by Theorem 3.5. Every non-trivial element of E has order p, and so if p is not invertible then E is contained in H by the Moore condition. Therefore M is projective over R_E because R_H is projective over R_E by Lemma 4.1.

Corollary 4.5. *(Serre) Let G be a torsion free group of finite virtual cohomological dimension. Then G has finite cohomological dimension.*

Proof Let

$$\cdots \longrightarrow P_2 \longrightarrow P_1 \longrightarrow P_0 \longrightarrow \mathbb{Z} \longrightarrow 0$$

be an augmented projective resolution of $\mathbb{Z}$ as a $\mathbb{Z}G$-module, and suppose that H is a normal subgroup of G with $|G : H| < \infty$ and $cd(H) = n$, so the kernel K of the map $P_{n-1} \longrightarrow P_{n-2}$ is projective as a $\mathbb{Z}H$-module. The triple $(\mathbb{Z}, G, H)$ clearly satisfies the Moore condition and $G \in \mathbf{H}_1 \mathfrak{F}$ since it has finite virtual cohomological dimension. It follows from the theorem that K is a projective $\mathbb{Z}G$-module.

5. Complete Resolutions

Let S be a ring and let M and N be S-modules. The complete cohomology groups $\widehat{\mathrm{Ext}}^*_S(M, N)$ can be defined in various ways. For example if

$$\cdots \longrightarrow Q_2 \longrightarrow Q_1 \longrightarrow Q_0 \longrightarrow N \longrightarrow 0$$

is a projective resolution of N and K_i is the kernel of $Q_i \longrightarrow Q_{i-1}$ then

$$\widehat{\mathrm{Ext}}^n_S(M, N) = \varinjlim_{i \geq n+1} \mathrm{Ext}^i_S(M, K_{i-(n+1)})$$

where the homomorphisms in the direct limit system are the connecting homomorphisms in the long exact Ext-sequences induced by the short exact sequences $K_i \longrightarrow Q_i \longrightarrow K_{i-1}$ [29]. See also [22], [8], and [24] where the terminology 'complete cohomology' was introduced.

One can see that this definition does not lend itself easily to computation, so with the special case of Tate-Farrell cohomology in mind we introduced our alternative Definition 1.9 of complete resolutions in Section 1. In this section we outline some results which will appear in [16], ignoring many of the technical details such as homotopy equivalence of complete resolutions. Of course we need to know the following result otherwise the definition is of no use to us.

Lemma 5.1. *Let $P_\bullet = (P_\bullet, \delta)$ be a complete resolution of the S-module M and let N be an arbitrary S-module. Then*

$$\widehat{\mathrm{Ext}}_S^*(M, N) \cong H^*(\mathrm{Hom}_S(P_\bullet, N)).$$

Proof The functor $H^*(\mathrm{Hom}_S(P_\bullet, \_))$ is a $(-\infty, \infty)$-cohomological functor from S-modules to abelian groups [21], which disappears on projective modules. Suppose that $Q_\bullet = (Q_\bullet, \delta)$ is an ordinary projective resolution of M which agrees with $P_\bullet$ in dimensions $i \geq n$ for some positive integer n. Thus there is a partial chain map $\phi = (\phi_{i\geq n}) : P_\bullet \longrightarrow Q_\bullet$ and every ϕ_i an isomorphism
We first extend ϕ to a chain map in all dimensions. By induction we need only construct a map $\phi_{n-1} : Q_{n-1} \longrightarrow P_{n-1}$ such that $\phi_{n-1}\delta_n = d_n\phi_n$. By hypothesis the sequence

$$\mathrm{Hom}_S(P_{n-1}, Q_{n-1}) \longrightarrow \mathrm{Hom}_S(P_n, Q_{n-1}) \longrightarrow \mathrm{Hom}_S(P_{n+1}, Q_{n-1})$$

is exact at $\mathrm{Hom}_S(P_n, Q_{n-1})$, so the element $d_n\phi_n \in \mathrm{Hom}_S(P_n, Q_{n-1})$ maps to $d_n\phi_n\delta_{n+1} = d_n d_{n+1}\phi_{n+1} = 0 \in \mathrm{Hom}_S(P_{n+1}, Q_{n-1})$ and so there exists

$$\phi_{n-1} \in \mathrm{Hom}_S(P_{n-1}, Q_{n-1})$$

such that $\phi_{n-1}\delta_n = d_n\phi_n$ as required. One can show that the chain map constructed in this way is unique up to homotopy.
Therefore we have a morphism of $(-\infty, \infty)$-cohomological functors

$$\mathrm{Ext}_S^*(M, \_) \longrightarrow H^*(\mathrm{Hom}_S(P_\bullet, \_))$$

where $\mathrm{Ext}_S^i(M, \_)$ is defined to be zero if $i < 0$. The maps are isomorphisms in dimensions $i \geq n$ and thus there is an induced morphism

$$\widehat{\mathrm{Ext}}_S^*(M, \_) \longrightarrow H^*(\mathrm{Hom}_S(P_\bullet, \_))$$

which is an equivalence by [29, Lemma 2.5].

Of course there is no point in making a definition if one does not have examples. Indeed we construct modules in [16] which cannot have a complete resolution. However we do have a positive result for strongly graded rings.

Definition 5.2. Let G be a group. Then $B(G, \mathbb{Z})$ is the ring of bounded functions from G to $\mathbb{Z}$.

Lemma 5.3. *(20, Corollary 97.4) $B(G, \mathbb{Z})$ is a $\mathbb{Z}G$-module which is free abelian as an additive group.*

The ring $B = B(G, \mathbb{Z})$ is used in several of the proofs in [14] and [15] and we shall have little to say about it here quoting results when we need them. For example the next proposition follows easily from the proof of [15, Proposition 9.2].

Proposition 5.4. *Let G be a group in the class* $\mathbf{H}_1\mathfrak{F}$, *let R be a strongly G-graded ring and let M be an R-module such that* $proj.dim_{R_1} M < \infty$. *Then*

$$proj.dim_{R} M \otimes B < \infty$$

where R acts semi-diagonally.

Proof of Theorem 1.10 By the previous proposition it is enough to show that M has a complete resolution if $proj.dim_R M \otimes B < \infty$. Since $\_ \otimes B$ is exact and R is projective over R_1 we may replace M by a suitable kernel in a projective resolution of M over R and assume that M is projective over R_1 and $M \otimes B$ is projective over R.

The inclusion $\mathbb{Z} \longrightarrow B$ as constant functions is $\mathbb{Z}$-split by $f \mapsto f(1)$ and thus we have a short exact sequence of $\mathbb{Z}G$-modules

$$\mathbb{Z} \longrightarrow B \longrightarrow \overline{B} \ (\dagger)$$

and $\overline{B}$ is $\mathbb{Z}$-free. Let V_i denote the $\mathbb{Z}G$-module $\overline{B}^{\otimes i} \otimes B$ for $i \geq 0$ (where $\overline{B}^{\otimes 0} = \mathbb{Z}$). Tensoring $(\dagger)$ with $\overline{B}^{\otimes i}$ for each $i \geq 0$ we have short exact sequences of $\mathbb{Z}G$-modules

$$\overline{B}^{\otimes i} \longrightarrow V_i \longrightarrow \overline{B}^{\otimes i+1}.$$

By splicing these short exact sequences together we have the long exact sequence $V_\bullet$ of $\mathbb{Z}G$-modules

$$0 \longrightarrow \mathbb{Z} \longrightarrow V_0 \longrightarrow V_1 \longrightarrow \cdots.$$

Now let $Q_\bullet$ be a projective resolution of $\mathbb{Z}$ over $\mathbb{Z}G$. By splicing $Q_\bullet$ and $V_\bullet$ together, tensoring with M and letting R act semi-diagonally we have an acyclic complex $P_\bullet$ of R-modules where $P_i = M \otimes Q_i$ for $i \geq 0$ and $P_{-i} = M \otimes V_i$ for $i \geq 1$. We claim that $P_\bullet$ is a complete resolution of M. It follows from the tensor identity $M \otimes \mathbb{Z}G \cong M \otimes_{R_1} R$, and so each $M \otimes Q_i$ is projective over R and $M \otimes Q_\bullet$ is a projective resolution of M. By Lemma 2.5,

$$\mathrm{Hom}_R(M \otimes V_i, \_) \cong \mathrm{Hom}_R(M \otimes B, \mathrm{Hom}_{\mathbb{Z}}(\overline{B}^{\otimes i}, \_)).$$

and so $M \otimes V_i$ is projective for all $i \geq 0$ because $M \otimes B$ is projective and $\overline{B}^{\otimes i}$ is $\mathbb{Z}$-free. Therefore $P_\bullet$ satisfies the first condition of the definition.

Now let X be a projective R-module. The inclusion $\mathbb{Z} \longrightarrow B$ induces an R-split epimorphism $\mathrm{Hom}_{\mathbb{Z}}(B, X) \longrightarrow X$ and thus it suffices to show that $\mathrm{Hom}_R(P_*, \mathrm{Hom}_{\mathbb{Z}}(B, X))$ is acyclic. Using Lemma 2.5 again we have

$$\mathrm{Hom}_R(P_*, \mathrm{Hom}_{\mathbb{Z}}(B, X)) \cong \mathrm{Hom}_R(P_* \otimes B, X).$$

The kernels of the complex $P_* \otimes B$ are projective in dimensions ≥ 0 because $M \otimes B$ is projective. In negative dimensions they have the form $M \otimes V_i$ which we have already shown are projective. Therefore $P_* \otimes B$ is split, so $\mathrm{Hom}_R(P_* \otimes B, X)$ is acyclic and the second condition of the definition is satisfied.

Corollary 5.5. *(Shapiro's Lemma) Let G, R and M be as in the theorem, let H be a subgroup of G and let N be an R_H-module. Then for all integers n, there are natural isomorphisms*

$$\widehat{\mathrm{Ext}}^n_{R_H}(M, N) \cong \widehat{\mathrm{Ext}}^n_R(M, \mathrm{Hom}_{R_H}(R, N)).$$

Proof Since $\mathrm{Hom}_{R_H}(P_*, N) \cong \mathrm{Hom}_R(P_*, \mathrm{Hom}_{R_H}(R, N))$ it suffices to show that the complete resolution P_* constructed in the theorem is also a complete resolution of M as an R_H-module. The first condition for a complete resolution is clearly satisfied since R is a projective R_H-module. Let Y be a projective R_H-module. The same argument as above shows that Y is a direct summand of $\mathrm{Hom}_\mathbb{Z}(B(G, \mathbb{Z}), Y)$ as R_H-modules and the result follows.

Remark 5.6. One can also construct spectral sequences in the standard way, (see [16]).

References

[1] J. L. Alperin and L. Evens, *Representations, resolutions, and Quillen's dimension theorem*, J. Pure Appl. Algebra **26** (1982), pp. 221–227.

[2] E. Aljadeff and Y. Ginosar, *Induction from elemntary abelian subgroups*, J. Algebra **179** (1996), pp. 599–606.

[3] E. Aljadeff, J. Cornick, Y. Ginosar and P. H. Kropholler, *On a conjecture of Moore*, J. Pure Appl. Algebra **110** (1996), pp. 109–112.

[4] H. Bass, *Algebraic K-Theory*, Benjamin, New York, (1968).

[5] H. Bass, *Euler characteristics and characters of discrete groups*, Inventiones Math. **35** (1976), pp. 155–196.

[6] D. J. Benson, *Representations and Cohomology I: Basic representation theory of finite groups and associative algebras*, Cambridge Studies in Advanced Mathematics, **30** Cambridge Uuniversity Press, (1991).

[7] D. J. Benson, *Representations and Cohomology 2: Cohomology of groups and modules*, Cambridge Studies in Advanced Mathematics, **31** Cambridge Uuniversity Press, (1991).

[8] D. J. Benson and J. Carlson, *Products in negative cohomology*, J. Pure Appl. Algebra **82** (1992), pp. 107–130.

[9] K. S. Brown, *Cohomology of Groups*, Graduate Texts in Mathematics, **87** Springer, Berlin, (1982).

[10] D. Burghelea, *The cyclic homology of group rings*, Comment. Math. Helvetici **60** (1985), pp. 354–365.

[11] L. Chouinard, *Projectivity and relative projectivity over group rings*, J. Pure Appl. Algebra **7** pp. 278–302.

[12] M. Cohen and S. Montgomery, *Group-graded rings, smash products and group actions*, Trans. Amer. Math. Soc. **282** (1984), pp. 237–257.

[13] J. Cornick, *On the homology of group graded algebras*, J. Algebra **174** (1995), pp. 999–1023.

[14] J. Cornick and P. H. Kropholler, *Homological finiteness conditions for modules over group algebras*, To appear in J. London Math. Soc.

[15] J. Cornick and P. H. Kropholler, *Homological finiteness conditions for modules over strongly group-graded rings*, Math. Proc. Camb. Philos. Soc. **120** (1996), pp. 43–54.

[16] J. Cornick and P. H. Kropholler, *On complete resolutions*, Topology Appl. **78** (1997), pp. 235–250.

[17] E. C. Dade, *Group graded rings and modules*, Math. Z. **174** (1980), pp. 241–262.

[18] B. Eckmann, *Cyclic homology of groups and the Bass conjecture*, Comment. Math. Helvetici **61** (1986), pp. 193–202.

[19] B. Feigin and B. Tsygan, *Cyclic homology of algebras with quadratic relations, universal enveloping algebras*, in: K-Theory, Arithmetic and Geometry Lect. Notes in Math, **1289** Springer, Berlin, (1987).

[20] L. Fuchs, *Infinite Abelian Groups, Volume II*, Academic Press, New York, (1973).

[21] T. V. Gedrich and K. W. Gruenberg, *Complete cohomological functors on groups*, Topology and its Applications **25** (1987), pp. 203–223.

[22] F. Goichot, *Homologie de Tate-Vogel équivariante*, J. Pure Appl. Algebra **82** (1992), pp. 39–64.

[23] P. H. Kropholler, *On groups of type FP_∞*, J. Pure Appl. Algebra **90** (1993), pp. 55–67.

[24] P. H. Kropholler, *Hierarchical decompositions, generalized Tate cohomology, and groups of type FP_∞*, in: Proceedings of the Edinburgh Conference on Geometric Group Theory, 1993 (A. Duncan, N. Gilbert, and J. Howie, eds.), Cambridge University Press, (1994).

[25] P. H. Kropholler, *Cohomological finiteness conditions*, in: Groups '93 Galway/St. Andrews Vol.I (C. M. Campbell et al., eds.), LMS Lecture Notes, **211** (1995), pp. 274–304.

[26] J. Loday and D. Quillen, *Cyclic homology and the Lie algebra homology of matrices*, Comment. Math. Helvetici **59** (1984), pp. 565–591.

[27] M. Lorenz, *On the homology of graded algebras*, Communications in Algebra **20** (1992), pp. 489–507.

[28] S. MacLane, *Homology*, Springer, Berlin, (1975).

[29] G. Mislin, *Tate cohomology for arbitrary groups via satellites*, Topology Appl. **56** (1994), pp. 293–300.

[30] D. Passman, *Infinite crossed products*, Academic Press, San Diego, (1989).

[31] D. Quinn, *Group-graded rings and duality*, Trans. Amer. Math. Soc. **292** (1985), pp. 155–167.

Buildings are $CAT(0)$

Michael W. Davis

Department of Mathematics, The Ohio State University, 231 W. 18th Avenue, Columbus, Ohio 43210

email:mdavis@math.ohio-state.edu

0. Introduction

Given a finitely generated Coxeter group W, I described, in [D, Section 14], a certain contractible simplicial complex, here denoted $|\mathbf{W}|$, on which W acts properly with compact quotient. After writing [D], I realized that there was a similarly defined, contractible simplicial complex associated to any building C. This complex is here denoted $|C|$ and called the "geometric realization" of C. The definition is such that the geometric realization of each apartment is isomorphic to $|\mathbf{W}|$. (N. B. Our terminology does not agree with standard usage. For example, if W is finite, then the usual Coxeter complex of W is homeomorphic to a sphere, while our $|\mathbf{W}|$ is homeomorphic to the cone on this sphere.)

There is a natural piecewise Euclidean metric on $|\mathbf{W}|$ (described in §9) so that W acts as a group of isometries. Following an idea of Gromov ([G, pp. 131-132]), Gabor Moussong proved in his Ph.D. thesis [M] that with this metric $|\mathbf{W}|$ is "$CAT(0)$" (in the sense of [G]). This is equivalent to saying that it is simply connected and "nonpositively curved". Moussong's result implies, via a standard argument, the following theorem.

Theorem. *The (correctly defined) geometric realization of any building is* $CAT(0)$.

Although this theorem was known to Moussong, it is not included in [M].

The theorem implies, for example, that the Bruhat Tits Fixed Point Theorem can be applied to any building. (See Corollary 11.9.)

Partially supported by NSF Grant DMS9208071

One of the purposes of this paper is to provide the "correct definition" of the geometric realization $|C|$ and to give the "standard argument" for deducing the above theorem from Moussong's result.

Another purpose is to explain and expand the results of [HM] and [Mei] on graph products of groups. In fact, I was inspired to write this paper after listening to J. Harlander's lecture at the Durham Conference on the results of [HM]. I realized that graph products of groups provide nice examples of buildings where the associated W is a right-angled Coxeter group.

Part I of this paper concerns the combinatorial theory of buildings. In §1, §2, §3 and §6 we recall some relevant definitions and results from Ronan's book [R]. In §4 we state a theorem of Tits [T] and use it in §5 to prove that the chamber system associated to a graph product of groups is a building (Theorem 5.1).

Part II concerns the geometric and topological properties of buildings. In §7, we recall some basic definitions concerning $CAT(0)$ spaces from [G], [Bri], [Bro] and [CD1]. The definitions of $|\mathbf{W}|$ and $|C|$ are given in §9 and §10, respectively. In §11 we prove the main theorem and deduce some consequences. In particular, in Corollary 11.7, we apply the theorem in the case of a graph product of groups.

I. The combinatorial theory of buildings

1. Chamber systems

A *chamber system over a set* I is a set C together with a family of partitions of C indexed by I. The elements of C are *chambers*. Two chambers are *i-adjacent* if they belong to the same subset in the partition corresponding to i.

Example 1.1. Let G be a group, B a subgroup and $(P_i)_{i \in I}$ a family of subgroups containing B. Define a chamber system $C = C(G, B, (P_i)_{i \in I})$ as follows: $C = G/B$ and the chambers gB and $g'B$ are *i-adjacent* if they have the same image in G/P_i.

Let C be a chamber system over I. Let I^* denote the free monoid on I. (An element of I^* is a word $\mathbf{i} = i_1 \cdots i_k$, where each $i_j \in I$.) A *gallery* in C is a finite sequence of chambers $(c_0, c_1, \cdots, c_k)$ such that c_{j-1} is adjacent but not equal to $c_j, 1 \leq j \leq k$. The gallery has *type* $\mathbf{i} = i_1 \cdots i_k$ if c_{j-1} is i_j-adjacent to c_j. If each i_j belongs to a given subset J of I, then it is a *J-gallery*. A chamber system is *connected* (or *J-connected*) if any two chambers can be joined by a gallery (or a J-gallery). The J-connected components of a chamber system C are called its *J-residues*. The *rank* of a chamber system over I is the cardinality of I. A *morphism* $\varphi : C \to C'$ of chamber systems over the same set I is a map which preserves *i*-adjacency.

Suppose that G is a group of automorphisms of C which acts transitively on C. Fix a chamber $c \in C$ and for each $i \in I$ choose an $\{i\}$-residue containing c. Let B denote the stabilizer of c and P_i the stabilizer of the $\{i\}$-residue. Then, clearly, C is isomorphic to the chamber system $C(G, B, (P_i)_{i \in I})$ of Example 1.1.

Suppose $C_1, \cdots, C_k$ are chamber systems over $I_1, \cdots, I_k$. Their *direct product*, $C_1 \times \cdots \times C_k$, is a chamber system over the disjoint union $I_1 \amalg \cdots \amalg I_k$. Its elements are k-tuples $(c_1, \cdots, c_k)$ where $c_t \in C_t$. For $i \in I_t$, $(c_1, \cdots, c_k)$ is *i-adjacent* to $(c_1', \cdots, c_k')$ if $c_j = c_j'$ for $j \neq t$ and c_t and c_t' are *i*-adjacent.

2. Coxeter groups

A *Coxeter matrix* M over a set I is a symmetric matrix $(m_{ij}), (i,j) \in I \times I$, with entries in $\mathbf{N} \cup \{\infty\}$ such that for all $i, j \in I, m_{ii} = 1$ and $m_{ij} \geq 2$ for $i \neq j$. If J is a subset of I, then M_J denotes the restriction of M to J, i.e., it is the matrix formed by the entries of M which are indexed by $J \times J$.

For each $i \in I$ introduce a symbol s_i and let $S = \{s_i | i \in I\}$. The *Coxeter group determined by* M is the group W (or $W(M)$) given by the presentation, $W = \langle S | (s_i s_j)^{m_{ij}} = 1$, for all $(i,j) \in I \times I$ with $m_{ij} \neq \infty \rangle$. The natural map $S \to W$ is an injection ([B, p. 92]) and henceforth, we identify S with its image in W. For any subset J of I, denote by W_J the subgroup of W generated by $\{s_j | j \in J\}$. The case $J = \emptyset$ corresponds to the trivial subgroup of W.

The Coxeter matrix M (and the group W) are *right-angled* if each off-diagonal entry of M is 2 or ∞.

The matrix M is *spherical* if W is finite. A subset J of I is *spherical* if $W_J (\cong W(M_J))$ is finite. Associated to M we have the poset $\mathcal{S}^f$ (or $\mathcal{S}^f(M)$) of all spherical subsets of I.

Suppose the $\mathbf{i} = i_1 \cdots i_k$ is an element in the free monoid I^*. Its *value* $s(\mathbf{i})$ is the element of W defined by $s(\mathbf{i}) = s_{i_1} \cdots s_{i_k}$. Two words $\mathbf{i}$ and $\mathbf{i}'$ are *equivalent* (with respect of M) if $s(\mathbf{i}) = s(\mathbf{i}')$. The word $\mathbf{i}$ is *reduced* (with respect to M) if the word length of $s(\mathbf{i})$ is k. (Alternative definitions of these concepts in terms of "homotopies of words" can be found in [R, p. 17].)

3. Buildings

Let I, I^*, M and W be as in §2.

A *building of type M* is a chamber system C over I such that

(1) for each $i \in I$, each subset of the partition corresponding to i contains at least two chambers, and

(2) there exists a "W-valued distance function" $\delta : C \times C \to W$ such that if **i** is a reduced word in I^* (with respect to M), then chambers c and c' can be joined by a gallery of type **i** if and only if $\delta(c, c') = s(\mathbf{i})$

Example 3.1. Let **W** be the chamber system $C(W, \{1\}, (W_{\{i\}})_{i \in I})$ where the notation is as in Example 1.1. In other words, the set of chambers is W and two chambers w and w' are i-adjacent if and only if $w' = w$ or $w' = ws_i$. There is a W-valued distance function $\delta : W \times W \to W$ defined by $\delta(w, w') = w^{-1}w'$. Thus, **W** is a building, called the *abstract Coxeter complex* of W.

Example 3.2. Suppose that C is a building of rank 1. Then W is the cyclic group of order 2. There is only one possibility for $\delta : C \times C \to W$; it must map the diagonal of $C \times C$ to the identity element and the complement of the diagonal to the nontrivial element. Thus, any two chambers are adjacent.

Example 3.3. Suppose that $M_1, \cdots, M_k$ are Coxeter matrices over $I_1, \cdots, I_k$, respectively. Let I denote the disjoint union $I_1 \amalg \cdots \amalg I_k$ and define a Coxeter matrix M over I by setting m_{ij} equal to the corresponding entry of M_t whenever i, j belong to the same component I_t of I and $m_{ij} = 2$ when they belong to different components. Then $W = W_1 \times \cdots \times W_k$ where $W = W(M)$ and $W_t = W(M_t)$. Suppose that $C_1, \cdots, C_k$ are buildings over $I_1, \cdots, I_k$. As in §1, their direct product $C = C_1 \times \cdots \times C_k$ is a chamber system over I. Moreover, the direct product of the W_t-valued distance functions gives a W-valued distance function on C. Hence, C is a building of type M.

4. A theorem of Tits

I, M, W and $\mathcal{S}^f$ are as in §2. Suppose that G is a chamber-transitive group of automorphisms on a building C of type M. Choose a chamber $c \in C$ and for each $J \in \mathcal{S}^f$ a J-residue containing c. Let P_J denote the stabilizer of this J-residue and B ($= P_\emptyset$) the stabilizer of c. Set $P_i = P_{\{i\}}$, so that $C \cong C(G, B, (P_i)_{i \in I})$. We note that P_J is just the subgroup generated by $(P_j)_{j \in J}$.

Thus, the chamber-transitive automorphism group G gives the following data: for each $J \in \mathcal{S}^f$, there is a group P_J, and whenever $J, J' \in \mathcal{S}^f$ with $J \leq J'$, there is an injective homomorphism $\varphi_{JJ'} : P_J \to P_{J'}$ such that $\varphi_{JJ} = id$ and $\varphi_{JJ''} = \varphi_{J'J''} \circ \varphi_{JJ'}$ when $J \leq J' \leq J''$. (This is the data for a "complex of groups" in the sense of [H] or more precisely a "simple complex of groups " in the sense of [CD2].)

For a natural number m, let $\mathcal{S}^f_m$ denote the subposet of $\mathcal{S}^f$ consisting of all J with $\mathrm{Card}\,(J) \leq m$.

Proposition 4.1. (Tits [T, Proposition 2]). *Suppose that $C(G, B, (P_i)_{i \in I})$ is a building of type M. Then G is the direct limit of the system of groups $\{P_J | J \in \mathcal{S}_2^f\}$.*

The "direct limit" of a system of groups is defined in [S, p. 1]; the term "amalgamated sum" is used in [T] for this concept.

There is a converse result to Proposition 4.1. Suppose we are given a simple complex of groups (P_J) over $\mathcal{S}_3^f$. (In other words, a system of groups as in the paragraph preceding Proposition 4.1, where the data is only specified for those J with $\mathrm{Card}\,(J) \leq 3$.) For any subset K of I, let G_K denote the direct limit of the system of groups $\{P_J | J \in \mathcal{S}_2^f \cap K\}$. Set $B = P_\emptyset, P_i = P_{\{i\}}$ and $G = G_I$.

Theorem 4.2. (Tits [T, Theorem 1]). *With notation as above, suppose that for all $J \in \mathcal{S}_3^f - \{\emptyset\}, C(P_J, B, (P_j)_{j \in J})$ is a building of type M_J. Then $C(G, B, (P_i)_{i \in I})$ is a building of type M.*

5. Graph products

Let Γ be a simplicial graph with vertex set I. Define a Coxeter matrix $M(= M(\Gamma))$ by setting

$$m_{ij} = \begin{cases} 1; & \text{if } i = j \\ 2; & \text{if } \{i, j\} \text{ spans an edge of } \Gamma \\ \infty; & \text{otherwise.} \end{cases}$$

Let W be the associated right-angled Coxeter group. Note that a subset J of I is spherical if and only if any two elements of J span and edge; furthermore, if this is the case, then W_J is the direct product of $|J|$ copies of the cyclic group of order two.

Suppose we are given a family of groups $(P_i)_{i \in I}$. For each $J \in \mathcal{S}^f$, let P_J denote the direct product

$$P_J = \prod_{j \in J} P_j$$

($P_\emptyset = \{1\}$). If $J \leq J' \in \mathcal{S}^f$, then $\varphi_{JJ'} : P_J \to P_{J'}$ is the natural inclusion. The direct limit G of $\{P_J | J \in \mathcal{S}_2^f\}$ is called the *graph product* of the $(P_i)_{i \in I}$ (with respect to Γ).

Alternatively, G could have been defined as the quotient of the free product of the $(P_i)_{i \in I}$ by the normal subgroup generated by all commutators of the form $[g_i, g_j]$, where $g_i \in P_i, g_j \in P_j$ and $m_{ij} = 2$.

Notice that for all $J \in \mathcal{S}^f, J \neq \emptyset, C(P_J, \{1\}, (P_j)_{j \in J})$ is a building of type M_J (it is a direct product of the rank one buildings $C(P_j, \{1\}, P_j)$ as in Example 3.3). Hence, Tits' Theorem 4.2 implies the following.

Theorem 5.1. *Let G be a graph product of groups $(P_i)_{i \in I}$, as above. Then $C(G, \{1\}, (P_i)_{i \in I})$ is a building of type M.*

Actually, it is not difficult to give a direct proof of this, without invoking Tits' Theorem, by producing the W-valued distance function $\delta : G \times G \to W$. It is clear that any g in G can be written in the form $g = g_{i_1} \cdots g_{i_k}$, where $g_{i_t} \in P_{i_t} - \{1\}$ and where $\mathbf{i} = i_1 \cdots i_k$ is a reduced word in I^* (with respect to M). Moreover, if $g' = g_{i'_1} \cdots g_{i'_k}$ is another such representation, then $g = g'$ if and only if we can get from one representation to the other by a sequence of replacements of the form, $g_i g_j \to g_j g_i$ where $m_{ij} = 2$. In particular, the words $\mathbf{i} = i_1 \cdots i_k$ and $\mathbf{i'} = i'_1 \cdots i'_k$ must have the same image in W. If $g = g_{i_1} \cdots g_{i_k}$, then define $\delta(1, g) = s(\mathbf{i})$ and then extend this to $G \times G$ by $\delta(g, g') = \delta(1, g^{-1}g')$. This δ has the desired properties.

6. Apartments and retractions

Suppose that C is a building of type M. Let $\mathbf{W}$ be the abstract Coxeter complex of W (defined in Example 3.1). A W-isometry of $\mathbf{W}$ into C is a map $\alpha : \mathbf{W} \to C$ which preserves W-distances, i.e.,

$$\delta_C(\alpha(w), \alpha(w')) = \delta_{\mathbf{W}}(w, w')$$

for all $w, w' \in W$. (Here δ_C is the W-valued distance function on C and $\delta_{\mathbf{W}}(w, w') = w^{-1}w'$.)

An *apartment* in C is an isometric image $\alpha(\mathbf{W})$ of $\mathbf{W}$ in C.

The set of W-isometries of $\mathbf{W}$ with itself is bijective with W. Indeed, given $w \in W$, there is a unique isometry $\alpha_w : \mathbf{W} \to \mathbf{W}$ sending 1 to w. It is defined by $\alpha_w(w') = ww'$. It follows that an isometry $\alpha : \mathbf{W} \to C$ is uniquely determined by its image $A = \alpha(\mathbf{W})$ together with a chamber $c = \alpha(1)$.

Fix an apartment A in C and a chamber c in A. Define a map

$$\rho_{c,A} : C \to A,$$

called the *retraction of C onto A with center c*, as follows. Let $A = \alpha(\mathbf{W})$ with $\alpha(1) = c$. Set $\rho_{c,A}(c') = \alpha(\delta(c, c'))$.

The following lemma is an easy exercise which we leave for the reader.

Lemma 6.1. *The map $\rho_{c,A} : C \to A$ is a morphism of chamber systems.*

In other words, if $\delta(c', c'') = s_i$, then either $\rho(c') = \rho(c'')$ or $\delta(\rho(c'), \rho(c'')) = s_i$, where $\rho = \rho_{c,A}$.

Corollary 6.2. *Let $\rho = \rho_{c,A}$. If $\delta(c', c'') \in W_J$, then $\delta(\rho(c'), \rho(c'')) \in W_J$.*

Lemma 6.3. *Let $\rho = \rho_{c,A} : C \to A$. If A' is another apartment containing c, then $\rho|_{A'} : A' \to A$ is a W-isometry (i.e., $\rho|_{A'}$ is an isomorphism).*

Proof Let $\alpha : \mathbf{W} \to A$ and $\beta : \mathbf{W} \to A'$ be W-isometries such that $\alpha(1) = c = \beta(1)$. If $c' \in A'$, then $\rho(c') = \alpha(\delta(c, c'))$ and $\beta(\delta(c, c')) = c'$. Therefore, $\rho(c') = \alpha \circ \beta^{-1}(c')$, i.e., $\rho|_{A'} = \alpha \circ \beta^{-1}$.

II. Geometric properties of buildings

7. $CAT(0)$ spaces

A metric space (X, d) is a *geodesic space* if any two points x and y can be joined by a path of length $d(x, y)$; such a path, parametrized by arc length, is a *geodesic*. A *triangle* in a geodesic space X is a configuration of three points and three geodesic segments connecting them. For each real number ϵ, let M_ϵ^2 denote the plane of curvature ϵ. (If $\epsilon > 0$, $M_\epsilon^2 = S_\epsilon^2$, the sphere of radius $1/\sqrt{\epsilon}$; if $\epsilon = 0$, $M_0^2 = \mathbb{E}^2$, the Euclidean plane; if $\epsilon < 0$, $M_\epsilon^2 = \mathbb{H}_\epsilon^2$, the hyperbolic plane of curvature ϵ.) Let T be a triangle in X and T^* a "comparison triangle" in M_ϵ^2. This means that the edge lengths of T^* are the same as those of T. (If $\epsilon > 0$, assume that the perimeter of T is $\leq 2\pi/\sqrt{\epsilon}$.) Denote the canonical isometry $T \to T^*$ by $p \to p^*$. Suppose x, y, z are the vertices of T and that, for $t \in [0, 1]$, p_t is the point on the edge from x to y of distance $td(x, y)$ from x. The $CAT(\epsilon)$-*inequality* is the inequality,

$$d(z, p_t) \leq d^*(z^*, p_t^*),$$

where d^* denotes distance in M_ϵ^2. The space X is $CAT(\epsilon)$ if this inequality holds for all triangles T (of perimeter $\leq 2\pi/\sqrt{\epsilon}$ when $\epsilon > 0$) and for all $t \in [0, 1]$.

If x, y, z are points in $\mathbb{E}^2$, then a simple argument ([Bro, p. 153]) shows

$$d^2(z, p_t) = (1 - t)d^2(z, x) + td^2(z, y) - t(1 - t)d^2(x, y)$$

Here d denotes Euclidean distance and $d^2(x, y) = d(x, y)^2$.

We return to the situation where T is a triangle in X with vertices x, y, z. Since the edge lengths of T are the same as those in the Euclidean comparison triangle, the equation in the previous paragraph immediately yields the following well-known lemma.

Lemma 7.1. *With notation as above, the $CAT(0)$-inequality for the triangle T in X is equivalent to*

$$d^2(z, p_t) \leq (1 - t)d^2(z, x) + td^2(z, y) - t(1 - t)d^2(x, y)$$

A *piecewise Euclidean polyhedron* X is a space formed by gluing together convex cells in Euclidean space via isometries of their faces. Each cell in X can then be identified with a Euclidean cell, well-defined up to isometry. It follows that arc-length makes sense in X. Thus, X has a natural "path metric": $d(x, y)$ is the infimum of the lengths of all piecewise linear paths joining x to y. One says that X has *finitely many shapes of cells* if there are only a finite number of isometry types of cells in the given cell structure on X. If X has finitely many shapes of cells, then the path metric gives it the structure of a complete geodesic space ([Bri]). We are interested in the question of when such piecewise Euclidean polyhedra are $CAT(0)$.

It also makes sense to speak of geodesically convex cells in S^n. This leads to the analogous notion of a piecewise spherical polyhedron. Piecewise spherical polyhedra play a distinguished role in this theory since they arise naturally as "links" of cells in piecewise constant curvature polyhedra. A basic result is that if X is a piecewise constant curvature polyhedron (of curvature ϵ), then X satisfies $CAT(\epsilon)$ locally if and only if the link of each cell is $CAT(1)$. (See [Bri] or Ballman's article in [GH, Ch. 10].)

8. The geometric realization of a poset

Let P be a poset. Its *derived complex*, denoted by P', is the poset of finite chains in P (partially ordered by inclusion). It is an abstract simplicial complex: the vertex set is P, the simplices are the elements of P'. The geometric realization of this simplicial complex is denoted by $geom(P)$ and called the *geometric realization* of P.

There are two different decompositions of $geom(P)$ into closed subspaces. Both decompositions are indexed by P. Given $p \in P$, let $geom(P)_{\leq p}$ (respectively, $geom(P)_{\geq p}$) denote the union of all simplices with maximal vertex p (respectively, minimal vertex p). The subcomplex $geom(P)_{\leq p}$ is a *face*, while $geom(P)_{\geq p}$ is a *coface*. Thus, the poset of faces of $geom(P)$ (partially ordered by inclusion) is naturally identified with P, while the poset of cofaces is identified with P^{op} (where P^{op} denotes the same set as P but with the order relations reversed).

9. The geometric realization of the Coxeter complex

Let I, M, W and $\mathcal{S}^f$ be as in §2 and $\mathbf{W}$ be the abstract Coxeter complex of Example 3.1. Consider the poset

$$W\mathcal{S}^f = \coprod_{J \in \mathcal{S}^f} W/W_J,$$

where the partial ordering is inclusion of cosets. We remark that $W\mathcal{S}^f$ can be identified with the poset of all residues in $\mathbf{W}$ of spherical type (where a residue of type $\emptyset$ is interpreted to be a chamber).

The *geometric realization of* $\mathbf{W}$, denoted $|\mathbf{W}|$, is defined to be the geometric realization of the poset $W\mathcal{S}^f$, i.e.,

$$|\mathbf{W}| = geom(W\mathcal{S}^f).$$

Also, we will use the notation,

$$|1| = geom(\mathcal{S}^f).$$

Subscripts will be used to denote the cofaces of $|\mathbf{W}|$ and $|1|$. Thus, if $J \in \mathcal{S}^f$ and $wW_J \in W\mathcal{S}^f$, then

$$|1|_J = geom(\mathcal{S}^f)_{\geq J}$$
$$|\mathbf{W}|_{wW_J} = geom(W\mathcal{S}^f)_{\geq wW_J}.$$

The maximal cofaces (when $J = \emptyset$) correspond to chambers in $\mathbf{W}$. If $w \in W$ is such a chamber, then the corresponding maximal coface is also called a *chamber* and will be denoted simply by $|w|$ (instead of $|\mathbf{W}|_{wW_\emptyset}$). Similarly, we will write $|w|_J$ instead of $|\mathbf{W}|_{wW_J}$.

Our next goal is to show that the faces of $|\mathbf{W}|$ are cells.

First suppose that W is finite. Then it has a representation as an orthogonal reflection group on $\mathbb{R}^n, n = \text{Card}(I)$. The reflection hyperplanes divide $\mathbb{R}^n$ into simplicial cones each of which is a fundamental domain for the W-action. Two of these simplicial cones (which are images of each other under the antipodal map) are bounded by the hyperplanes corresponding to the $s_i, i \in I$. Choose one and call it the *fundamental simplicial cone*. (See [B, Ch. V §4].) For each $\lambda \in (0,\infty)^I$ there is a unique point p_λ in the interior of the fundamental simplicial cone so that the distance from p_λ to the hyperplane fixed by s_i is λ_i. The convex hull of the W-orbit of p_λ is a convex cell in $\mathbb{R}^n$ called a *Coxeter cell of type W* and denoted by $P_W(\lambda)$. The intersection of $P_W(\lambda)$ with a fundamental simplicial cone is also a convex cell in $\mathbb{R}^n$ called a *Coxeter block* and denoted by $B_W(\lambda)$.

For example, if W is a direct product of cyclic groups of order two, then $P_W(\lambda)$ is the Cartesian product of intervals $([-\lambda_i, \lambda_i])_{i \in I}$, while $B_W(\lambda)$ is the Cartesian product of $([0, \lambda_i])_{i \in I}$.

The next lemma is an easy exercise. A proof is found in [CD3, Lemma 2.1.3].

Lemma 9.1. *Suppose W is finite. The vertex set of $P_W(\lambda)$ is Wp_λ (the W-orbit of p_λ). Let $\theta : W \to Wp_\lambda$ denote the bijection $w \to wp_\lambda$. Given a subset V of W, $\theta(V)$ is the vertex set of a face of $P_W(\lambda)$ if and only if $V = wW_J$ for some $wW_J \in W\mathcal{S}^f$. Thus, θ induces an isomorphism from $W\mathcal{S}^f$ to the poset of faces of $P_W(\lambda)$.*

The map θ of Lemma 9.1 induces a simplicial isomorphism from $|\mathbf{W}|$ to the barycentric subdivision of $P_W(\lambda)$ taking faces to faces. Moreover, the restriction of this identification to $|1|$ yields an identification of $|1|$ with $B_W(\lambda)$. Thus, when W is finite each chamber of $|\mathbf{W}|$ is identified with a Coxeter block. We also note that the face of $P_W(\lambda)$ corresponding to wW_J is isometric to $P_{W_J}(\lambda_J)$ where λ_J denotes the image of λ under the projection $(0,\infty)^I \to (0,\infty)^J$.

We return to the general situation where W may be infinite. As before, choose $\lambda \in (0, \infty)^I$. By Lemma 9.1 we can identify the face of $|\mathbf{W}|$ corresponding to wW_J with the Coxeter cell $P_{W_J}(\lambda_J)$. Moreover, this identification is well-defined up to an element of W_J (which acts by isometries on $P_{W_J}(\lambda_J)$). Hence, we have given $|\mathbf{W}|$ the structure of a piecewise Euclidean cell complex in which each cell is isometric to a Coxeter cell.

We make a few observations.

(1) The natural W-action on $|\mathbf{W}|$ is by isometries.

(2) Any chamber $|w|$ is a fundamental domain for the W-action.

(3) The projection $W\mathcal{S}^J \to \mathcal{S}^J$ induces a projection $|\mathbf{W}| \to |1|$ which is constant on W-orbits. The induced map $|\mathbf{W}|/W \to |1|$ is a homeomorphism

(4) If I is finite (a mild assumption), then $|1|$ is a finite complex and $|\mathbf{W}|$ has only finitely many shapes of cells.

We henceforth assume that I is finite.

Theorem 9.2. (Moussong [M]). *For any finitely generated Coxeter group W, the space $|\mathbf{W}|$, with the piecewise Euclidean structure defined above, is a complete $CAT(0)$ geodesic space.*

Remark 9.3. The choice of $\lambda \in (0, \infty)^I$ plays no role in Moussong's Theorem (or anywhere else). Henceforth, we normalize the situation by setting $\lambda_i = 1$, for all $i \in I$.

Remark 9.4. Suppose W is right-angled. Then (with the above normalization) each Coxeter cell is a regular Euclidean cube of edge length 2. In this situation $|\mathbf{W}|$ is a cubical complex and Theorem 9.2 was proved by Gromov, [G, p. 122], by a relatively easy argument.

10. The geometric realization of a building

Let C be a building of type M. The quickest way to define the geometric realization of C is as follows. Let $\mathcal{C}$ denote the poset of all J-residues in C with $J \in \mathcal{S}^J$. Then $|C|$ is defined to be the geometric realization of $\mathcal{C}$. For each $c \in C$ let $|c|$ denote the maximal coface $geom(\mathcal{C})_{\geq c}$. The map Type: $\mathcal{C} \to \mathcal{S}^J$ which associates to each residue its type induces a map of geometric realizations $|C| \to |1|$. Moreover, the restriction of this map to each chamber $|c|$ is a homeomorphism. Since we showed in §9 how to put a piecewise Euclidean structure on $|1|$ (a union of Coxeter blocks), this defines a piecewise Euclidean structure on $|C|$.

We shall now describe an alternate approach to this definition which is probably more illuminating. Let X be a space and $(X_i)_{i \in I}$ a family of closed subspaces. For each $x \in X$, set $J(x) = \{j \in I | x \in X_j\}$. Define an equivalence relation $\sim$ on $C \times X$ by $(c, x) \sim (c', x')$ if and only if $x = x'$ and

$\delta(c, c') \in W_{J(x)}$. The *X-realization* of C, denoted by $X(C)$, is defined to be the quotient space $(C \times X)/ \sim$. (Here C has the discrete topology.)
If c is a chamber in C, then $X(c)$ denotes the image of $c \times X$ in $X(C)$. If A is an apartment, then

$$X(A) = \bigcup_{c \in A} X(c)$$
$$\cong (A \times X)/ \sim$$

Let $\rho = \rho_{c,A} : C \to A$ be the retraction onto A with center c. By Corollary 6.2, the map $\rho \times id : C \times X \to A \times X$ is compatible with the equivalence relation $\sim$. Hence, there is an induced map $\overline{\rho} : X(C) \to X(A)$. The map $\overline{\rho}$ is a retraction in the usual topological sense.

We shall apply this construction in the following two special cases.

Case 1. $X = |1|$, the geometric realization of $\mathcal{S}^f$, and for each $i \in I, |1|_i = |1|_{\{i\}}$, the coface corresponding to $\{i\}$.

Case 2. W is finite (so that it acts as an orthogonal reflection group on $\mathbb{R}^n, n = \mathrm{Card}\,(I)$) and $X = \Delta$, the spherical $(n-1)$-simplex which is the intersection the fundamental simplicial cone with S^{n-1}. The subspace Δ_i is the codimension-one face of Δ which is the intersection of the hyperplane corresponding to s_i and Δ.

In Case 1, we will use the notation $|C|$ for $|1|(C)$ and call it the *geometric realization of C*. It is a simple matter to check that this agrees with the definition in the initial paragraph of this section. Similarly, $|c| = |1|(c)$ and $|A| = |1|(A)$.

In Case 2 (where C is a building of spherical type), $\Delta(C)$ will be called the *spherical realization of C*.

As explained in §9, $|1|$ has a natural piecewise Euclidean cell structure: the cells are Coxeter blocks. Thus,

$$|1| = \bigcup_{J \in \mathcal{S}^f} B_{W_J}.$$

Here the Coxeter block B_{W_J} is identified with the face of $|1|$ corresponding to J, i.e., with $geom(\mathcal{S}^f)_{\leq J}$. This induces a piecewise Euclidean cell structure (and a resulting path metric) on $|C|$. Thus, if A is an apartment of C, then $|A|$ is isometric to $|\mathbf{W}|$.

Similarly, if C is of spherical type, then $\Delta(C)$ inherits a piecewise spherical simplicial structure from the spherical simplex Δ. If A is an apartment, then $\Delta(A)$ is isometric to the round sphere S^{n-1}. These spherical realizations of buildings of spherical type arise naturally as links of certain cells in $|C|$ for a general building C.

Example 10.1. (A continuation of Example 3.2). Suppose that C is a building of rank one. Then $|1| = [0,1]$. Hence, $|C| = (C \times [0,1])/ \sim$ where $(c,0) \sim (c',0)$ for all $c, c' \in C$, i.e., $|C|$ is the cone on C. Similarly, $\Delta(C) = C$ (with the discrete topology).

Example 10.2. (A continuation of Example 3.3). Suppose that $C = C_1 \times \cdots \times C_k$ is a direct product of buildings. We use the notation of Example 3.3 Let $|1_t|$ denote the geometric realization of $\mathcal{S}^f(M_t), 1 \leq t \leq k$, and let $|1|$ denote the geometric realization of $\mathcal{S}^f(M)$. Then $|1| = |1_1| \times \cdots \times |1_k|$, where the Cartesian product has the product piecewise Euclidean structure and the product metric. It follows that $|C| = |C_1| \times \cdots \times |C_k|$. Similarly, if C is of spherical type, then $\Delta(C)$ is the "orthogonal join" (defined in [CD1, p. 1001]) of the $\Delta(C_i)$.

In particular suppose that each C_i is a rank one building. Then W is the direct product of k copies of the cyclic group of order two; $|C|$ is a Cartesian product of k cones; each chamber in $|C|$ is isometric to the k-cube $[0,1]^k$ and each apartment is isometric to $[-1,1]^k$. Similarly, $\Delta(C)$ is the orthogonal join of the C_i; each chamber is an "all right" spherical $(k-1)$-simplex and each apartment is isometric to S^{k-1}, triangulated as the boundary of a k-dimensional octahedron.

Remark 10.3. The usual definition of the geometric realization of a building C, say in [Bro] or [R], is as $X(C)$ where X is a simplex of dimension $\text{Card}(I) - 1$ and $(X_i)_{i \in I}$ is the set of codimension-one faces. If C is spherical, then $X(C) = \Delta(C)$, and if C is irreducible and of affine type, then $X(C) = |C|$. But, if, for example, C is the direct product of two buildings of affine type, then $X(C)$ is not a Cartesian product (it is a join). In [R, p. 184], Ronan comments to the effect that in the general case, the geometric realization of a building should be defined as above.

Remark 10.4. Another possibility is to take X and $(X_i)_{i \in I}$ to be a "Bestvina complex" as in [Bes] or [HM]. This means that X is a CW-complex, each X_i is a subcomplex and that if, for J a subset of I, we set

$$X_J = \bigcap_{i \in J} X_j$$

(and $X_\emptyset = X$), then X_J is nonempty and acyclic if and only if $J \in \mathcal{S}^f$. Furthermore, X is required to have the smallest possible dimension among all such complexes with this property. The argument in [D] then shows that $X(C)$ is acyclic (and contractible if X is contractible). In the case where C admits a chamber-transitive automorphism group $G, X(C)$ can be used to determine information about the cohomological dimension (or virtual cohomological dimension) of G.

11. The main theorem and some of its consequences

Theorem 11.1. *The geometric realization of any building is a complete, $CAT(0)$ geodesic space.*

In the case where the building C is irreducible and of affine type, this result is well-known. A proof can be found in [Bro, Ch. VI §3]. Our proof follows the argument given there.

Suppose $\rho = \rho_{c,A} : C \rightarrow A$ is the retraction onto A with center c. Its geometric realization $\bar{\rho} : |C| \rightarrow |A|$ (defined in §10) takes each chamber of $|C|$ isometrically onto a chamber of $|A|$. Hence, $\bar{\rho}$ maps a geodesic segment in $|C|$ to a piecewise geodesic segment in $|A|$ of the same length. From this observation we conclude the following.

Lemma 11.2. *The retraction $\bar{\rho} : |C| \rightarrow |A|$ is distance decreasing, i.e., for all $x, y \in |C|$,*

$$d_{|A|}(\bar{\rho}(x), \bar{\rho}(y)) \leq d_{|C|}(x, y),$$

where $d_{|A|}$ and $d_{|C|}$ denote distance in $|A|$ and $|C|$, respectively. In particular, if $x, y \in |A|$, then $d_{|A|}(x, y) = d_{|C|}(x, y)$.

Lemma 11.3. *There is a unique geodesic between any two points in $|C|$.*

Proof One of the basic facts about buildings is any two chambers are contained in a common apartment ([R, Theorem 3.11, p. 34]). This implies that, given $x, y \in |C|$, there is an apartment A such that $x, y \in |A|$. (Choose chambers $c, c' \in C$ so that $x \in |c|, y \in |c'|$ and an apartment A containing c and c'; then $x, y \in |A|$.) A basic fact about $CAT(0)$ spaces is that any two points are connected by a unique geodesic segment. Since $|A|$ is $CAT(0)$ (Moussong's Theorem), there is a unique geodesic segment in $|A|$ from x to y. Let $\gamma : [0, d] \rightarrow |A|$ be a parametrization of this segment by arc length, where $d = d(x, y)$. By the last sentence of the previous lemma, γ is also a geodesic in $|C|$. Let $\gamma' : [0, d] \rightarrow |C|$ be another geodesic from x to y. If $\bar{\rho} : |C| \rightarrow |A|$ is the geometric realization of any retraction onto A, then $\bar{\rho} \circ \gamma'$ is a geodesic in $|A|$ from x to y (since it is a piecewise geodesic of length d). Hence, $\bar{\rho} \circ \gamma' = \gamma$. Let $t_0 = \sup\{t | \gamma|_{[0,t]} = \gamma'|_{[0,t]}\}$. Suppose that $t_0 < d$. Then, for small positive values of ϵ, $\gamma(t_0 + \epsilon)$ lies in the relative interior of some coface $|c|_J$, with $c \in A$, while $\gamma'(t_0 + \epsilon)$ lies in the relative interior of a different coface $|c'|_J$, where $|c'|_J \not\subset |A|$. Set $\rho = \rho_{c,A}$. Since $|c|_J \neq |c'|_J, \delta(c, c') \notin W_J$. Hence, $\bar{\rho}(\gamma'(t_0 + \epsilon)) \neq \gamma(t_0 + \epsilon)$, a contradiction. Therefore, $t_0 = d$ and $\gamma = \gamma'$.

Lemma 11.4. *Suppose $\rho = \rho_{c,A}$. If $x \in |c|$, then $d(x, \bar{\rho}(y)) = d(x, y)$ for all $y \in |C|$.*

Proof Choose an apartment A' so that $|A'|$ contains both x and y. By the previous lemma, the image of the geodesic γ from x to y is contained in $|A'|$. By Lemma 6.3, $\rho|_{A'} : A' \to A$ is a W-isometry. It follows that $\overline{\rho}|_{|A'|} : |A'| \to |A|$ is an isometry. Hence, $\overline{\rho} \circ \gamma$ is actually a geodesic (of the same length as γ).

Proof of Theorem 11.1 Suppose $x, y, z \in |C|$. For $t \in [0,1]$, let p_t be the point on the geodesic segment from x to y such that $d(x, p_t) = td(x,y)$. By Lemma 7.1, to prove that $|C|$ is $CAT(0)$ we must show that

$$d^2(z, p_t) \le (1 - t)d^2(z, x) + td(z, y) - t(1 - t)d^2(x, y).$$

Choose an apartment A so that $x, y \in |A|$. Since the geodesic segment from x to y lies in $|A|$, $p_t \in |A|$. Hence, we can choose a chamber c in A so that $p_t \in |c|$. Let $\rho = \rho_{c,A}$. By Lemma 11.4, $d(z, p_t) = d(\overline{\rho}(z), p_t)$. Hence,

$$
\begin{aligned}
d^2(z, p_t) &= d^2(\overline{\rho}(z), p_t) \\
&\le (1 - t)d^2(\overline{\rho}(z), x) + td^2(\overline{\rho}(z), y) - t(1 - t)d^2(x, y) \\
&\le (1 - t)d^2(z, x) + td^2(z, y) - t(1 - t)d^2(x, y).
\end{aligned}
$$

The first inequality holds since $\overline{\rho}(z), x, y$ all lie in $|A|$ and since $|A|$ is $CAT(0)$. The second inequality follows from Lemma 11.2. Therefore, $|C|$ is $CAT(0)$. $\quad\blacksquare$

Since $CAT(0)$ spaces are contractible (via geodesic contraction), we have the following corollary.

Corollary 11.5. $|C|$ *is contractible.*

As mentioned previously (in Remark 10.4) this can also be proved as in [D]. Suppose, for the moment, that C is spherical. Then $|C|$ has a distinguished vertex v, namely the coface $|C|_{W_I}(=|c|_I$ for any $c \in C$). The link of v in C is $\Delta(C)$. Since the link of a vertex in a $CAT(0)$ space is $CAT(1)$ ([G, p. 120]), this gives the following corollary.

Corollary 11.6. *Suppose C is spherical, then $\Delta(C)$ is $CAT(1)$.*

Of course, one could also give a direct argument for this by proving the analogs of Lemmas 11.2 and 11.3 for $\Delta(C)$ (in the case of 11.3 one shows the uniqueness of geodesics only in the case where the endpoints are of distance less than π).

Corollary 11.7. (Meier [Mei]) *With notation as in §5, let G be a graph product of groups $(P_i)_{i \in I}$ and let C be the building $C(G, \{1\}, (P_i)_{i \in I})$ (cf. Theorem 5.1). Then $|C|$ is $CAT(0)$.*

Remark 11.8. Since the Coxeter block associated to a direct product of cyclic groups of order two is a Euclidean cube, $|C|$ is a cubical complex. In the proof of Theorem B in [Mei], in the case of graph products, the complex $|C|$ is defined (without mentioning buildings) and proved to be $CAT(0)$. There is a simple proof of this by induction on $\mathrm{Card}\,(I)$. One first notes that for any subset J of I, $|C_J|$ is a totally geodesic subcomplex of $|C|$, where $C_J = C(G_J, \{I\}, (P_j)_{j \in J})$. If I is spherical, then $|C|$ is a direct product of cones on the P_i and hence, is $CAT(0)$. If I is not spherical, then there exist $i, j \in I$ with $m_{ij} = \infty$. Then G is the amalgamated product of $G_{I-\{i\}}$ and $G_{I-\{j\}}$ along $G_{I-\{i,j\}}$ and $|C|$ is a union of components each of which is a translate of $|C_{I-\{i\}}|$ or $|C_{I-\{j\}}|$. Furthermore, the intersection of two such components is a translate of the totally geodesic subcomplex $|C_{I-\{i,j\}}|$. The result follows from a gluing lemma of Gromov ([G, p. 124] and [GH, p. 192]) and induction.

The Bruhat-Tits Fixed Point Theorem (see [Bro, p. 157]) states that if a group of isometries of a complete, $CAT(0)$ space has a bounded orbit, then it has a fixed point. Applying this result to the case of a building with a chamber-transitive automorphism group, we get the following.

Corollary 11.9. *With notation as in §4, suppose that $C = C(G, B, (P_i)_{i \in I})$ is a building and that H is a subgroup of G which has a bounded orbit in $|C|$. Then H is conjugate to a subgroup of P_J, for some $J \in \mathcal{S}^f$.*

Remark 11.10. First suppose W is finite. Then it can be represented as a reflection group on hyperbolic n-space. Given $\lambda \in (0, \infty)^I$, one defines a "hyperbolic Coxeter cell" $P_W^h(\lambda)$ and a "hyperbolic Coxeter block" $B_W^h(\lambda)$, exactly as in §9. Returning to the situation where W and C are arbitrary, given $\lambda \in (0, \infty)^I$, we get a piecewise hyperbolic structure on $|C|$. Let us denote it $|C|^h(\lambda)$. If, for the link of each cell in the $CAT(0)$ structure on $|W|$, the length of the shortest closed geodesic is strictly greater than 2π, then for sufficiently small values of λ, $|W|^h(\lambda)$ will be $CAT(-1)$ (and hence, W will be word hyperbolic). If this is the case, then the proof of Theorem 11.1 shows that $|C|^h(\lambda)$ is also $CAT(-1)$. In [M] Moussong determined exactly when this holds. His condition is that for each subset J of I neither of the following occur:

a) W_J is of affine type, with $\mathrm{Card}\,(J) \geq 3$,

b) W_J is a direct product $W_{J_1} \times W_{J_2}$ where both J_1 and J_2 are infinite.

Thus, if neither condition holds, then $|C|$ can be given a $CAT(-1)$ structure.

References

[Bes] M. Bestvina, *The virtual cohomological dimension of Coxeter groups*, in: Geometric Group Theory I London Maths. Soc. Lecture Notes, **181** (G. A. Niblo and M. A. Roller, eds.), Cambridge University Press, Cambridge, (1993) pp. 19–23.

[B] N. Bourbaki, *Groupes et Algebrès de Lie,* Chapters 4–6 Masson, Paris (1981).

[Bri] M. R. Bridson, *Geodesics and curvature in metric simplicial complexes,* in: Group Theory from a Geometrical Viewpoint (E. Ghys, A. Haefliger and A. Verjovsky, eds.), World Scientific, Singapore (1991), pp. 373–463.

[Bro] K. Brown, *Buildings,* Springer-Verlag, New York (1989).

[CD1] R. Charney and M. W. Davis, *Singular metrics of nonpositive curvature on branched covers of Riemannian manifolds,* Amer. J. Math. **115** (1993), pp. 929–1009.

[CD2] R. Charney and M. W. Davis, *The $K(\pi,1)$-problem for hyperplane complements associated to infinite reflection groups,* To appear in J. Amer. Math. Soc.

[CD3] R. Charney and M. W. Davis, *Finite $K(\pi,1)$'s for Artin groups,* in: Prospects in Topology (F. Quinn, ed.), Annals of Math. Studies, **138** Princeton Univ. Press, Princeton (1995), pp. 110–124.

[D] M. W. Davis, *Groups generated by reflections and aspherical manifolds not covered by Euclidean space,* Ann. of Math. **117** (1983), pp. 293–324.

[GH] E. Ghys and P. de la Harpe, (eds.), *Sur les Groupes Hyperboliques d'apres Mikhael Gromov,* Prog. Math. **83** Birkhauser, Boston (1990).

[G] M. Gromov, *Hyperbolic groups,* in: Essays in Group Theory (S. M. Gersten, ed.), M.S.R.I. Publ., 8 Springer-Verlag, New York (1987), pp. 75–264.

[H] A. Haefliger, *Complexes of groups and oribhedra,* in: Group Theory from a Geometrical Viewpoint (E. Ghys, A. Haefliger and A. Verjosky, eds.), World Scientific Singapore (1991), pp. 504–540.

[HM] J. Harlander and H. Meinert, *Higher generation subgroup sets and the virtual coholmological dimension of graph products,* J. London Math. Soc. **53** (1996), pp. 99–117.

[Mei] J. Meier, *When is the graph product of hyperbolic groups hyperbolic?,* Geom. Dedicata **61** (1996), pp. 29–41.

[M] G. Moussong, *Hyperbolic Coxeter groups,* Ph.D. thesis, Ohio State University (1988).

[R] M. Ronan, *Lectures on Buildings,* Academic Press, San Diego (1989).

[S] J. P. Serre, *Trees,* Springer-Verlag, New York (1989).

[T] J. Tits, *Buildings and group amalgamations,* in: Proceedings of Groups - St. Andrews 1985 London Maths, Soc. Lecture Notes, **121** Cambridge University Press, Cambridge (1986), pp. 110–127.

On Subgroups of Coxeter Groups

Warren Dicks and Ian J. Leary

Departament de Matemàtiques, Universitat Autònoma de Barcelona, E 08193, Bellaterra (Barcelona), Spain.
Faculty of Mathematical Studies, University of Southampton, Southampton SO17 1BJ, England.
Dedicated to the memory of Brian Hartley

1. Outline

For a finitely generated Coxeter group Γ, its virtual cohomological dimension over a (non-zero, associative) ring R, denoted $\mathrm{vcd}_R\Gamma$, is finite and has been described [8,1,11,13]. In [8], M. Davis introduced a contractible Γ-simplicial complex with finite stabilisers. The dimension of such a complex gives an upper bound for $\mathrm{vcd}_R\Gamma$. In [1], M. Bestvina gave an algorithm for constructing an R-acyclic Γ-simplicial complex with finite stabilisers of dimension exactly $\mathrm{vcd}_R\Gamma$, for R the integers or a prime field; he used this to exhibit a group whose cohomological dimension over the integers is finite but strictly greater than its cohomological dimension over the rationals. For the same rings, and for right-angled Coxeter groups, J. Harlander and H. Meinert [13] have shown that $\mathrm{vcd}_R\Gamma$ is determined by the local structure of Davis' complex and that Davis' construction can be generalised to graph products of finite groups.

Our contribution splits into three parts. Firstly, Davis' complex may be defined for infinitely generated Coxeter groups (and infinite graph products of finite groups). We determine which such groups Γ have finite virtual cohomological dimension over the integers, and give partial information concerning $\mathrm{vcd}_{\mathbf{Z}}\Gamma$. We discuss a form of Poincaré duality for simplicial complexes that are like manifolds from the point of view of R-homology, and give conditions for a (finite-index subgroup of a) Coxeter group to be a Poincaré duality group over R. We give three classes of examples: we recover Bestvina's examples (and give more information about their cohomology); we exhibit a group whose virtual cohomological dimension over the integers is finite but strictly greater than its virtual cohomological dimension over any field; we exhibit a torsion-free rational Poincaré duality group which is not an integral Poincaré duality group.

Secondly, we discuss presentations for torsion-free subgroups of low index in right-angled Coxeter groups. In some cases (depending on the local structure of Davis' complex) we determine the minimum number of generators for any torsion-free normal subgroup of minimal index. Using the computer package GAP [17] we find good presentations for one of Bestvina's examples, where 'good' means having as few generators and relations as possible.

Finally, we give an 8-generator 12-relator presentation of a group Δ and a construction of Δ as a tower of amalgamated free products, which allows us to describe a good CW-structure for an Eilenberg-Mac Lane space $K(\Delta, 1)$ and explicitly show that Δ has cohomological dimension three over the integers and cohomological dimension two over the rationals. (In fact Δ is isomorphic to a finite-index subgroup of a Coxeter group, but our proofs do not rely on this.) The starting point of the work contained in this paper was the desire to see an explicit Eilenberg-Mac Lane space for an example like Δ.

2. Introduction

A *Coxeter system* (Γ, V) is a group Γ and a set of generators V for Γ such that Γ has a presentation of the form

$$\Gamma = \langle V \mid (vw)^{m(v,w)} = 1 \ (v, w \in V) \rangle,$$

where $m(v, v) = 1$, and if $v \neq w$ then $m(v, w) = m(w, v)$ is either an integer greater than or equal to 2, or is infinity (in which case this relation has no significance and may be omitted). Note that we do not require that V should be finite. The group Γ is called a *Coxeter group*, and in the special case when each $m(v, w)$ is either 1, 2 or ∞, Γ is called a *right-angled Coxeter group*.

Remark. Let (Γ, V) be a Coxeter system, and let $m : V \times V \to \mathbf{N} \cup \{\infty\}$ be the function occurring in the Coxeter presentation for Γ. If W is any subset of V and Δ the subgroup of Γ generated by W, then it may be shown that (Δ, W) is a Coxeter system, with m_W being the restriction of m_V to $W \times W$ [5]. The function m is determined by (Γ, V) because $m(v, w)$ is the order of vw (which is half the order of the subgroup of Γ generated by v and w).

Definition. A *graph* is a 1-dimensional simplicial complex (i.e., our graphs contain no loops or multiple edges). A *labelled graph* is a graph with a function from its edge set to a set of 'labels'. A *morphism of graphs* is a simplicial map which does not collapse any edges. A *morphism of labelled graphs* is a graph morphism such that the image of each edge is an edge having the same label. A *colouring* of a graph X is a function from its vertex set to a set of 'colours' such that the two ends of any edge have different images. Colourings of a graph X with colour set C are in 1-1-correspondence with graph morphisms from X to the complete graph with vertex set C.

Definition. For a Coxeter system (Γ, V), the simplicial complex $K(\Gamma, V)$ is defined to have as n-simplices the $(n+1)$-element subsets of V that generate finite subgroups of Γ. Note that our $K(\Gamma, V)$ is Davis' $K_0(\Gamma, V)$ in [8]. The graph $K^1(\Gamma, V)$ is by definition the 1-skeleton of this complex. The graph $K^1(\Gamma, V)$ has a labelling with labels the integers greater than or equal to 2, which takes the edge $\{v, w\}$ to $m(v, w)$. This labelled graph is different from, but carries the same information as, the Coxeter diagram. The labelled graph $K^1(\Gamma, V)$ determines the Coxeter system (Γ, V) up to isomorphism, and any graph labelled by the integers greater than or equal to 2 may arise in this way. A morphism of labelled graphs from $K^1(\Gamma, V)$ to $K^1(\Delta, W)$ gives rise to a group homomorphism from Γ to Δ.

Call a subgroup of Γ *special* if it is generated by a (possibly empty) subset of V. Thus the simplices of $K(\Gamma, V)$ are in bijective correspondence with the non-trivial finite special subgroups of Γ. Let $D(\Gamma, V)$ be the simplicial complex associated to the poset of (left) cosets of finite special subgroups of Γ. By construction Γ acts on D, and the stabiliser of each simplex is conjugate to a finite special subgroup of Γ. In [8], Davis showed that $D(\Gamma, V)$ is contractible if V is finite, and the general case follows easily (for example because any cycle (resp. based loop) in $D(\Gamma, V)$ is contained in a subcomplex isomorphic to $D(\langle V' \rangle, V')$ for some finite subset V' of V, so *a fortiori* bounds (resp. bounds a disc) in $D(\Gamma, V)$). Note that $K(\Gamma, V)$ is finite-dimensional if and only if $D(\Gamma, V)$ is, and in this case the dimension of $D(\Gamma, V)$ is one more than the dimension of $K(\Gamma, V)$.

A *graph product* Γ of finite groups in the sense of E. R. Green [12] is the quotient of the free product of a family $\{G_v \mid v \in V\}$ of finite groups by the normal subgroup generated by the sets $\{[g, h] \mid g \in G_v, h \in G_w\}$ for some pairs $v \neq w$ of elements of V. A graph product of groups of order two is a right-angled Coxeter group. If a *special subgroup* of a graph product is defined to be a subgroup generated by some subset of the given family of finite groups, then the above definitions of $K(\Gamma, V)$ and $D(\Gamma, V)$ go through unchanged. In [13] it is proved that for a graph product of finite groups, $D(\Gamma, V)$ is contractible. (As in [8] only the case when V is finite is considered, but the general case follows easily.) The group algebra for a graph product $\mathbb{Z}\Gamma$ is isomorphic to the quotient of the ring coproduct of the $\mathbb{Z}G_v$ by relations that ensure that the pairs $\mathbb{Z}G_v$ and $\mathbb{Z}G_w$ generate their tensor product whenever G_v and G_w commute. Theorem 4.1 of [11] is a result for algebras formed in this way which in the case of the group algebra of a graph product is equivalent to the acyclicity of $D(\Gamma, V)$.

3. Virtual cohomology of Coxeter groups

Henceforth we shall make use of the abbreviations vcd and cd to denote the phrases 'virtual cohomological dimension'and 'cohomological dimension', respectively, and when no ring is specified, these dimensions are understood to be over the ring of integers.

Theorem 1. *The Coxeter group Γ has finite* vcd *if and only if there is a labelled graph morphism from $K^1(\Gamma, V)$ to some finite labelled graph.*

Proof The complex $K = K(\Gamma, V)$ has simplices of arbitrarily large dimension if and only if V contains arbitrarily large finite subsets generating finite subgroups of Γ. In this case Γ cannot have a torsion-free subgroup of finite index, and there can be no graph morphism from K^1 to any finite graph. Thus we may assume that K and hence also D are finite-dimensional. Any torsion-free subgroup of Γ acts freely on D, and so it remains to show that if D is finite-dimensional then there is a labelled graph morphism from K^1 to a finite graph if and only if Γ has a finite-index torsion-free subgroup. As remarked above, a morphism from $K^1(\Gamma, V)$ to $K^1(\Delta, W)$ gives rise to a group homomorphism from Γ to Δ in an obvious way. Moreover, if v, v' have product of order $m(v, v')$, then so do their images in W, because the edge $\{v, v'\}$ and its image in $K^1(\Delta, W)$ are both labelled by $m(v, v')$. Now if V' is a finite subset of V generating a finite subgroup of Γ, and W' is its image in W, then it follows that $\langle V' \rangle$ and $\langle W' \rangle$ have identical Coxeter presentations, so are isomorphic. Thus a morphism from $K^1(\Gamma, V)$ to $K^1(\Delta, W)$ gives rise to a homomorphism from Γ to Δ which is injective on every finite special subgroup of Γ. Now suppose that there is a morphism from $K^1(\Gamma, V)$ to $K^1(\Delta, W)$ for some finite W. The finitely generated Coxeter group Δ has a finite-index torsion-free subgroup Δ_1, so let Γ_1 be the inverse image of this subgroup in Γ. Since Γ_1 intersects any conjugate of any finite special subgroup trivially, it follows that Γ_1 acts freely on $D(\Gamma, V)$ and is torsion-free.

Conversely, if Γ has a finite-index torsion-free subgroup Γ_1, which we may assume to be normal, let Q be the quotient Γ/Γ_1, and build a labelled graph X with vertices the elements of Q of order two and all possible edges between them. Label the edge $\{q, q'\}$ by the order of qq'. Now the homomorphism from Γ onto Q induces a simplicial map from $K^1(\Gamma, V)$ to X which is a labelled graph morphism because if vv' has finite order then its image in Q has the same order.

Remarks. 1) If we are interested only in right-angled Coxeter groups then all the edges of K^1 have the same label, 2, and we may replace the condition that there is a morphism from K^1 to a finite labelled graph by the equivalent condition that K^1 admits a finite colouring. The above proof can be simplified slightly in this case, because the right-angled Coxeter group corresponding to a finite complete graph is a finite direct product of cyclic groups of order two.

2) An easy modification of the proof of Theorem 1 shows that a graph product Γ of finite groups has finite vcd if and only if there are only finitely many isomorphism types among the vertex groups G_v, and the graph $K^1(\Gamma, V)$ admits a finite colouring.

3) Let (Γ, V) be the Coxeter system corresponding to the complete graph on an infinite set, where each edge is labelled n for some fixed $n \geq 3$. Then $K(\Gamma, V)$ is one-dimensional (since the Coxeter group on three generators such that the product of any two has order n is infinite), Γ has an action on a 2-dimensional contractible complex with stabilisers of orders 1, 2, and $2n$, but by the above theorem Γ does not have finite vcd. Similarly, if we take a triangle-free graph which cannot be finitely coloured, then the corresponding right-angled Coxeter group acts on a contractible 2-dimensional complex with stabilisers of orders 1, 2, and 4, but does not have finite vcd. In contrast, any group acting on a tree with finite stabilisers of bounded order has finite vcd; see for example [10], Theorem I.7.4.

If a Coxeter group Γ has finite vcd then $D(\Gamma, V)$ is finite-dimensional and the dimension of D gives an upper bound for $\text{vcd}\,\Gamma$. Parts a) and c) of the following theorem determine when this upper bound is attained. The information concerning the right Γ-module structure on various cohomology groups will be used only during the construction (in example 3 of the next section) of a torsion-free rational Poincaré duality group that is not a Poincaré duality group over the integers. To avoid cluttering the statement unnecessarily we first give some definitions that are used in it.

Definition. For a Coxeter system (Γ, V) and an abelian group A, let A° denote the Γ-bimodule with underlying additive group A and Γ-actions given by $va = av = -a$ for all $v \in V$. This does define compatible actions of Γ because each of the relators in the Coxeter presentation for Γ has even length as a word in V. For a Γ-module M, let M_a denote the underlying abelian group. For a simplicial complex D, let $C_*(D)$ denote the simplicial chain complex of D, let $C_*^+(D)$ denote the augmented simplicial chain complex (having a -1-simplex equal to the boundary of every 0-simplex) and let $\tilde{H}^*(D; A)$ denote the reduced cohomology of D with coefficients in A, i.e., the homology of the cochain complex $\text{Hom}(C_*^+(D), A)$. All our Γ-modules (in particular, all our chain complexes of Γ-modules) are left modules unless otherwise stated.

Theorem 2. *Let (Γ, V) be a Coxeter system such that Γ has finite vcd, let $K = K(\Gamma, V)$ have dimension n (which implies that $\text{vcd}\,\Gamma \leq n+1$), let $D = D(\Gamma, V)$, let Γ_1 be a finite-index torsion-free subgroup of Γ, and let A be an abelian group containing no elements of order two. Then*
a) For any Γ_1-module M, $H^{n+1}(\Gamma_1; M)$ is a quotient of a finite direct sum of

copies of $\tilde{H}^n(K; M_a)$.

b) For each j, there is an isomorphism of right Γ-modules as follows.

$$H^{j+1}\mathrm{Hom}_\Gamma(C_*(D), A^\circ) \cong \tilde{H}^j(K; A)^\circ$$

c) The right Γ-module $H^{n+1}(\Gamma; A\Gamma)$ (which is isomorphic to $H^{n+1}(\Gamma_1; A\Gamma_1)$ as a right Γ_1-module) admits a surjective homomorphism onto $\tilde{H}^n(K; A)^\circ$.

d) If multiplication by the order of each finite special subgroup of Γ induces an isomorphism of A, then for each j the right Γ-modules $H^{j+1}(\Gamma; A^\circ)$ and $\tilde{H}^j(K; A)^\circ$ are isomorphic.

Proof Let K' be the simplicial complex associated to the poset of non-trivial finite special subgroups of Γ, so that K' is the barycentric subdivision of K. Let D' be the complex associated to the poset of cosets of non-trivial finite special subgroups of Γ. Then D' is a subcomplex of D, and consists of all the simplices of D whose stabiliser is non-trivial. We obtain a short exact sequence of chain complexes of $\mathbb{Z}\Gamma$-modules

$$0 \to C_*(D') \to C_*(D) \to C_*(D, D') \to 0, \qquad (*)$$

such that for each n the corresponding short exact sequence of $\mathbb{Z}\Gamma$-modules is split.

There is a chain complex isomorphism as shown below.

$$C_*(D, D') \cong \mathbb{Z}\Gamma \otimes_{\mathbb{Z}} C^+_{*-1}(K')$$

Topologically this is because the quotient semi-simplicial complex D/D' is isomorphic to a wedge of copies of the suspension of K', with Γ acting by permuting the copies freely and transitively. More explicitly, one may identify m-simplices of D with equivalence classes of $(m + 2)$-tuples $(\gamma, V_0, \ldots, V_m)$, where $V_0 \subseteq \cdots \subseteq V_m$ are subsets of V generating finite subgroups of Γ, γ is an element of Γ, and two such expressions $(\gamma, V_0, \ldots, V_m)$ and $(\gamma', V_0', \ldots, V_m')$ are equivalent if $V_i = V_i'$ for all i and the cosets $\gamma\langle V_0\rangle$ and $\gamma'\langle V_0\rangle$ are equal. A map from $C_*(D)$ to $\mathbb{Z}\Gamma \otimes_{\mathbb{Z}} C^+_{*-1}(K')$ may be defined by

$$(\gamma, V_0, \ldots, V_m) \mapsto \begin{cases} 0 & \text{if } V_0 \neq \emptyset, \\ \gamma \otimes (V_1, \ldots, V_m) & \text{if } V_0 = \emptyset, \end{cases}$$

and it may be checked that this is a surjective chain map with kernel $C_*(D')$. The claim of part a) now follows easily. Applying $\mathrm{Hom}_{\Gamma_1}(\,\cdot\,, M)$ to the sequence $(*)$ and taking the cohomology long exact sequence for this short exact sequence of cochain complexes, one obtains the following sequence.

$$H^{n+1}\mathrm{Hom}_{\Gamma_1}(C_*(D, D'), M) \to H^{n+1}\mathrm{Hom}_{\Gamma_1}(C_*(D), M) \to 0$$

Now $H^{n+1}\mathrm{Hom}_{\Gamma_1}(C_*(D), M) = H^{n+1}(\Gamma_1; M)$, and there is a chain of isomorphisms as below.

$$H^{n+1}\mathrm{Hom}_{\Gamma_1}(C_*(D, D'), M) \cong H^n\mathrm{Hom}_{\Gamma_1}(\mathbb{Z}\Gamma \otimes C_*^+(K'), M)$$
$$\cong \bigoplus_{\Gamma/\Gamma_1} H^n\mathrm{Hom}(C_*^+(K'), M_a)$$
$$\cong \bigoplus_{\Gamma/\Gamma_1} \tilde{H}^n(K; M_a)$$

To prove b), note that since A has no elements of order two, there are no non-trivial Γ-module homomorphisms from the permutation module $\mathbb{Z}\Gamma/\langle V'\rangle$ to A° for any non-empty subset V' of V. Hence applying $\mathrm{Hom}_\Gamma(\,\cdot\,, A^\circ)$ to the sequence (∗) one obtains an isomorphism of cochain complexes of right Γ-modules

$$\mathrm{Hom}_\Gamma(C_*(D), A^\circ) \cong \mathrm{Hom}_\Gamma(C_*(D, D'), A^\circ).$$

Taking homology gives the following chain of isomorphisms.

$$H^{j+1}\mathrm{Hom}_{\mathbb{Z}\Gamma}(C_*(D), A^\circ) \cong H^{j+1}\mathrm{Hom}_{\mathbb{Z}\Gamma}(C_*(D, D'), A^\circ)$$
$$\cong H^j\mathrm{Hom}_{\mathbb{Z}\Gamma}(\mathbb{Z}\Gamma \otimes C_*^+(K'), A^\circ)$$
$$\cong H^j\mathrm{Hom}_{\mathbb{Z}}(C_*^+(K'), A^\circ)$$
$$\cong \tilde{H}^j(K; A)^\circ$$

Now d) follows easily. Let R be the subring of $\mathbb{Q}$ generated by the inverses of the orders of the finite special subgroups of Γ. Now $\mathrm{Hom}_{R\Gamma}(R \otimes C_*(D), A^\circ)$ is isomorphic to $\mathrm{Hom}_{\mathbb{Z}\Gamma}(C_*(D), A^\circ)$, and $R \otimes C_*(D)$ is a projective resolution for R over $R\Gamma$, so d) follows from b).

For c), note that there is an equivalence of functors (defined on Γ-modules) between $\mathrm{Hom}_\Gamma(\,\cdot\,, A\Gamma)$ and $\mathrm{Hom}_{\Gamma_1}(\,\cdot\,, A\Gamma_1)$. In particular, $H^*(\Gamma_1; A\Gamma_1)$ and $H^*(\Gamma; A\Gamma)$ are both isomorphic to the homology of the cochain complex $\mathrm{Hom}_\Gamma(C_*(D), A\Gamma)$.

There is a Γ-bimodule map ϕ from $A\Gamma$ to A° sending $a.w$ to $(-1)^l a$, where w is any element of Γ representable by a word of length l in the elements of V. Consider the following commutative diagram of cochain complexes, where the vertical maps are induced by ϕ:

$$
\begin{array}{ccc}
\mathrm{Hom}_\Gamma(C_*(D), A\Gamma) & \to & \mathrm{Hom}_\Gamma(C_*(D, D'), A\Gamma) \\
\downarrow & & \downarrow \\
\mathrm{Hom}_\Gamma(C_*(D), A^\circ) & \to & \mathrm{Hom}_\Gamma(C_*(D, D'), A^\circ).
\end{array}
$$

The horizontal maps are surjective because $C_i(D, D')$ is a direct summand of $C_i(D)$ for each i, and the lower horizontal map is an isomorphism as in the proof of b). The right-hand vertical map is surjective because $C_*(D, D')$ is $\mathbb{Z}\Gamma$-free, and hence the left-hand vertical map is surjective.

Since each of the cochain complexes is trivial in degrees greater than $n + 1$, one obtains a surjection

$$H^{n+1}\mathrm{Hom}_\Gamma(C_*(D), A\Gamma) \to H^{n+1}\mathrm{Hom}_\Gamma(C_*(D), A^\circ),$$

and hence by b) a surjection of right Γ-modules

$$H^{n+1}(\Gamma_1; A\Gamma_1) \longrightarrow \tilde{H}^n(K; A)^\circ.$$

Remark. Parts b) and d) of Theorem 2 do not generalise easily to graph products of finite groups having finite virtual cohomological dimension, and we have no application for these statements except in the Coxeter group case. We outline the generalisation of a) and a weaker version of c) below.

The statement and proof of part a) carry over verbatim, and there is a generalisation of part c). If Γ is a graph product of finite groups with l distinct isomorphism types of vertex group such that the graph $K^1(\Gamma, V)$ can be m-coloured, then the graph product version of Theorem 1 implies that Γ admits a finite quotient $G = G_1 \times \cdots \times G_k$ for some $k \leq lm$, where each G_i is isomorphic to a vertex group of Γ and each finite special subgroup of Γ is mapped injectively to G with image of the form $G_{i(1)} \times \cdots \times G_{i(j)}$ for some subset $\{i(1), \ldots, i(j)\}$ of $\{1, \ldots, k\}$. Now for $1 \leq i \leq k$, let $x_i \in \mathbb{Z}G$ be the sum of all the elements of G_i, and let Z be the $\mathbb{Z}\Gamma$-module defined as the quotient of $\mathbb{Z}G$ by the ideal generated by the x_i. This Z is the appropriate generalisation of $\mathbb{Z}^\circ$ to the case of a graph product, because it is a quotient of $\mathbb{Z}\Gamma$ of finite $\mathbb{Z}$-rank and contains no non-zero element fixed by any vertex group. To see this, note that

$$Z \cong \mathbb{Z}G_1/(x_1) \otimes \cdots \otimes \mathbb{Z}G_k/(x_k),$$

where Γ acts on the ith factor via its quotient G_i. Each factor is $\mathbb{Z}$-free, and the action of G_i on $\mathbb{Z}G_i/(x_i)$ has no fixed points, because for example $\mathbb{C} \otimes (\mathbb{Z}G_i/(x_i))$ does not contain the trivial $\mathbb{C}G_i$-module.

The arguments used in the proof of Theorem 2 may be adapted to prove a statement like that of part b) for the module Z, namely that for any j,

$$H^{j+1}\mathrm{Hom}_\Gamma(C_*(D), Z) \cong \tilde{H}^j(K; Z_a).$$

From this it may be deduced that if Γ_1 is a torsion-free finite-index subgroup of Γ, then $H^{n+1}(\Gamma_1; \mathbb{Z}\Gamma_1)$ admits $\tilde{H}^n(K; Z_a)$ as a quotient. A similar result could then be deduced for any torsion-free abelian group A. A similar result could also be proved for A an $\mathbb{F}_p$-vector space, for p a prime not dividing the order of any of the vertex groups, by using the fact that $\mathbb{F}_p G$ is semisimple to deduce that $\mathbb{F}_p \otimes Z$ has no fixed points for the action of any G_i.

Corollary 3. *If* (Γ, V) *is a finite Coxeter system such that the topological realisation* $|K|$ *of* $K = K(\Gamma, V)$ *is the closure of a subspace which is a connected n-manifold, then for any finite-index torsion-free subgroup* Γ_1 *of* Γ,

$$H^{n+1}(\Gamma_1; \mathbb{Z}\Gamma_1) \cong \tilde{H}^n(K; \mathbb{Z}).$$

Proof We shall apply the condition on $|K|$ in the following equivalent form: Every simplex of the barycentric subdivision K' of K is contained in an n-simplex, and any two n-simplices of K' may be joined by a path consisting of alternate n-simplices and $(n-1)$-simplices, each $(n-1)$-simplex being a face of its two neighbours in the path and of no other n-simplex. It suffices to show that under this hypothesis, $H^{n+1}(\Gamma_1; \mathbb{Z}\Gamma_1)$ is a cyclic group, because by Theorem 2 it admits $\tilde{H}^n(K; \mathbb{Z})$ as a quotient and has the same exponent as $\tilde{H}^n(K; \mathbb{Z})$.

Recall the description of the m-simplices of $D = D(\Gamma, V)$ as $(m+2)$-tuples as in the proof of Theorem 2. The boundary of the simplex $\sigma = (\gamma, V_0, \ldots, V_m)$ is given by

$$d(\sigma) = \sum_{i=0}^{m}(-1)^i(\gamma, V_0, \ldots, V_{i-1}, V_{i+1}, \ldots, V_m),$$

and the action of Γ by

$$\gamma'\sigma = (\gamma'\gamma, V_0, \ldots, V_m).$$

The stabiliser of σ is $\gamma\langle V_0\rangle\gamma^{-1}$. In the case when $m = n+1$, V_i must be a subset of V of cardinality i, and σ is therefore in a free Γ-orbit. For σ an $(n+1)$-simplex, define $f_\sigma \in \mathrm{Hom}_\Gamma(C_{n+1}(D), \mathbb{Z}\Gamma)$ by the equations

$$f_\sigma(\sigma') = \begin{cases} \gamma' & \text{if } \sigma' = \gamma'\sigma \text{ for some } \gamma' \in \mathbb{Z}\Gamma, \\ 0 & \text{otherwise.} \end{cases}$$

The f_σ form a $\mathbb{Z}$-basis for $\mathrm{Hom}_\Gamma(C_{n+1}(D), \mathbb{Z}\Gamma)$, so it suffices to show that for each σ and σ', $f_\sigma \pm f_{\sigma'}$ is a coboundary.

From now on we shall fix $\sigma = (\gamma, V_0, \ldots, V_{n+1})$, and show that $f_\sigma \pm f_{\sigma'}$ is a coboundary for various choices of σ'. If

$$\sigma' = (\gamma, V_0, \ldots, V_{i-1}, V_i', V_{i+1}, \ldots, V_{n+1})$$

for some $i > 0$, let τ be the n-simplex $(\gamma, V_0, \ldots, V_{i-1}, V_{i+1}, \ldots, V_{n+1})$. There are exactly two i-element subsets of V_{i+1} containing V_{i-1}, so σ and σ' are the only $(n+1)$-simplices of D having τ as a face. Defining f_τ in the same way as f_σ and $f_{\sigma'}$ (which we can do because τ is in a free Γ-orbit), we see that the coboundary of f_τ is $(-1)^i(f_\sigma + f_{\sigma'})$.

If $\sigma' = (\gamma, W_0, \ldots, W_{n+1})$, take a path in K', of the form guaranteed by the hypothesis, joining the simplices $(V_1, \ldots, V_{n+1})$ and $(W_1, \ldots, W_{n+1})$, and use this to make a similar path of $(n + 1)$- and n-simplices between σ and σ', and use induction on the length of this path to reduce to the case considered above.

It will suffice now to consider the case when $\sigma' = (\gamma\gamma', V_0, \ldots, V_{n+1})$. By induction on the length of γ' as a word in V, it suffices to consider the case when $\gamma' = v$. Using the cases done above and the fact that $\{v\}$ is a vertex of some n-simplex of K', we may assume that $V_1 = \{v\}$. Now the n-simplex $\tau = (\gamma, V_1, \ldots, V_{n+1})$ of D is a face of only σ and σ'. The simplex τ has stabiliser in Γ the subgroup $\gamma\langle v \rangle\gamma^{-1}$, so we may define

$$g_\tau(\tau') = \begin{cases} \gamma'\gamma(1 + v)\gamma^{-1} & \text{if } \tau' = \gamma'\tau, \\ 0 & \text{otherwise.} \end{cases}$$

It is easy to check that $d(g_\tau) = f_\sigma + f_{\sigma'}$, using the fact that $f_{\sigma'}(\gamma'\sigma) = \gamma'\gamma v\gamma^{-1}$.

In the same vein we have the following.

Proposition 4. *If (Γ, V) is a finite Coxeter system such that the topological realisation $|K|$ of $K = K(\Gamma, V)$ is the closure of a subspace which is a connected n-manifold, and Γ_1 is a finite-index torsion-free subgroup of Γ, then the topological space $|D(\Gamma, V)|/\Gamma_1$ (which is an Eilenberg-Mac Lane space for Γ_1) is homeomorphic to a CW-complex with exactly one $(n + 1)$-cell.*

Proof We shall give only a sketch. The complex $D(\Gamma, V)/\Gamma_1$ consists of copies of the cone on $K(\Gamma, V)'$ indexed by the cosets of Γ_1 in Γ, where each n-simplex not containing a cone point is a face of exactly two $(n + 1)$-simplices. (The simplex $(\Gamma_1\gamma, V_1, \ldots, V_{n+1})$, where $V_1 = \{v\}$, is a face of $(\Gamma_1\gamma, \emptyset, V_1, \ldots, V_{n+1})$ and $(\Gamma_1\gamma v, \emptyset, V_1, \ldots, V_{n+1})$.) By hypothesis and this observation there exists a tree whose vertices consist of all the $(n + 1)$-simplices of $D(\Gamma, V)/\Gamma_1$ and whose edges are n-simplices of $D(\Gamma, V)/\Gamma_1$ which are faces of exactly two $(n + 1)$-simplices. The ends of an edge of the tree are of course the two $(n + 1)$-simplices containing it. The union of the (topological realisations of the) open simplices of such a tree is homeomorphic to an open $(n + 1)$-cell. The required CW-complex has n-skeleton the simplices of $D(\Gamma, V)/\Gamma_1$ not in the tree, with a single $(n + 1)$-cell whose interior consists of the union of the open simplices of the tree.

Remark. The condition on $K(\Gamma, V)$ occurring in the statements of Corollary 3 and Proposition 4 is equivalent to '$K(\Gamma, V)$ is a *pseudo-manifold*' in the sense of [15]. Neither Corollary 3 nor Proposition 4 has a good analogue for graph products, because both rely on the fact that the n-simplices of $D(\Gamma, V)$ in non-free Γ-orbits are faces of exactly two $(n + 1)$-simplices.

In Theorem 5 we summarize a version of Poincaré duality for simplicial complexes that look like manifolds from the point of view of R-homology for a commutative ring R. Our treatment is an extension of that of J. R. Munkres book [15], which covers the case when $R = \mathbb{Z}$. We also generalise the account in [15] by allowing a group to act on our 'manifolds'. The proofs are very similar to those in [15] however, so we shall only sketch them here.

Definition. Let R be a commutative ring. An *R-homology n-manifold* is a locally finite simplicial complex L such that the link of every i-simplex of L has the same R-homology as an $(n - i - 1)$-sphere, where a sphere of negative dimension is empty. From this definition it follows that L is an n-dimensional complex, and that every $(n - 1)$-simplex of L is a face of exactly two n-simplices. Thus (the topological realisation of) every open $(n - 1)$-simplex of L has an open neighbourhood in $|L|$ homeomorphic to a ball in $\mathbb{R}^n$. Say that L is *orientable* if the n-simplices of L may be oriented consistently across every $(n - 1)$-simplex. Call such a choice of orientations for the n-simplices an *orientation* for L. If L is connected and orientable then a choice of orientation for one of the n-simplices of L, together with the consistency condition, determines a unique orientation for L. In particular, a connected L has either two or zero orientations, and a simply connected L has two.

For any locally finite simplicial complex L, the *cohomology with compact supports of L with coefficients in R*, written $H_c^*(L; R)$, is the cohomology of the subcomplex of the R-valued simplicial cochains on L consisting of the functions which vanish on all but finitely many simplices of L. (This graded R-submodule may be defined for any L, but is a subcomplex only when L is locally finite.)

Theorem 5. *Fix a commutative ring R, and let L be a connected R-homology n-manifold. Let Γ be a group acting freely and simplicially on L. If L is orientable, let R° stand for the right $R\Gamma$-module upon which an element γ of Γ acts as multiplication by -1 if it exchanges the two orientations of L and as the identity if it preserves the orientations of L. If R has characteristic two, let R° be R with the trivial right Γ-action. Then if either L is orientable or R has characteristic two, there is for each i an isomorphism of right $R\Gamma$-modules*

$$H_c^i(L; R) \otimes R^\circ \cong H_{n-i}(L; R).$$

Proof The statements and proofs contained in sections 63–65 of [15] hold for R-homology manifolds provided that all of the (co)chain complexes and (co)homology are taken with coefficients in R. For each simplex σ of L, one defines the dual block $D(\sigma)$ and its boundary exactly as in section 64 of [15]. From the point of view of R-homology, the dual block to an i-simplex of L looks like an $(n - i)$-cell, and its boundary looks like the boundary of an $(n - i)$-cell. Thus as in Theorem 64.1 of [15], the homology of the dual block complex $D_*(L; R)$ is isomorphic to the R-homology of L. There is a natural bijection between the dual blocks of L and the simplices of L. This is clearly preserved by the action of Γ. Each choice of orientation on L gives rise to homomorphisms

$$\psi : D_{n-i}(L;R) \otimes_R C_i(L;R) \to R$$

which behave well with respect to the boundary maps, and have the property that for any simplex σ with dual block $D(\sigma)$, and simplex σ', $\psi(D(\sigma) \otimes \sigma') = \pm 1$ if $\sigma = \sigma'$, and 0 otherwise. This allows one to identify $D_{n-*}(L;R)$ with $C_c^*(L;R)$. With the diagonal action of Γ on $D_{n-i}(L;R) \otimes C_i(L;R)$, ψ is not Γ-equivariant, but gives rise to a Γ-equivariant map

$$\psi' : D_{n-i}(L;R) \otimes_R C_i(L;R) \to R^\circ,$$

and hence an $R\Gamma$-isomorphism between $D_{n-*}(L;R)$ and $C_c^*(L;R) \otimes R^\circ$.

Remark. The referee pointed out that the sheaf-theoretic proof of Poincaré duality in G. E. Bredon's book ([3], 207–211) also affords a proof of Theorem 5.

Corollary 6. *Let R be a commutative ring and let L be a contractible R-homology n-manifold. Let Γ be a group and assume that Γ admits a free simplicial action on L with finitely many orbits of simplices. Then Γ is a Poincaré duality group of dimension n over R, with orientation module the module R° defined in the statement of Theorem 5. The same result holds if L is assumed only to be orientable and R-acyclic rather than contractible.*

Proof The simplicial R-chain complex for L is a finite free $R\Gamma$-resolution for R, and hence Γ is FP over R. Since L has only finitely many Γ-orbits of simplices, the cochain complexes (of right $R\Gamma$-modules)

$$\mathrm{Hom}_{R\Gamma}(C_*(L), R\Gamma) \quad \text{and} \quad C_c^*(L;R)$$

are isomorphic. Hence by Theorem 5, the graded right $R\Gamma$-module $H^*(\Gamma; R\Gamma)$ is isomorphic to R° concentrated in degree n. Thus Γ satisfies condition d) of Definition V.3.3 of [10] and is a Poincaré duality group over R as claimed.

Corollary 7. *Let (Γ, V) be a Coxeter system, and let R be a commutative ring. If $K(\Gamma, V)$ is an R-homology n-sphere (i.e., $K(\Gamma, V)$ is an R-homology n-manifold and $H_*(K(\Gamma, V); R)$ is isomorphic to the R-homology of an n-sphere), then any finite-index torsion-free subgroup of Γ is a Poincaré duality group over R, of dimension $n + 1$.*

Proof It suffices to show that whenever $K = K(\Gamma, V)$ is an R-homology n-sphere, $D = D(\Gamma, V)$ is an R-homology $(n + 1)$-manifold, because then Corollary 6 may be applied to the free action of the finite-index torsion-free subgroup of Γ on D. We shall show that the link of any simplex in D is isomorphic to either K' or a suspension (of the correct dimension) of the link of some simplex in K'. This implies that under the hypothesis on K, the link of every simplex of D is an R-homology n-sphere.

Let $\sigma = (\gamma, V_0, \ldots, V_m)$ be an m-simplex of D, and without loss of generality we assume that $\gamma = 1$. Then the link of σ is the collection of simplices σ' of D having no vertex in common with σ but such that the union of the vertex sets of σ and σ' is the vertex set of some simplex of D. Thus the link of the simplex σ as above consists of those simplices $(\gamma', U_0, \ldots, U_l)$ of D such that the finite subsets $U_0, \ldots, U_l, V_0, \ldots, V_m$ of V are all distinct, generate finite subgroups of Γ, and are linearly ordered by inclusion, where γ' is an element of the subgroup Γ_0 of Γ generated by V_0. The link of σ decomposes as a join of pieces corresponding to posets of the three types listed below, where we adopt the convention that the join of a complex X with an empty complex is isomorphic to X, and spheres of dimension -1 are empty.

1) The poset of all subsets U of V such that $\langle U \rangle$ is finite and U properly contains V_m. This is isomorphic to the poset of faces of the link in K' of any simplex of K' of dimension $|V_m| - 1$ of the form $(V_1', \ldots, V_i', \ldots, V_{m'}')$, where $m' = |V_m|$ and $V_{m'}' = V_m$. By the hypothesis on K, this is an R-homology sphere of dimension $n - |V_m|$.

2) For each i such that $0 \le i < m$, the poset of all subsets of V properly containing V_i and properly contained in V_{i+1}. This is isomorphic to the poset of faces of the boundary of a simplex with vertex set $V_{i+1} - V_i$, so is a triangulation of a sphere of dimension $|V_{i+1}| - |V_i| - 2$.

3) The poset of all cosets in Γ_0 of proper special subgroups of (Γ_0, V_0). (Recall that we defined $\Gamma_0 = \langle V_0 \rangle$.) This is a triangulation of a sphere of dimension $|V_0| - 1$ on which the group Γ_0 acts with each $v \in V_0$ acting as a reflection in a hyperplane (see [5], I.5, especially I.5H).

The link of the m-simplex σ consists of a join of one piece of type 1), m pieces of type 2), and one piece of type 3). All of these are spheres except that the piece of type 1) is only an R-homology sphere. It follows that the link of σ is an R-homology sphere, whose dimension is equal to the sum

$$(n-|V_m|) + 1 + (|V_m| - |V_{m-1}| - 2) + 1 + \cdots$$
$$+ 1 + (|V_2| - |V_1| - 2) + 1 + (|V_1| - |V_0| - 2) + 1 + (|V_0| - 1) = n - m.$$

This sum is obtained from the fact that the dimension of the join of two simplicial complexes is equal to the sum of their dimensions plus one, which is correct in all cases, provided that the empty complex is deemed to have dimension equal to -1.

Remarks. 1) Theorem 10 may also be used to prove Corollary 7.

2) A $\mathbb{Z}$-homology sphere is a *generalised homology sphere* in the sense of [8]. Davis shows that if (Γ, V) is a Coxeter group such that $K(\Gamma, V)$ is a manifold and a $\mathbb{Z}$-homology sphere, then Γ acts on an acyclic manifold with finite stabilisers (see Sections 12 and 17 of [8]).

4. Unusual cohomological behaviour

Not every simplicial complex may be $K(\Gamma, V)$ for some Coxeter system (Γ, V); for example the 2-skeleton of a 6-simplex cannot occur. To see this, note that any labelling of the edges of a 6-simplex by the labelling set {red, blue} that contains no red triangle must contain a vertex incident with at least three blue edges. Now recall that the Coxeter group corresponding to a labelled triangle can be finite only if one of the edges has label two. Writing the integer two in blue, and other integers in red, one sees that any 7-generator Coxeter group with all 3-generator special subgroups finite has a 3-generator special subgroup which commutes with a fourth member of the generating set.

A condition equivalent to a complex K being equal to $K(\Gamma, V)$ for some right-angled Coxeter system (Γ, V) is that whenever K contains all possible edges between a finite set of vertices, this set should be the vertex set of some simplex of K. Complexes satisfying this condition are called 'full simplicial complexes' or 'flag complexes' [1], [5]. The barycentric subdivision of any complex satisfies this condition. The barycentric subdivision of an n-dimensional complex admits a colouring with $n + 1$ colours, where the barycentre $\hat{\sigma}$ of an i-simplex σ is given the colour $i \in \{0, \ldots, n\}$. This proves the following (see 11.3 of [8]).

Proposition 8. *The barycentric subdivision of any n-dimensional simplicial complex is isomorphic to $K(\Gamma, V)$ for some right-angled Coxeter system (Γ, V) such that* vcdΓ *is finite.*

$\square$

We refer the reader to [14] for a statement of the universal coefficient theorem and a calculation of the Ext-groups arising in the following examples.

Example 1 (Bestvina). (A group of finite cohomological dimension over the integers whose rational cohomological dimension is strictly less than its integral cohomological dimension.) Let X be the space obtained by attaching a disc to a circle by wrapping its edge around the circle n times, so that $H_1(X) \cong \mathbb{Z}/(n)$ and $H_2(X) = 0$. Now let (Γ, V) be any Coxeter system such that $K(\Gamma, V)$ is a triangulation of X. The generating set V will be finite since X is compact, so any such Γ will have finite vcd. Now if Γ_1 is a finite-index torsion-free subgroup of Γ, cdΓ_1 is at most 3, and for any $\mathbb{Z}\Gamma_1$-module M, $H^3(\Gamma_1; M)$ is a quotient of a finite sum of copies of $H^2(X; M_a)$, which is in turn isomorphic to $\mathrm{Ext}(\mathbb{Z}/(n), M_a)$ by the universal coefficient theorem. In particular, $nH^3(\Gamma_1; M) = 0$ for any M, and $H^3(\Gamma_1; \mathbb{Z}\Gamma_1) \cong \mathrm{Ext}(\mathbb{Z}/(n), \mathbb{Z}) \cong \mathbb{Z}/(n)$ by Corollary 3. Note that the methods used by Bestvina [1] and by Harlander and Meinert [13] seem to show only that $H^3(\Gamma_1; \mathbb{Z}\Gamma_1)$ contains elements of order p for each prime p dividing n, whereas our argument gives elements of order exactly n.

Example 2. (A group whose cohomological dimension over the integers is finite but strictly greater than its cohomological dimension over any field.) Let X be a 2-dimensional CW-complex with $H_1(X) \cong \mathbb{Q}$ and $H_2(X) = 0$, for example X could be an Eilenberg-Mac Lane space $K(\mathbb{Q}, 1)$ built from a sequence $C_1, C_2, \ldots$ of cylinders, where the end of the ith cylinder is attached to the start of the $(i + 1)$st cylinder by a map of degree i. Now let (Γ, V) be a Coxeter system such that $K(\Gamma, V)$ is a 2-dimensional simplicial complex homotopy equivalent to X, and Γ has finite vcd. We shall show that $\mathrm{vcd}\,\Gamma = 3$, and that for any field $\mathbb{F}$, $\mathrm{vcd}_{\mathbb{F}}\Gamma = 2$. Let Γ_1 be a finite-index torsion-free subgroup of Γ. Then $\mathrm{cd}\,\Gamma_1$ is at most 3, and for any M, $H^3(\Gamma_1; M)$ is a quotient of a finite sum of copies of $H^2(X; M_a) \cong \mathrm{Ext}(\mathbb{Q}, M_a)$. If M is an $\mathbb{F}\Gamma_1$-module for $\mathbb{F}$ a field of non-zero characteristic p, then $\mathrm{Ext}(\mathbb{Q}, M_a)$ is an abelian group which is both divisible and annihilated by p, so is trivial. If M is an $\mathbb{F}\Gamma_1$-module for $\mathbb{F}$ a field of characteristic zero, then M_a is a divisible abelian group, so is $\mathbb{Z}$-injective, and so once again $\mathrm{Ext}(\mathbb{Q}, M_a) = 0$. On the other hand, $H^3(\Gamma_1; \mathbb{Z}\Gamma_1)$ is non-zero, because it admits $\mathrm{Ext}(\mathbb{Q}, \mathbb{Z})$ as a quotient, and $\mathrm{Ext}(\mathbb{Q}, \mathbb{Z})$ is a $\mathbb{Q}$-vector space of uncountable dimension.

The group Γ_1 requires infinitely many generators, but a 2-generator example may be constructed from Γ_1 using an embedding theorem of Higman Neumann and Neumann ([16], Theorem 6.4.7). They show that any countable group G may be embedded in a 2-generator group $\widehat{G}$ constructed as an HNN-extension with base group the free product of G and a free group of rank two, and associated subgroups free of infinite rank. An easy Mayer-Vietoris argument shows that for any ring R,

$$\mathrm{cd}_R G \leq \mathrm{cd}_R \widehat{G} \leq \max\{2, \mathrm{cd}_R G\}.$$

Thus $\widehat{\Gamma}_1$ is a 2-generator group with $\mathrm{cd}\,\widehat{\Gamma}_1 = 3$ but $\mathrm{cd}_{\mathbb{F}}\widehat{\Gamma}_1 = 2$ for any field $\mathbb{F}$. We do not know whether there is a finitely presented group with this property, but the referee showed us the following Proposition (see also [2], 9.12).

Proposition 9. *Let G be a group of type FP. Then there is a prime field $\mathbb{F}$ such that $\mathrm{cd}_{\mathbb{F}}G = \mathrm{cd}\,G$.*

Proof Recall that if G is of type FP, then for any ring R, $\mathrm{cd}_R G$ is equal to the maximum n such that $H^n(G; RG)$ is non-zero, and that if $\mathrm{cd}\,G = n$, then for any R, $H^n(G; RG)$ is isomorphic to $H^n(G; \mathbb{Z}G) \otimes R$ ([4], p199–203). Let M stand for $H^n(G; \mathbb{Z}G)$, where $n = \mathrm{cd}\,G$. Since $\mathrm{Hom}_G(P, \mathbb{Z}G)$ is a finitely generated right $\mathbb{Z}G$-module for any finitely generated projective P, it follows that M is a finitely generated right $\mathbb{Z}G$-module.

If $M \otimes \mathbb{F}_p$ is non-zero for some prime p we may take $\mathbb{F} = \mathbb{F}_p$. If not, then M is divisible and hence, as an abelian group, M is a direct sum of a number of copies of $\mathbb{Q}$ and a divisible torsion group ([16], Theorem 4.1.5). If $M \otimes \mathbb{Q}$

is non-zero then we may take $\mathbb{F} = \mathbb{Q}$. It remains to show that M cannot be a divisible torsion abelian group. Suppose that this is the case, and let $m_1, \ldots, m_r$ be a generating set for M as a right $\mathbb{Z}G$-module. If N is the least common multiple of the additive orders of the elements $m_1, \ldots, m_r$, multiplication by N annihilates M, contradicting the divisibility of M.

Example 3. (A torsion-free rational Poincaré duality group of dimension four which is not an integral Poincaré duality group.) Fix an odd prime q, and let X be a lens space with fundamental group of order q, i.e., X is a quotient of the 3-sphere by a free linear action of the cyclic group of order q. It is easy to see that X is triangulable. The homology groups of X are (in ascending order) $\mathbb{Z}$, $\mathbb{Z}/(q)$, $\{0\}$ and $\mathbb{Z}$. From the universal coefficient theorem it is easy to see that X has the same R-homology as the 3-sphere for any commutative ring R in which q is a unit. Now let (Γ, V) be a right-angled Coxeter system such that $K(\Gamma, V)$ is a triangulation of X. By Corollary 7, any finite-index torsion-free subgroup Γ_1 of Γ is a Poincaré duality group of dimension four over any R in which q is a unit.

We claim however, that Γ_1 is not a Poincaré duality group (or PD-group for short) over the field $\mathbb{F}_q$, which implies that it cannot be a PD-group over the integers. Since all finite subgroups of Γ have order a power of two and q is an odd prime, it follows from Theorem V.5.5 of [10] that Γ_1 is a PD-group over $\mathbb{F}_q$ if and only if Γ is. We shall assume that Γ is a PD-group over $\mathbb{F}_q$ and obtain a contradiction.

Firstly, note that the $\mathbb{F}_q$-cohomology groups $H^0, \ldots, H^3$ of X are all isomorphic to $\mathbb{F}_q$. Now it follows from Theorem 2 part c) that the right Γ-module $H^4(\Gamma; \mathbb{F}_q \Gamma)$ admits $\mathbb{F}_q^\circ$ (as defined just above the statement of Theorem 2) as a quotient. Thus Γ has cohomological dimension four over $\mathbb{F}_q$, and if Γ is a PD-group over $\mathbb{F}_q$, its orientation module must be $\mathbb{F}_q^\circ$. In particular, for any $\mathbb{F}_q \Gamma$-module M, there should be an isomorphism for each i

$$H^i(\Gamma; M) = \operatorname{Ext}^i_{\mathbb{F}_q \Gamma}(\mathbb{F}_q, M) \cong \operatorname{Tor}^{\mathbb{F}_q \Gamma}_{4-i}(\mathbb{F}_q^\circ, M).$$

Now let M be the Γ-bimodule $\mathbb{F}_q^\circ$, viewed as a left $\mathbb{F}_q \Gamma$-module. There is an $\mathbb{F}_q$-algebra automorphism ϕ of $\mathbb{F}_q \Gamma$ defined by $\phi(v) = -v$ for each $v \in V$, because the relators in Γ have even length as words in V. The Γ-bimodule obtained from $\mathbb{F}_q^\circ$ by letting $\mathbb{F}_q \Gamma$ act via ϕ is the trivial bimodule $\mathbb{F}_q$. Thus for each i, ϕ induces an isomorphism

$$\operatorname{Tor}^{\mathbb{F}_q \Gamma}_i(\mathbb{F}_q^\circ, \mathbb{F}_q^\circ) \cong \operatorname{Tor}^{\mathbb{F}_q \Gamma}_i(\mathbb{F}_q, \mathbb{F}_q).$$

(Here we are viewing the bimodules as left modules when they appear as the right-hand argument in Tor, and as right modules when they appear as the left-hand argument.) Putting this together with the isomorphism obtained earlier, it follows that if Γ is a PD-group over $\mathbb{F}_q$, then for each i,

$$H^i(\Gamma; \mathbb{F}_q^\circ) = \operatorname{Ext}^i_{\mathbb{F}_q \Gamma}(\mathbb{F}_q, \mathbb{F}_q^\circ) \cong \operatorname{Tor}^{\mathbb{F}_q \Gamma}_{4-i}(\mathbb{F}_q, \mathbb{F}_q) = H_{4-i}(\Gamma; \mathbb{F}_q). \qquad (*)$$

The cohomology groups $H^0, \ldots, H^4$ of Γ with coefficients in $\mathbb{F}_q^c$ are calculated in Theorem 2 part d); as vector spaces over $\mathbb{F}_q$ they have dimensions 0, 0, 1, 1, and 1 respectively. We claim now that any finitely generated right-angled Coxeter group is $\mathbb{F}_q$-acyclic, i.e., its homology with coefficients in the trivial module $\mathbb{F}_q$ is 1-dimensional and concentrated in degree zero. This leads to a contradiction because given the claim, the isomorphism (*) for $i = 2$ or 3 is between a 1-dimensional vector space and a 0-dimensional vector space. The claim follows from Theorem 4.11 of [9], which is proved using an elegant spectral sequence argument. It is also possible to provide a direct proof by induction on the number of generators using the fact (see [6] or [12]) that a finitely generated right-angled Coxeter group which is not a finite 2-group is a free product with amalgamation of two of its proper special subgroups, and applying the Mayer-Vietoris sequence.

Remarks. 1) Note that the only properties of the space X used in Example 3 are that X be a compact manifold which is triangulable (in the weak sense that it is homeomorphic to the realisation of some simplicial complex), and that for some rings R, X be an R-homology sphere, but that there be a prime field $\mathbb{F}_q$ for $q \neq 2$ such that X is not an $\mathbb{F}_q$-homology sphere. These examples show that being a GD-group over R (in the sense of [10], V.3.8) is not equivalent to being a PD-group over R.

2) If Γ is a Coxeter group such that $K(\Gamma, V)$ is a triangulation of 3-dimensional real projective space, then Corollary 7 shows that any torsion-free finite-index subgroup of Γ is a PD-group over any R in which 2 is a unit. The methods used above do not show that such a group is not a PD-group over R when $2R \neq R$ however. The results of the next section show that this is indeed the case.

5. Cohomology of Coxeter groups with free coefficients

The results of this section were shown to us by the referee, although we are responsible for the proofs given here. The main result is Theorem 10, which computes $H^*(\Gamma, R\Gamma)$ for any Coxeter group Γ such that $K(\Gamma, V)$ is a (triangulation of a) closed compact oriented manifold. This should be contrasted with Theorem 2, which applies to any Coxeter group, but gives only partial information concerning cohomology with free coefficients. Theorem 10 may be used to prove Corollary 7 and to give an alternative proof of the existence of Example 3 of the previous section.

Let K be a locally finite simplicial complex, so that the cohomology of K with compact supports, $H_c^*(K)$ is defined, and suppose that we are given a sequence

$$K_1 \supseteq K_2 \supseteq K_3 \supseteq \cdots$$

of subcomplexes such that each K_i is cofinite (i.e., each K_i contains all but finitely many simplices of K), and the intersection of the K_i's is trivial. For $j > i$, the inclusion of the pair (K, K_j) in (K, K_i) gives a map from $H^*(K, K_i)$ to $H^*(K, K_j)$, and the direct limit is isomorphic to $H_c^*(K)$:

$$H_c^*(K) \cong \lim_{\rightarrow}(H^*(K, K_1) \to H^*(K, K_2) \to \cdots). \qquad (**)$$

(To see this, it suffices to check that a similar isomorphism holds at the cochain level.)

Theorem 10. *Let R be a ring, let (Γ, V) be a Coxeter system such that $K(\Gamma, V)$ is a triangulation of a closed compact connected n-manifold X, and suppose that either X is orientable or $2R = 0$. Then as right $R\Gamma$-module, the cohomology of Γ with free coefficients is:*

$$H^i(\Gamma, R\Gamma) = \begin{cases} 0 & \text{for } i = 0 \text{ or } 1, \\ H^{i-1}(X; R) \otimes_R R\Gamma & \text{for } 2 \leq i \leq n, \\ R^\circ & \text{for } i = n+1. \end{cases}$$

(Here R° is the Γ-module defined above Theorem 2.)

Proof By hypothesis the complex $D = D(\Gamma, V)$ is locally finite, and so $H^*(\Gamma, R\Gamma)$ is isomorphic to $H_c^*(D; R)$. Recall that the complex D may be built up from a union of a collection of cones on the barycentric subdivision K' of $K(\Gamma, V)$ indexed by the elements of Γ. For $\gamma \in \Gamma$, let $C(\gamma)$ be the cone with apex the coset $\gamma\{1\}$. In terms of the description of D in the proof of Theorem 2, $C(\gamma)$ consists of all simplices of D which may be represented in the form $(\gamma, V_0, \ldots, V_m)$ for some m and subsets $V_0 \ldots, V_m$ of V. For any enumeration $1 = \gamma_1, \gamma_2, \gamma_3, \ldots$ of the elements of Γ, define subcomplexes E_i and D_i of D by

$$E_i = \bigcup_{j \leq i} C(\gamma_j), \qquad D_i = \bigcup_{j > i} C(\gamma_j).$$

Davis' original proof that D is contractible [8] uses the following argument. For W a subset of V, define $K_\sigma(W)$ to be the subcomplex of $K' = K(\Gamma, V)'$ consisting of the simplices having a face in common with $K(\langle W\rangle, W)' \subseteq K(\Gamma, V)$ and all of their faces. The simplicial interior, $\text{int}K_\sigma(W)$, of $K_\sigma(W)$ (i.e., the union of the topological realisations of the open simplices of $K_\sigma(W)$ which are not faces of any simplex of $K' - K_\sigma(W)$) deformation retracts onto $K(\langle W\rangle, W)$ by a linear homotopy. In particular, this interior is contractible if $\langle W\rangle$ is finite, and it may also be shown that in this case $K_\sigma(W)$ is itself contractible. In [8] an enumeration of the elements of Γ is given such that for each i there exists $W \subseteq V$ with $\langle W\rangle$ a finite subgroup of Γ, and $E_i \cap C(\gamma_{i+1})$ is isomorphic to $K_\sigma(W) \subseteq K'$ by the restriction of the natural isomorphism $C(\gamma_{i+1}) \cong CK'$. By induction it follows that each E_i is contractible, and hence that D is.

Throughout the remainder of the proof fix an enumeration of Γ as in the previous paragraph. From $(**)$ it follows that

$$H^*(\Gamma; R\Gamma) \cong H^*_c(D; R) \cong \varinjlim(H^*(D, D_i; R)).$$

Let $F_i = D_i \cap E_i$, which could be thought of as the boundary of E_i. By excision, $H^*(D, D_i; R)$ is isomorphic to $H^*(E_i, F_i; R)$, and since E_i is contractible, this is in turn isomorphic to the reduced cohomology group $\tilde{H}^{*-1}(F_i; R)$. So far we have used none of the conditions on $K(\Gamma, V)$ except that V be finite. The hypothesis that $K(\Gamma, V)$ be a closed R-oriented n-manifold is used to compute the limit of the groups $\tilde{H}^{*-1}(F_i; R)$. In effect, F_i is a connected sum of i copies of K'. More precisely, if $W_i \subseteq V$ is such that $E_i \cap C(\gamma_{i+1}) \cong K_\sigma(W_i)$, then F_{i+1} is obtained from $F_i - \operatorname{int}K_\sigma(W_i)$ and $K' - \operatorname{int}K_\sigma(W_i)$ by identifying the two copies of $K_\sigma(W_i) - \operatorname{int}K_\sigma(W_i)$. (We defined the simplicial interior $\operatorname{int}L$ of a subcomplex L of K' to be a topological space, but it is easy to see how to define a subcomplex $M - \operatorname{int}L$ of M for any $L \subseteq M \subseteq K'$.) Given our hypotheses on $K(\Gamma, V)$, there are equalities

$$H^j(K' - \operatorname{int}K_\sigma(W_i); R) = \begin{cases} H^j(K'; R) & \text{for } i \neq n, \\ 0 & \text{for } i = n, \end{cases}$$

while $K_\sigma(W_i) - \operatorname{int}K_\sigma(W_i)$ is a homology $(n-1)$-sphere by Poincaré-Lefschetz duality for $K_\sigma(W_i)$. From the usual argument used to compute the cohomology of a connected sum and induction it follows that

$$H^j(F_i; R) \cong \begin{cases} \bigoplus_{k=1}^i H^j(K'; R) & \text{for } 0 < j < n, \\ R & \text{for } j = n. \end{cases}$$

Moreover, the map from $\tilde{H}^j(F_i; R)$ to $\tilde{H}^j(F_{i+1}; R)$ given by

$$\tilde{H}^j(F_i; R) \cong H^{j+1}(D, D_i; R) \to H^{j+1}(D, D_{i+1}; R) \cong \tilde{H}^j(F_{i+1}; R)$$

is the inclusion of the first i direct summands for $j < n$ and the identity for $j = n$.

As a right Γ-module, $H^{n+1}_c(D; \mathbb{Z})$ is isomorphic to $\mathbb{Z}^\circ$ by Theorem 2c), and the claim for general R follows by the universal coefficient theorem. To verify the claimed right Γ-module structure for $H^j_c(D; R)$ for $j \leq n$, note that for each i, the images of D_i under translation by each of $\gamma_1^{-1} = 1, \ldots, \gamma_i^{-1}$ are contained in D_1, and check that the i corresponding maps

$$\tilde{H}^{j-1}(K'; R) \quad \cong \quad H^j(D, D_1; R)$$
$$\downarrow$$
$$H^j(D, D_i; R) \quad \cong \quad \tilde{H}^{j-1}(K'; R) \oplus \cdots \oplus \tilde{H}^{j-1}(K'; R)$$

are the inclusions of the i distinct direct summands.

Remark. A group G is a PD-group over R if and only if it is of type FP over R and $H^*(G; RG)$ is isomorphic to R concentrated in a single degree. It follows that if Γ is such that $K(\Gamma, V)$ satisfies the hypotheses of Theorem 10, then a finite-index torsion-free subgroup of Γ is a PD-group over R if and only if $K(\Gamma, V)$ is an R-homology sphere. For example, if $K(\Gamma, V)$ is a triangulation of 3-dimensional real projective space, then a finite-index torsion-free subgroup of Γ is a PD-group of dimension four over R if and only if $2R = R$.

6. Generating sets for torsion-free subgroups

In this section we shall consider only right-angled Coxeter groups, so (Γ, V) shall be a right-angled Coxeter system, and $K(\Gamma, V)$ a full simplicial complex. The numerical information that we give will not be very useful if V is infinite. Recall from Section 3 that a colouring $c : V \to W$ of the graph $K^1(\Gamma, V)$ with colour set W gives rise to a homomorphism from Γ to a product of copies of the cyclic group C_2 indexed by the elements of W, such that the kernel Γ_1 is torsion-free.

Proposition 11. *Let (Γ, V) be a right-angled Coxeter system, and let $c : V \to W$ be a colouring of $K^1(\Gamma, V)$. Let Γ_1 be the kernel of the induced map from Γ to C_2^W. For $w, w' \in W$, define $K^1(\Gamma, V)(w, w')$ to be the largest subgraph of $K^1(\Gamma, V)$ all of whose vertices have colour w or w'. Let S be the following set of elements of Γ_1.*

$$S = \{vv' \mid v, v' \in V, \quad c(v) = c(v')\}$$

i) If c is such that any two colours w, w' are adjacent in $K^1(\Gamma, V)$, then S generates Γ_1 as a normal subgroup of Γ.
ii) If c is such that for each $w, w' \in W$ the graph $K^1(\Gamma, V)(w, w')$ is connected, then S generates Γ_1 as a group.

Proof Let Q be the quotient of Γ by the normal subgroup generated by S. Then Q has C_2^W as a quotient, and the images of v and v' in Q are equal if $c(v) = c(v')$, so Q is generated by a set of elements of order two bijective with W. If the colours w, w' are adjacent in $K^1(\Gamma, V)$, there exist v, v' with $c(v) = w$, $c(v') = w'$ which commute in Γ. Thus under the hypothesis in i), the relations of Q include relations saying that all pairs of generators commute, and so Q is isomorphic to C_2^W. The hypothesis in ii) implies that in i), so it remains to prove that when each $K^1(\Gamma, V)(w, w')$ is connected, the subgroup generated by S is normal. For this it suffices to show that if u, v, v' are elements of V with $c(v) = c(v')$, then $uvv'u$ is in the subgroup generated by S. Let

$$v = v_1, \; u_1, \; v_2, \; u_2, \ldots, v_{n-1}, \; u_{n-1}, \; v_n = v'$$

be the sequence of vertices on a path in the graph $K^1(\Gamma, V)(c(v), c(u))$ between v and v'. Thus $c(v_i) = c(v)$, $c(u_i) = c(u)$, and for all $1 \leq i \leq n - 1$, u_i commutes with v_i and v_{i+1}. These commutation relations imply that $uvv'u$ is expressible as the following word in elements of S.

$$(uu_1)(v_1v_2)(u_1u_2)(v_2v_3) \cdots (v_{n-2}v_{n-1})(u_{n-2}u_{n-1})(v_{n-1}v_n)(u_{n-1}u) = uvv'u$$

Remark. The hypothesis in i) is not very strong. Given a colouring of a graph in which there exist colours w and w' which are not adjacent, it is possible to identify the colours w and w' to produce a new colouring of the same graph using fewer colours.

Corollary 12. *Let L be an n-dimensional simplicial complex having N simplices in total, such that every simplex of L is a face of an n-simplex and $|L| - |L^{n-2}|$ is connected. If (Γ, V) is such that $K(\Gamma, V)$ is the barycentric subdivision of L, then Γ has a torsion-free normal subgroup Γ_1 of index 2^{n+1}, which may be generated by $N-n-1$ elements. Γ has no torsion-free subgroups of lower index, and any normal subgroup of this index requires at least this number of generators.*

Proof Vertices of $K(\Gamma, V)$ correspond bijectively with simplices of L, and we may colour $K^1(\Gamma, V)$ with the set $\{0, \ldots, n\}$ by sending a vertex to the dimension of the corresponding simplex of L. Let Γ_1 be the kernel of the induced homomorphism onto C_2^{n+1}. Even without the extra conditions on L, Γ_1 is a torsion-free subgroup of Γ of index 2^{n+1}. Since the vertices of an n-simplex of $K(\Gamma, V)$ generate a subgroup of Γ isomorphic to C_2^{n+1}, Γ cannot have a torsion-free subgroup of lower index. The abelianisation of Γ is isomorphic to C_2^N, and so Γ cannot be generated by fewer than N elements. If Γ_2 is any normal subgroup of Γ of index 2^{n+1}, then Γ/Γ_2 can be generated by $n + 1$ elements, so Γ_2 cannot be generated by fewer than $N - n - 1$ elements. Now we claim that the extra conditions on L are equivalent to the condition that for any $i, j \in \{0, \ldots, n\}$, the graph $K^1(\Gamma, V)(i, j)$ (as defined in the statement of Proposition 11) is connected. Firstly, $|L| - |L^{n-2}|$ is connected if and only if $K^1(\Gamma, V)(n, n-1)$ is connected. Now if $K^1(\Gamma, V)(n, i)$ is connected, every i-simplex of L is a face of some n-simplex. For the converse, note first that in the special case when L is a single n-simplex, $K^1(\Gamma, V)(i, j)$ is connected for each $i < j \leq n$. In the case when $j = n$ this is trivial, and for general n follows by an easy induction. For the case of arbitrary L, to find a path between any two vertices v, v' of $K^1(\Gamma, V)(i, j)$, first pick n-simplices u, u' of L such that v is a face of u and v' is a face of u'. Then pick a path in $K^1(\Gamma, V)(n, n - 1)$ between u and u'. For each $(n - 1)$-simplex occurring on this path, pick one of its i-simplices. This gives a sequence $w_1, \ldots, w_m$ of i-simplices such that w_l and w_{l+1} are faces of the same n-simplex for each l, and similarly for the pairs v, w_1 and v', w_n. By the special case already proved, there are paths in $K^1(\Gamma, V)(i, j)$ between each of these pairs, which concatenate to give a path from v to v'.

Hence by Proposition 11, Γ_1 can be generated by the set of pairs vv', where v and v' correspond to simplices of L of the same dimension. If we fix for each dimension i one generator v_i, then $vv' = (v_i v)^{-1}(v_i v')$, so we really need only the pairs $v_i v$ to generate Γ_1, and there are exactly $N - n - 1$ of these.

Remarks. 1) The conditions imposed on the simplicial complex L are satisfied if L is a triangulation of a connected n-manifold, or more generally if $|L|$ is the closure of a subset which is a connected n-manifold. These conditions cannot be omitted. Let L be the simplicial bow tie, consisting of two triangles joined at a point, and let Γ, Γ_1 be the Coxeter group and subgroup of index 8 constructed from L as in Corollary 12. Note that $|L| - |L^0|$ is not connected. Using GAP [17] it may be shown that the abelianisation of Γ_1 is free of rank 11, and so Γ_1 requires at least 11 generators, rather than the 10 which would suffice if Corollary 12 applied.

2) When (as in Corollary 12) the Coxeter group Γ has a torsion-free normal subgroup Γ_1 of index equal to the order of the largest finite subgroup of Γ, this torsion-free subgroup will not usually be unique.

There are other ways to construct a torsion-free group having similar homological properties to a given right-angled Coxeter group. One generalisation of the construction given above is as follows. Given a right-angled Coxeter group Γ, and a homomorphism ψ from Γ onto a finite group Q with torsion-free kernel Γ_1, let Δ be any torsion-free group and ϕ any homomorphism from Δ to Q. The group P defined as the pullback of the following diagram is torsion-free and has finite index in $\Gamma \times \Delta$.

$$
\begin{array}{ccc}
P & \longrightarrow & \Delta \\
\downarrow & & \downarrow{\scriptstyle\phi} \\
\Gamma & \overset{\psi}{\longrightarrow} & Q
\end{array}
$$

The group Γ_1 occurs as such a pullback in the case when Δ is the trivial group. The point about taking different choices of Δ is that the resulting group may have a simpler presentation than Γ_1. One such result is the following.

Proposition 13. *Let (Γ, V) be a right-angled Coxeter system, and fix an $(n+1)$-colouring c of $K^1(\Gamma, V)$. Let $K(\Gamma, V)$ have M edges, and N vertices. Suppose that the colouring of $K^1(\Gamma, V)$ has the property that for every $v \in V$, the star of v contains vertices of all colours. Then there is a torsion-free group, P, of finite index in $\Gamma \times \mathbb{Z}^{n+1}$ having a presentation with N generators and $M + N - n - 1$ relators, all of length four. Identifying the generators of P with the set V, the relators are the following words.*

i) For every edge with ends v, v' in $K(\Gamma, V)$, the commutator $[v, v']$.

ii) For each colour $w \in W$, *for some fixed* v_w *with* $c(v_w) = w$, *and for each* $v \neq v_w$ *such that* $c(v) = w$, *the word* $v^2 v_w^{-2}$.

Proof Let P_1 be the group presented by the above generators and relations. It suffices to show that P_1 is isomorphic to the pullback P in the diagram below, where W is the $(n+1)$-element set of colours, ϕ is the natural projection and ψ is the homomorphism induced by the colouring c.

$$
\begin{array}{ccc}
P & \longrightarrow & \mathbb{Z}^W \\
\downarrow & & \downarrow{\phi} \\
\Gamma & \xrightarrow{\psi} & C_2^W
\end{array}
$$

Identify the standard basis for $\mathbb{Z}^W$ with the set W. The elements $(v, c(v))$ of $\Gamma \times \mathbb{Z}^W$ are naturally bijective with V, and satisfy the relations given in the statement. Thus there is a homomorphism from P_1 to $P \leq \Gamma \times \mathbb{Z}^W$ which sends v to $(v, c(v))$. It remains to show that this homomorphism is injective and is onto P. The relations given between the elements of V suffice to show that each v^2 is central in P_1, because given v, v', there exists v'' such that $c(v) = c(v'')$ and there is an edge in $K(\Gamma, V)$ between v' and v''. By applying the relations given, one obtains $[v^2, v'] = [(v'')^2, v'] = 1$. Let P_2 be the subgroup of P_1 generated by the elements v^2. It is now easy to see that P_2 is central in P_1, and is free abelian of rank $n+1$. Under the map from P_1 to P, P_2 is mapped isomorphically to the kernel of the map from P to Γ. The quotient P_1/P_2 has the same presentation as Γ. It follows that P_1 is mapped isomorphically to P.

Remark. The condition on the colouring in the statement of Proposition 13 is weaker than the condition of part ii) of Proposition 11. It could be restated as saying that 'every component of each $K(\Gamma, V)(w, w')$ contains vertices of both colours w and w''. In the case when $K(\Gamma, V)$ is the barycentric subdivision of an n-dimensional simplicial complex L and the colouring taken is the usual 'colouring by dimension', the condition is equivalent to the statement that every simplex of L should be contained in an n-simplex.

7. Presentations of some of Bestvina's examples

In this section and the next we shall give some explicit presentations of groups whose cohomological dimensions differ over $\mathbb{Z}$ and $\mathbb{Q}$. To simplify notation slightly we shall adopt the convention that if x is a generator in a group presentation, $\bar{x}$ denotes x^{-1}. Much of the section will be based around the eleven vertex full triangulation of the projective plane given in figure 1, where of course vertices and edges around the boundary are to be identified in pairs. Proposition 14 shows that this triangulation is minimal in some sense.

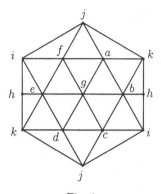

Fig. 1

Proposition 14. *There is no full triangulation of the projective plane having fewer than 11 vertices.*

Proof The following statements are either trivial or followed by their proof. A triangulation of the projective plane with N vertices has $3(N-1)$ edges and $2(N-1)$ faces. A full triangulation of a closed 2-manifold can have no vertex of valency 3 or less. The only 2-manifold having a full N-vertex triangulation with a vertex of valency $N-1$ is the disc. The only closed 2-manifold having a full triangulation with a vertex of valency $N-2$ is the 2-sphere. There is no triangulation of the projective plane having 7, 8, or 9 vertices, each of which has valency 4 or 5. (Write an equation for the numbers of vertices of each valency and obtain a negative number of vertices of valency 4.) There are triangulations of the projective plane having 6 vertices, all of valency 5, but they are not full.

There is no 9 vertex full triangulation of the projective plane: Assume that there is such a triangulation. Then by the above we know that it has a vertex of valency 6. This vertex and its neighbours form a hexagon containing twelve edges. There are no further edges between the vertices of this hexagon, so all the remaining twelve edges contain at least one of the remaining two vertices. Hence at least one of these two vertices is joined to each of the boundary vertices of the hexagon, giving an eight vertex triangulation of the 2-sphere before adding the final vertex.

Any 10 vertex full triangulation of the projective plane has no vertex of valency seven: Assume that there is such a vertex. Then it and its neighbours form a heptagon containing 14 edges. There are 13 other edges, each of which must contain at least one of the other two vertices. Thus either one of these vertices is joined to all of the boundary vertices of the heptagon and this gives a 9 vertex triangulation of the 2-sphere before adding the final vertex, or the final two vertices are joined to each other and to six each of the boundary vertices of the heptagon, in which case the complex contains a tetrahedron (consisting of the two final vertices and any two of the boundary vertices of the heptagon adjacent to both of the final vertices).

Any 10-vertex full triangulation of the projective plane has at least four vertices of valency six (there are no vertices of valency higher than 6 by the above, and the sum of the 10 valencies is 54). If no two of these are adjacent, then they all have the same set of neighbours, but now all the other vertices have valency 4 and the total is wrong. Hence we may assume that a pattern of edges as in figure 2a occurs in the triangulation.

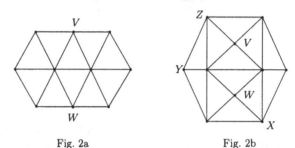

Fig. 2a   Fig. 2b

All of the vertices are already in the picture, and so the vertices marked V and W can have no other neighbours. Hence the triangulation contains the pattern of edges shown in figure 2b. With two vertices of valency 4, there must be at least six vertices of valency 6. Hence by symmetry it may be assumed that the vertex X is one such. The only possible new neighbours for X are the vertices Y and Z. Adding edges XY and XZ, together with the faces implied by fullness, a triangulated disc whose boundary consists of four edges is obtained. There is only one way to close up this surface without adding new vertices or violating fullness, and this gives a triangulation of the 2-sphere (as may be seen either by calculating its Euler characteristic, or simply from the fact that after removing one face it may be embedded in the plane).

Remark. The smallest triangulation of the projective plane which is a barycentric subdivision has 31 vertices.

It is easy to see that the triangulation of figure 1 cannot be 3-coloured, and also that it has no 4-colourings in which every vertex has a neighbour of each of the three other colours (see Proposition 13). It does have a 4-colouring in which all but one of the vertices has neighbours of three colours, and even in which all but one of the colour-pair subgraphs (as defined in the statement of Proposition 11) are connected. The colouring with vertex classes

$$\{a,c,e\}, \quad \{b,d,f\}, \quad \{g,h,j\}, \quad \{i,k\}$$

is one such. Let (Γ, V) be the right-angled Coxeter system with $K(\Gamma, V)$ as in figure 1. The above 4-colouring gives rise to an index 16 torsion-free subgroup of Γ. It is easy to see that as a normal subgroup this group is generated by the eight elements

$$ac, ae, bd, bf, gh, gj, ik, gigi = [g, i].$$

Moreover, the techniques of the proof of Proposition 11 can be used to show that the subgroup generated by these elements is already normal, and hence that these eight elements generate an index 16 subgroup Γ_2 of Γ. It is still possible to improve upon Γ_2. Let elements l, m, and n generate a product of three cyclic groups of order two, and define a homomorphism ψ from Γ to this group by the following equations.

$$\psi(a) = \psi(c) = \psi(e) = l \qquad \psi(b) = \psi(d) = \psi(f) = m$$
$$\psi(g) = \psi(h) = \psi(j) = n \qquad \psi(i) = \psi(k) = lmn$$

The homomorphism ψ maps each of the maximal special subgroups of Γ isomorphically to the group generated by l, m and n, and hence its kernel Γ_1 is a torsion-free normal subgroup of Γ of index eight, which contains the index sixteen subgroup Γ_2 of the previous paragraph. The element $bcgi$ is in Γ_1, but is not in Γ_2. However, since both b and c commute with g and i, its square is $(bcgi)^2 = gigi$. It follows that the eight elements

$$ac, ae, bd, bf, gh, gj, ik, bcgi$$

generate the group Γ_1. Any normal subgroup of Γ of index eight requires at least eight generators (see Proposition 11), so there is a sense in which Γ_1 is best possible.

The Euler characteristic $\chi(\Gamma)$ may be calculated using I. M. Chiswell's formula [7], and then $\chi(\Gamma_1)$ is equal to $|\Gamma : \Gamma_1| \chi(\Gamma)$. In fact, for a group having a finite Eilenberg-Mac Lane space (such as Γ_1), this Euler characteristic is just the usual (topological) Euler characteristic of the Eilenberg-Mac Lane space. Indeed, Chiswell's formula can be obtained by using the Davis complex to make a finite Eilenberg-Mac Lane space for a torsion-free subgroup of finite index in a Coxeter group, and then dividing by the index. Since the complex $K(\Gamma, V)$ has 11 vertices, 30 edges, and 20 2-simplices, while Γ_1 has index eight, the formula gives

$$\chi(\Gamma_1) = 8(1 - \frac{11}{2} + \frac{30}{4} - \frac{20}{8}) = 4.$$

Using the computer algebra package GAP [17], together with some adjustments suggested by V. Felsch, we were able to find the presentation for Γ_1 given below, which has 8 generators and 12 relations of total length 70 as words in the generators. (Recall our convention that $\bar{x}$ stands for x^{-1}.)

$$\Gamma_1 = \langle s,\, t,\, u,\, v, w,\, x,\, y,\, z \mid \bar{y}sy\bar{s},\, vxv\bar{x},\, \bar{z}^2wx^2w,$$
$$x^2wu\bar{w}u,\, \bar{y}uz\bar{y}z\bar{u},\, u\bar{w}uz^2\bar{w},\, \bar{z}y\bar{z}\bar{t}yt,\, uzsu\bar{s}\bar{z}$$
$$\bar{t}z\bar{w}t\bar{w}z,\, vzt\bar{z}\bar{v}t,\, vw\bar{y}vy\bar{w},\, \bar{v}w\bar{z}svzs\bar{w}\rangle$$

As words in the eleven generators of the Coxeter group Γ the above generators are:

$$s = ca,\ t = db,\ u = fb,\ v = ki,\ w = eigb,\ x = hbie,\ y = jg,\ z = cigb.$$

We know that Γ_1 needs eight generators, and also that an Eilenberg-Mac Lane space $K(\Gamma_1, 1)$ must have at least one 3-cell (because $H^3(\Gamma_1; \mathbb{Z}\Gamma_1)$ is non-zero by Theorem 2 or Corollary 3). We also know that the Euler characteristic of Γ_1 is 4, and it follows that the above presentation has the minimum possible numbers of generators and relations. Proposition 4 implies that there is a $K(\Gamma_1, 1)$ of dimension three having exactly one 3-cell, but it does not follow that one may make a $K(\Gamma_1, 1)$ by attaching one 3-cell to the 2-complex for a presentation with 8 generators and 12 relations.

In the next section we shall give another presentation for Γ_1 (although we shall not prove this), also having the minimum numbers of generators and relations, but with the total length of the relations slightly longer than here. The advantage of the other presentation is that it shows how the group presented (which is in fact Γ_1) can be built up using free products with amalgamation from surface groups, and gives an independent proof that the cohomological dimension of Γ_1 over a ring R depends on whether 2 is a unit in R. We also describe an attaching map for a 3-cell to make an Eilenberg-Mac Lane space $K(\Gamma_1, 1)$ from the 2-complex for the new presentation.

It is worth noting that the technique used in Proposition 13 may be applied to the group Γ to give an 11 generator group of cohomological dimension five over any ring R such that $2R = R$, and six over other rings, with thirty-seven relators of length four, and one relator of length eight. The relators are the thirty commutators corresponding to the edges in figure 1, together with the following words.

$$a^2\bar{c}^2,\ a^2\bar{e}^2,\ b^2\bar{d}^2,\ b^2\bar{f}^2,\ g^2\bar{h}^2,\ g^2\bar{j}^2,\ i^2\bar{k}^2,\ b^2c^2g^2\bar{i}^2$$

As in Proposition 13, one verifies that the subgroup generated by the squares of the eleven generators is central and free abelian of rank three, and that the quotient group is isomorphic to Γ.

8. A handmade Eilenberg-Mac Lane space

Theorem 15. *The group Δ given by the presentation below has cohomological dimension three over rings R such that $2R \neq R$, and cohomological dimension two over rings R such that $R = 2R$.*

$$\Delta = \langle s, t, u, v, w, x, y, z \mid \bar{s}v\bar{s}tu\bar{t}u\bar{v},\ w^2\bar{v}^2,\ \bar{s}v\bar{s}v\bar{w}s\bar{v}sv\bar{w},$$
$$x^2\bar{v}\bar{s}v\bar{s},\ \bar{x}w\bar{v}s\bar{v}w\bar{x}s,\ u\bar{v}w\bar{x}u\bar{x}\bar{w}v,\ \bar{w}yt\bar{v}yw\bar{v}t,$$
$$\bar{y}wu\bar{w}ytu\bar{t},\ y^2s\bar{v}s\bar{v},\ zsu\bar{v}su\bar{v}z,$$
$$\bar{z}v\bar{u}v\bar{u}x\bar{w}zx\bar{w},\ s\bar{w}yzs\bar{w}yx\bar{w}zx\bar{w}\rangle$$

There is an Eilenberg-Mac Lane space $K(\Delta, 1)$, having only one 3-cell and whose 2-skeleton is given by the above presentation. The 2-sphere forming the boundary of the 3-cell is formed from the hemispheres depicted in figures 3a and 3b.

Remark. In fact this group is isomorphic to the index eight subgroup Γ_1 of the 11 generator Coxeter group Γ of the previous section. The following function ϕ from the generating set for Δ given above to Γ extends to a homomorphism from Δ to Γ, because the image of each relator is the identity element. Moreover, it is easy to see that the image of ϕ is equal to Γ_1. We shall not prove that the kernel of ϕ is trivial.

$$\phi(s) = ca \qquad \phi(t) = hbia \qquad \phi(u) = akia \qquad \phi(v) = bgia$$
$$\phi(w) = gifa \qquad \phi(x) = fija \qquad \phi(y) = fgie \qquad \phi(z) = dgkc$$

Proof of Theorem 15 We shall build the group Δ in stages via a tower of free products with amalgamation, obtained by applying an algorithm due to Chiswell [6]. We shall use the same argument at each stage to justify this process, but shall give less detail as the steps become more complicated. Let Δ_0 be the group given by the following presentation.

$$\Delta_0 = \langle s,\, t,\, u,\, v \mid \bar{s}v\bar{s}tu\bar{t}u\bar{v} \rangle$$

Δ_0 is the fundamental group of a closed non-orientable surface of Euler characteristic -2, and the 2-complex corresponding to the above presentation is a CW-complex homeomorphic to this surface. In particular Δ_0 is torsion-free, and the 2-complex corresponding to the given presentation is an Eilenberg-Mac Lane space $K(\Delta_0, 1)$.

Now let F_0 be the free group on generators w and w'. Define automorphisms a and g of F_0 by the equations

$$w^a = w\bar{w}', \qquad w'^a = \bar{w}', \qquad w^g = w'\bar{w}, \qquad w'^g = ww'\bar{w}.$$

It is easy to check that a and g have order two and commute with each other, so that they generate a subgroup of $\mathrm{Aut}(F_0)$ isomorphic to the direct product of two cyclic groups of order two. Let G_0 be the subgroup of $\mathrm{Aut}(F_0)$ generated by F_0, a and g, which is isomorphic to the split extension with kernel F_0 and quotient of order four generated by a and g. Now define a homomorphism ψ_0 from Δ_0 to G_0 by

$$\psi_0(s) = a, \qquad \psi_0(t) = gw, \qquad \psi_0(u) = agw, \qquad \psi_0(v) = w.$$

This does define a homomorphism from Δ_0 to G_0, because the image of the relator is the identity element, as shown below.

$$\psi_0(\bar{s}v\bar{s}tu\bar{t}u\bar{v}) = awa(gw)(agw)(\bar{w}g)(agw)\bar{w}$$
$$= awagwg$$
$$= w\bar{w}'w'\bar{w}$$
$$= 1$$

Regions are to be read counter-clockwise.

Edge labels are on the left of the edge,
and on the right of the inverse edge.

Pins indicate base points for regions.

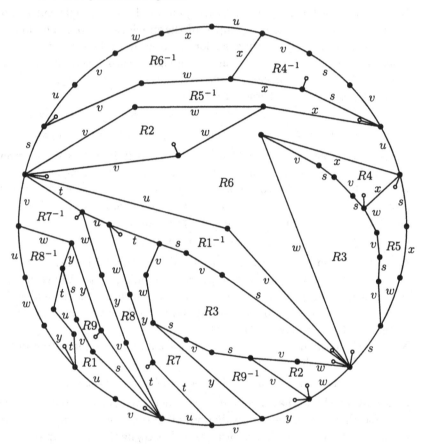

Fig. 3a

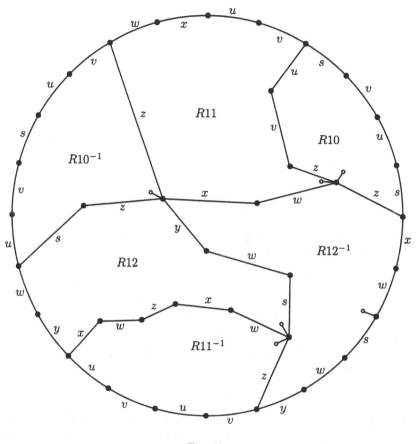

Fig. 3b

Now the images of $s\bar{v}sv$, v^2 and $\bar{v}s\bar{v}s$ under ψ_0 are w', w^2 and $\bar{w}w'\bar{w}$ respectively. These elements generate a normal subgroup of F_0 of index two, which is therefore free of rank three. (Subgroups of index two are always normal, but what one does is to check that the subgroup is normal, and then show that it has index two by calculating the order of the quotient.) It follows that the elements $s\bar{v}sv$, v^2 and $\bar{v}s\bar{v}s$ freely generate a free subgroup of Δ_0, and that this subgroup is mapped isomorphically by ψ_0 to the subgroup of F_0 generated by w', w^2 and $\bar{w}w'\bar{w}$. Hence a free product with amalgamation may be made from Δ_0 and F_0 by taking the free product and adding the relations $w' = s\bar{v}sv$, $w^2 = v^2$, and $\bar{w}w'\bar{w} = \bar{v}s\bar{v}s$. This gives a group Δ_1. Using the first of the three new relations to eliminate the generator w', it follows that Δ_1 has a presentation as below.

$$\Delta_1 = \langle s, t, u, v, w \mid \bar{s}v\bar{s}tu\bar{t}u\bar{v}, \ w^2\bar{v}^2, \ \bar{s}v\bar{s}v\bar{w}s\bar{v}svw \rangle$$

Moreover, the 2-complex corresponding to this presentation is a $K(\Delta_1, 1)$. Now take a free group F_1 of rank three generated by elements x, x' and x''. Define an automorphism f of F_1 by

$$x^f = x, \quad x'^f = x\bar{x}'x, \quad x''^f = x\bar{x}''x.$$

It may be seen that f has order two. Let G_1 be the split extension with kernel F_1 and quotient the group of order two generated by f, or equivalently the subgroup of $\mathrm{Aut}(F_1)$ generated by F_1 and f. Define a homomorphism ψ_1 from Δ_1 to G_1 as below.

$$\psi_1(s) = x', \quad \psi_1(t) = xf, \quad \psi_1(u) = x'', \quad \psi_1(v) = xf, \quad \psi_1(w) = x$$

As before, to check that this does define a homomorphism it suffices to verify that ψ_1 sends each relator to the identity in G_1. The images of s, u, $s\bar{v}sv$, $\bar{w}v\bar{s}v\bar{w}$, and $\bar{w}v\bar{u}v\bar{w}$ under ψ_1 are x', x'', x^2, $\bar{x}x'\bar{x}$, and $\bar{x}x''\bar{x}$ respectively. These five elements generate a normal subgroup of F_1 of index two, which is therefore a free group on five generators. It follows that the subgroup of Δ_1 generated by s, u, $s\bar{v}sv$, $\bar{w}v\bar{s}v\bar{w}$, and $\bar{w}v\bar{u}v\bar{w}$ is freely generated by these elements and is mapped isomorphically to a free subgroup of F_1 by ψ_1. Hence we may form an amalgamated free product of Δ_1 and F_1, identifying the two five generator free subgroups via ψ_1. Call the resulting group Δ_2. After eliminating the generators x' and x'' and the relations $x' = s$, $x'' = u$, the group Δ_2 has the presentation given below. Once again, because the amalgamating subgroup is free, the 2-complex for this presentation is a $K(\Delta_2, 1)$.

$$\Delta_2 = \langle s, t, u, v, w, x \mid \bar{s}v\bar{s}tu\bar{t}u\bar{v}, \ w^2\bar{v}^2,$$
$$\bar{s}v\bar{s}v\bar{w}s\bar{v}sv\bar{w}, \ x^2\bar{v}\bar{s}v\bar{s},$$
$$\bar{x}w\bar{v}s\bar{v}w\bar{x}s, \ u\bar{v}w\bar{x}u\bar{x}\bar{w}v \rangle$$

Take a free group F_2 of rank three with generators y, y', y'', and define automorphisms f and i of F_2 by the following equations.

$$y^f = y \qquad y'^f = \bar{y}\bar{y}'\bar{y} \qquad y''^f = \bar{y}\bar{y}'\bar{y}''y'y$$
$$y^i = \bar{y} \qquad y'^i = y'y^2 \qquad y''^i = \bar{y}''$$

It may be checked that i and f have order two and commute, so they generate a subgroup of $\text{Aut}(F_2)$ isomorphic to a direct product of two cyclic groups of order two. Let G_2 be the subgroup of $\text{Aut}(F_2)$ generated by F_2, f and i. Define a homomorphism ψ_2 from Δ_2 to G_2 by checking that the function defined as follows on the generators sends each relator to the identity in G_2.

$$\psi_2(s) = 1 \qquad \psi_2(t) = y'yf \qquad \psi_2(u) = y''$$
$$\psi_2(v) = yf \qquad \psi_2(w) = y \qquad \psi_2(x) = fi$$

Now ψ_2 sends the elements $t\bar{v}$, $t\bar{u}\bar{t}$, $v\bar{s}v\bar{s}$, $w\bar{t}v\bar{w}$, and $wu\bar{w}$ to y', y'', y^2, $yy'y$ and $yy''\bar{y}$ respectively, and these five elements generate a normal subgroup of F_2 of index two, which they therefore freely generate. Hence we may make an amalgamated free product Δ_3 from the free product of Δ_2 and F_2 by adding the five relations $t\bar{v} = y'$, ..., $wu\bar{w} = yy''\bar{y}$. After eliminating the generators y' and y'', we obtain the following presentation for Δ_3, such that the corresponding 2-complex is a $K(\Delta_3, 1)$.

$$\Delta_3 = \langle s, t, u, v, w, x, y, z \mid \bar{s}v\bar{s}tu\bar{t}u\bar{v}, \; w^2\bar{v}^2, \; \bar{s}v\bar{s}v\bar{w}s\bar{v}sv\bar{w},$$
$$x^2\bar{v}\bar{s}v\bar{s}, \; \bar{x}w\bar{v}s\bar{v}w\bar{x}s, \; u\bar{v}w\bar{x}u\bar{x}\bar{w}v, \; \bar{w}yt\bar{v}yw\bar{v}t,$$
$$\bar{y}wu\bar{w}ytu\bar{t}, \; y^2s\bar{v}s\bar{v}\rangle$$

Now take a 1-relator group F_3, with presentation

$$F_3 = \langle z, z', z'' \mid \bar{z}'z\bar{z}'z''zz''\rangle.$$

F_3 is the fundamental group of the closed nonorientable surface of Euler characteristic -1. The 2-complex corresponding to this presentation is a cellular decomposition of this surface, so in particular is a $K(F_3, 1)$. Define automorphisms c and g of F_3 by the following equations.

$$z^c = \bar{z}'z\bar{z}' \qquad z'^c = \bar{z}' \qquad z''^c = z''$$
$$z^g = z'\bar{z}z' \qquad z'^g = z \qquad z''^g = \bar{z}''$$

To check that c and g as defined above do extend to homomorphisms from F_3 to itself, note that

$$(\bar{z}'z\bar{z}'z''zz'')^c = z'\bar{z}'z\bar{z}'z'z''\bar{z}'z\bar{z}'z'' = zz''\bar{z}'z\bar{z}'z''$$

is equal (in the free group) to a conjugate of the relator, and similarly $(\bar{z}'z\bar{z}'z''zz'')^g$ is equal to a conjugate of the inverse of the relator. It is easy to check that c and g define commuting involutions in $\mathrm{Aut}(F_3)$ (and even on the free group with the same generating set). Now let G_3 be the split extension with kernel F_3 and quotient the subgroup of $\mathrm{Aut}(F_3)$ generated by c and g. (Since F_3 has trivial centre, G_3 is isomorphic to a subgroup of $\mathrm{Aut}(F_3)$.) Define a homomorphism ψ_3 from Δ_3 to G_3 by taking the following specification on the generators, and checking that the image of each relator is equal to the identity in G_3.

$$\psi_3(s) = c \qquad \psi_3(t) = 1 \qquad \psi_3(u) = gzc \qquad \psi_3(v) = g$$
$$\psi_3(w) = g \qquad \psi_3(x) = z''g \qquad \psi_3(y) = gz'c$$

Let H be the subgroup of Δ_3 generated by $\bar{y}w\bar{s}$, $x\bar{w}$, $v\bar{u}\bar{s}v\bar{u}\bar{s}$ and $w\bar{x}u\bar{v}u\bar{v}$. The images of these elements under ψ_3 are z', z'', z^2 and $zz''\bar{z}$ respectively. These four elements of F_3 generate a normal subgroup $\psi_3(H)$ of index two, which turns out to be an orientable surface group (necessarily of Euler characteristic -2). If we write $\alpha = z^2$ and $\beta = zz''\bar{z}$, then $\psi_3(H)$ may be presented as follows.

$$\psi_3(H) = \langle z', z'', \alpha, \beta \mid \bar{\alpha}\beta\alpha z''\bar{z}'\bar{\beta}\bar{z}''z' \rangle$$

We claim that ψ_3 restricted to H is injective. For this it suffices to show that the word in the four generators for H mapping to the relator in $\psi_3(H)$ is equal to the identity in Δ_3. Expressed in terms of the generators for Δ_3, the word is

$$s u \bar{v} s u \bar{v} w \bar{x} u \bar{v} u \bar{v} v \bar{u} \bar{s} v \bar{u} \bar{s} x \bar{w} s \bar{w} y v \bar{u} v \bar{u} x \bar{w} w \bar{x} \bar{y} w \bar{s},$$

so after cyclically reducing this word it suffices to show that in Δ_3,

$$u \bar{v} s u \bar{v} w \bar{x} u \bar{v} \bar{s} v \bar{u} \bar{s} x \bar{w} s \bar{w} y v \bar{u} v \bar{u} \bar{y} w = 1.$$

A Lyndon (or van Kampen) diagram whose boundary is this word, made from eighteen 2-cells each of which has boundary one of the nine relators in Δ_3, is shown in figure 3a. This diagram was found, with too much effort, using the normal form for elements of free products with amalgamation. It follows that we may form an amalgamated free product Δ of Δ_3 and F_3 amalgamating H and $\psi_3(H)$. The generators z' and z'' may be eliminated using the relations $z' = \bar{y}w\bar{s}$ and $z'' = x\bar{w}$, and the resulting presentation of Δ is the one given in the statement.

To make an Eilenberg-Mac Lane space for Δ it suffices to attach a 3-cell to the 2-complex for the above presentation corresponding to the relator in the amalgamating subgroup. The boundary of this 3-cell is made from two discs (Lyndon diagrams expressing the relator in H in terms of the relators in Δ_3 and the relator in $\psi_3(H)$ in terms of the relator in F_3) and a cylinder (whose sides represent the identification of the generators of H with their images under ψ_3). One of the discs is figure 3a. After eliminating z' and z'' as above, the rest of the sphere is as shown in figure 3b.

To verify the claim concerning $H^3(\Delta; M)$ we use the free resolution for $\mathbb{Z}$ over $\mathbb{Z}\Delta$ given by the cellular chain complex for the universal cover of the Eilenberg-Mac Lane space $K(\Delta, 1)$ constructed above. In the 1-skeleton of the sphere illustrated in figure 3, choose, for each of the twelve relators, an oriented path between the base points of the two occurrences of the relator. These paths determine elements $w_1, \ldots, w_{12}$ of Δ. Now for any Δ-module M, $H^3(\Delta; M)$ is isomorphic to M/IM, where I is the right ideal of $\mathbb{Z}\Delta$ generated by the twelve elements $1 \pm w_1, \ldots, 1 \pm w_{12}$, and the sign in $1 \pm w_i$ is positive if the ith relator appears in figure 3 with the same orientation each time, and is negative otherwise. In figure 3a, the two copies of the third relator meet at their base points and have the same orientation. It follows that $1 + w_3 = 2$, and hence that $H^3(\Delta; M)$ is a quotient of $M/2M$. This completes the proof of the statement. With a little more work it may be shown that I is equal to the ideal of $\mathbb{Z}\Delta$ generated by 2 and the augmentation ideal, which implies that for any M,

$$H^3(\Delta; M) \cong M_\Delta / 2M_\Delta.$$

We leave this as an exercise.

9. Further questions

1) We also used GAP [17] to try to find good presentations of various other finite-index torsion-free subgroups of Coxeter groups. The examples that we tried include:

a) The index sixteen subgroup Γ_2 of the right-angled Coxeter group Γ described in section 7. Comparison of the Euler characteristic (which is twice that of Γ_1, or eight) with the known minimum number of generators, together with the fact that any $K(\Gamma_2, 1)$ must have at least one 3-cell indicate that the minimum number of relators in a presentation of Γ_2 must be at least sixteen. Using GAP we were able to get down only to an 8-generator 17-relator presentation, but by hand (and then checking the result using GAP) we were able to eliminate one of the relators. The sum of the lengths of the relators in our presentation is 152. We found a CW-complex $K(\Gamma_2, 1)$ having eight 1-cells, sixteen 2-cells and one 3-cell.

b) Other index eight normal subgroups of the Coxeter group Γ of Section 7. If ψ' is any homomorphism from Γ onto a product of three cyclic groups of order two which restricts to an isomorphism on each maximal special subgroup of Γ, then the kernel of ψ' is a torsion-free index eight normal subgroup of Γ. One way to create such a ψ' is to take ψ (the homomorphism given earlier, with kernel Γ_1) and modify it slightly. We did not find a ψ' whose kernel had a smaller presentation than the one given for Γ_1.

c) Take figure 1, remove the vertex h and all edges leaving it, and add a new edge between vertices i and k. Label the three boundary edges with the label three, and give all other edges the label two. This gives a presentation of a ten-generator Coxeter system (Δ, V) such that $K(\Delta, V)$ is a triangulation of the projective plane. Δ has torsion-free normal subgroups of index 24, and clearly has no torsion-free subgroup of lower index, since it contains elements of order three and subgroups isomorphic to C_2^3. The subgroup we looked at required nine generators.

d) In [1] Bestvina pointed out that a finite-index torsion-free subgroup Γ_1 of a Coxeter group Γ such that $K(\Gamma, V)$ is an acyclic 2-complex would have cohomological dimension two over any ring, but might not have a 2-dimensional Eilenberg-Mac Lane space. (A famous conjecture of Eilenberg and Ganea asserts that any group of cohomological dimension two has a 2-dimensional Eilenberg-Mac Lane space [4].) Let K be the barycentric subdivision of the acyclic 2-complex having five vertices, ten edges corresponding to the ten pairs of vertices, and six pentagonal faces corresponding to a conjugacy class in A_5 of elements of order five. The simplicial complex K is full. If (Γ, V) is the corresponding right-angled Coxeter system, then the easy extension of Corollary 12 to polyhedral complexes shows that colouring $K(\Gamma, V)$ by dimension gives rise to a torsion-free index eight normal subgroup Γ_1 of Γ, requiring exactly eighteen generators. The complex $K(\Gamma, V)$ has 21 vertices, 80 edges and 60 2-simplices, so by Chiswell's formula [7], the Euler characteristic of Γ_1 is

$$\chi(\Gamma_1) = 8(1 - \frac{21}{2} + \frac{80}{4} - \frac{60}{8}) = 24.$$

Hence if a presentation for Γ_1 could be found having twenty-three more relators than generators, the corresponding 2-complex would be a $K(\Gamma_1, 1)$. An argument similar to that sketched in the proof of Proposition 4 shows that there is a 3-dimensional $K(\Gamma_1, 1)$ having exactly six 3-cells. Presentations of Γ_1 arising from this complex will have 29 more relators than generators. Using GAP we found an 18-generator 52-relator presentation for Γ_1, but were unable to reduce the number of relators any further. The problem of whether there exists a $K(\Gamma_1, 1)$ with less than six 3-cells remains open.

2) The groups exhibited in Section 4 Example 2 (whose cohomological dimension over the integers is three and whose cohomological dimension over any field is two) are finitely generated, but cannot be FP by Proposition 9. The question of whether there can be similar examples which are $FP(2)$ or even finitely presented remains open.

3) The result proved in Proposition 14 does not really prove that the examples of Sections 7 and 8 are minimal, even in the sense of being finite-index subgroups of Coxeter groups with the least possible number of generators. It may be true that there can be no full simplicial complex having ten vertices or fewer whose highest non-zero homology group is non-free, which would suggest that no right-angled Coxeter group on less than eleven generators can have different virtual cohomological dimensions over $\mathbb{Z}$ and $\mathbb{Q}$.

4) Is there a simpler example of a group whose cohomological dimension over $\mathbb{Z}$ is finite and strictly greater than its cohomological dimension over $\mathbb{Q}$ than the group $\Gamma_1 \cong \Delta$ given in Sections 7 and 8? Applying the embedding theorem of Higman-Neumann-Neumann to Δ we were able to construct a 2-generator 12-relator presentation of a group whose cohomological dimension over any ring is equal to that of Δ. An Euler characteristic argument shows that this group requires at least 12 relators. The total length of the 12 relators we found was 1,130, so this group can hardly be said to be simpler than Δ.

The two distinct 8-generator 12-relator presentations for Δ given in Sections 7 and 8 have short relators (i.e., simple attaching maps for the 2-cells) and a simple attaching map for the 3-cell respectively. Is it possible to make a presentation for Δ in which each 2-cell occurs only twice in the boundary of the 3-cell and such that the sum of the lengths of the relators is smaller than 96 (the sum of the lengths of the relators in the presentation given in Theorem 15)?

5) (P. H. Kropholler) Can there be a group Γ with $\mathrm{cd}\,\Gamma = 4$ and $\mathrm{cd}_{\mathbb{Q}}\Gamma = 2$? Notice that taking direct products of copies of Bestvina's examples gives groups with arbitrary finite differences between their cohomological dimensions over $\mathbb{Z}$ and over $\mathbb{Q}$.

6) What we call an R-homology manifold is really a *simplicial R-homology manifold*. One could give a similar definition and an analogue of Theorem 5 and Corollary 6 for general (locally compact Hausdorff) topological spaces. It may be the case that any torsion-free Poincaré duality group over R acts freely cocompactly and properly discontinuously on an orientable R-acyclic R-homology manifold.

Acknowledgements. We thank the referee for his comments on an earlier version of this paper, and especially for showing us Proposition 9 and Theorem 10. We thank Volkmar Felsch for informing us of how to modify GAP to suit our purposes. We gratefully acknowledge that this work was generously funded by the DGICYT, through grants PB90-0719 and PB93-0900 for the first-named author, and a post-doctoral fellowship held at the Centre de Recerca Matemàtica for the second-named author. The second-named author gratefully acknowledges the support of the Max Planck Institut für Mathematik and the Leibniz Fellowship Scheme during the period when this paper was being revised.

References

[1] M. Bestvina, *The virtual cohomological dimension of Coxeter groups*, in: Geometric Group Theory Volume I (G.A. Niblo and M.A. Roller, eds.), London Math. Soc. Lecture Notes, **181** Cambridge University Press (1993), pp. 19–23.

[2] R. Bieri, *Homological dimension of discrete groups*, Queen Mary College Mathematics Notes, London (1976).

[3] G.E. Bredon, *Sheaf theory*, McGraw-Hill (1967).

[4] K.S. Brown, *Cohomology of groups*, Graduate Texts in Mathematics, **87** Springer-Verlag (1982).

[5] K.S. Brown, *Buildings*, Springer-Verlag (1989).

[6] I.M. Chiswell, *Right-angled Coxeter groups*, in: Low Dimensional Topology and Kleinian Groups (D. B. A. Epstein, ed.), London Math. Soc. Lecture Notes, **112** (1986), pp. 297–304.

[7] I.M. Chiswell, *The Euler characteristic of graph products and of Coxeter groups*, in: Discrete Groups and Geometry (W.J. Harvey, C. MacLachlan, eds.), London Math. Soc. Lecture Notes, **173** (1992), pp. 36–46.

[8] M.W. Davis, *Groups generated by reflections*, Ann. of Math. **117** (1983), pp. 293–324.

[9] M.W. Davis and T. Januszkiewicz, *Convex polytopes, Coxeter orbifolds and torus actions*, Duke Math. Journal **62** (1991), pp. 417–451.

[10] W. Dicks and M.J. Dunwoody, *Groups Acting on Graphs*, Cambridge Studies in Advanced Mathematics, **17** Cambridge Univ. Press (1989).

[11] W. Dicks and I.J. Leary, *Exact sequences for mixed coproduct/tensor product ring constructions*, Publicacions Matemàtiques **38** (1994), pp. 89–126.

[12] E.R. Green, *Graph products of groups*, Ph. D. Thesis University of Leeds (1990).

[13] J. Harlander and H. Meinert, *Higher generation subgroup sets and the virtual cohomological dimension of graph products of finite groups*, J. London Math. Soc. **53** (1996), pp. 99–117.

[14] P.J. Hilton and U. Stammbach, *A course in homological algebra*, Graduate Texts in Mathematics, **4** Springer-Verlag (1971).

[15] J.R. Munkres, *Elements of algebraic topology*, Addison-Wesley (1984).

[16] D.S. Robinson, *A course in the theory of groups*, Graduate Texts in Mathematics, **80** Springer-Verlag (1982).

[17] M. Schönert et al., *GAP (Groups, Algorithms and Programming)*, Version 3 release 2 Lehrstuhl D für Mathematik, RWTH, Aachen (1993).

The p-primary Farrell Cohomology of Out(F_{p-1})

H.H. Glover, G. Mislin and S.N. Voon

Ohio State University, Columbus, Ohio
ETH Zürich, Switzerland and Ohio State University, Columbus, Ohio
Ohio State University, Newark, Ohio
email:glover@math.ohio-state.edu
email:mislin@math.ethz.ch and mislin@math.ohio-state.edu
email:snvoon@math.ohio-state.edu

Abstract. We compute for an odd prime p the p-primary part of the Farrell cohomology of $Out(F_{p-1})$, the group of outer automorphisms of a free group F_{p-1} of rank $p - 1$. We also determine the p-period, Yagita invariant and cohomological Krull dimension for automorphism groups of free groups.

0. Introduction

Let F_n denote a free group of rank n, $Aut(F_n)$ its automorphism group and $Out(F_n)$ the associated group of outer automorphisms. Recall that these groups are of finite virtual cohomological dimension. The main purpose of this note is to compute, for p a prime, the p-primary cohomology of $Aut(F_{p-1})$ and $Out(F_{p-1})$ above the virtual cohomological dimension. It is convenient to express our result in terms of Farrell cohomology. We shall write $\hat{H}^*(\Gamma; \mathbb{Z})_p$ for the p-primary part of the Farrell cohomology of a group Γ of finite vcd.

Theorem. *Let p be an odd prime. Then*

$$\hat{H}^*(Aut(F_{p-1}); \mathbb{Z})_p \cong \hat{H}^*(Out(F_{p-1}); \mathbb{Z})_p \cong \mathbb{F}_p[w, w^{-1}],$$

with w of degree $2(p - 1)$.

Note that the mod p Farrell cohomology of these groups is periodic, implying that the Krull dimension of the corresponding mod p cohomology rings equals one. In Section 1 we will determine for general n the Krull dimension of the mod p cohomology rings of $Aut(F_n)$ and $Out(F_n)$; the result can be read off from the work of Smillie and Vogtmann [SV2]. Section 2 is devoted to computing the Yagita invariant of these groups; it is interesting to compare the result with the corresponding one for the case of mapping class groups (cf. [GMX]). In the third Section we present the proof of the theorem stated above; the interested reader should compare this result with the analogous computation for the mapping class group $\Gamma_{(p-1)/2}$ (cf. [X]) and for $Gl_{p-1}(\mathbb{Z})$ (cf. [A], [BE]).

We thank the referee for many valuable suggestions.

1. The Krull dimension

Recall that for a group Γ of finite vcd the Krull dimension $\kappa_p(\Gamma)$ of the mod p cohomology of Γ equals the maximal rank of an elementary Abelian p-subgroup of Γ. Thus, if $\Lambda \longrightarrow \Gamma$ is a map of groups of finite vcd with p-torsion-free kernel, one has $\kappa_p(\Lambda) \leq \kappa_p(\Gamma)$.

The natural map $Out(F_n) \longrightarrow Gl_n(\mathbb{Z})$ has torsion-free kernel and thus

$$\kappa_p(Out(F_n)) \leq \kappa_p(Gl_n(\mathbb{Z})).$$

It is well-known that the rank of a maximal elementary Abelian p-subgroup of $Gl_n(\mathbb{Z})$ is $[\frac{n}{p-1}]$, where $[x]$ stands for the integral part of the real number x (cf. Minkowski [M]).

The following theorem then follows from [SV2].

Theorem 1.1. *For every prime p and $n \geq 1$ one has*

$$\kappa_p(Aut(F_n)) = \kappa_p(Out(F_n)) = \left[\frac{n}{p-1}\right].$$

Proof Because the kernel of the natural map $Aut(F_n) \longrightarrow Out(F_n)$ is torsion-free and in view of our discussion above, we have

$$\kappa_p(Aut(F_n)) \leq \kappa_p(Out(F_n)) \leq \kappa_p(Gl_n(\mathbb{Z})) = \left[\frac{n}{p-1}\right].$$

It suffices therefore to show that $(\mathbb{Z}/p)^m \subset Aut(F_{m(p-1)})$. For this, we consider the graph $G(p)$ with 2 vertices, connected by p edges, so that $\pi_1 G(p) \cong F_{p-1}$. Let $\mathbb{Z}/p$ act on $G(p)$ by permuting the edges and fixing the vertices. This action extends in an obvious way to an action of $(\mathbb{Z}/p)^m$ on $\vee_m G(p)$, yielding an injection

$$(\mathbb{Z}/p)^m \longrightarrow Aut(\pi_1(\underset{m}{\vee} G(p))) \cong Aut(F_{m(p-1)}),$$

and the result follows.

Remark. The whole symmetric group Σ_p acts by automorphisms on the graph $G(p)$, fixing the two vertices. Furthermore, the symmetric group Σ_m acts on $\vee_m G(p)$ by permuting the copies of $G(p)$ and fixing the basepoint. One concludes that the wreath product $\Sigma_p \int \Sigma_m := (\Sigma_p \times \cdots \times \Sigma_p) \rtimes \Sigma_m$ acts on $\vee_m G(p)$ fixing the basepoint so that $\Sigma_p \int \Sigma_m \subset Aut(F_{m(p-1)})$. Of course, m and p can here be arbitrary integers ≥ 1.

2. The Yagita invariant

Recall (see [Y]) that for a group Γ of finite *vcd* and prime p the Yagita invariant $p(\Gamma)$ is defined as the least common multiple of numbers $2m(\pi)$, where $m(\pi)$ is the largest number such that

$$H^*(\Gamma; \mathbb{Z}) \xrightarrow{\text{res}} \text{im}\,(res) \subset \mathbb{Z}/p[u^{m(\pi)}] \subset H^*(\pi; \mathbb{Z}/p)$$

where $\pi \subset \Gamma$ is a subgroup of order p and $u \in H^2(\pi; \mathbb{Z}/p)$ is a generator. (If no such $\pi \subset \Gamma$ exists, one puts $p(\Gamma) = 1$). It is obvious from the definition that for any map $\Lambda \longrightarrow \Gamma$ of groups of finite *vcd* with p-torsion-free kernel one has $p(\Lambda) \mid p(\Gamma)$. We also recall the well-known and elementary fact that in case $\kappa_p(\Gamma) = 1$, the Yagita invariant $p(\Gamma)$ agrees with the p-period, that is, it is equal to the smallest positive integer k such that

$$H^i(\Gamma; \mathbb{Z})_p \cong H^{i+k}(\Gamma; \mathbb{Z})_p\,, \qquad \forall i > vcd(\Gamma).$$

We call such a group p-periodic; it has periodic Farrell cohomology of period $p(\Gamma)$.

Definition. For $x \in \mathbb{R}^{>0}$ and p a prime we shall write $[x]_p$ for the largest power of p less than or equal to x; thus $x = [x]_p + y$ with $[x]_p = p^m$ and $0 \leq y < (p-1)p^m$.

Theorem 2.1. *Let p be a prime and $n \geq 1$. Then*

$$p(Aut(F_n)) = p(Out(F_n)) = 2(p-1)\left[\frac{n}{p-1}\right]_p.$$

Proof The natural maps $Aut(F_n) \longrightarrow Out(F_n) \longrightarrow Gl_n(\mathbb{Z})$ with torsion-free kernels show that

$$p(Aut(F_n)) \leq p(Out(F_n)) \leq p(Gl_n(\mathbb{Z})).$$

From [GMX, Lemma 1.2] we infer that $p(Gl_n(\mathbb{Z}))$ divides $2(p-1)\left[\frac{n}{p-1}\right]_p$ (see also [GLT] for more information concerning the Yagita invariant of a general linear group). It suffices therefore to find a finite subgroup $\sigma \subset Aut(F_n)$ with $p(\sigma) \geq 2(p-1) \left[\frac{n}{p-1}\right]_p$. Let $p^m = \left[\frac{n}{p-1}\right]_p$. Note that $(p-1)p^m \leq n$ and

$$\Sigma_p \int \Sigma_{p^m} \subset Aut(F_{(p-1)p^m}) \subset Aut(F_n)$$

according to our remark at the end of section 1. We claim that $p(\Sigma_p \int \Sigma_{p^m})$ is larger or equal to $2(p-1)p^m$. Indeed, the normalizer of $\mathbb{Z}/p \subset \Sigma_p$ contains the holomorph $Hol(\mathbb{Z}/p)$ of $\mathbb{Z}/p$ (i.e., the split extension of $\mathbb{Z}/p$ by its automorphism group $Aut(\mathbb{Z}/p) \cong \mathbb{Z}/(p-1)$). Clearly,

$$H^*(Hol(\mathbb{Z}/p); \mathbb{Z}) = \mathbb{Z}[w]/pw \quad \text{with} \quad |w| = 2(p-1).$$

Now consider $Hol(\mathbb{Z}/p) \int \Sigma_{p^m} \subset \Sigma_p \int \Sigma_{p^m}$. The diagonal embedding

$$\Delta : \mathbb{Z}/p \longrightarrow Hol(\mathbb{Z}/p) \times \cdots \times Hol(\mathbb{Z}/p) \subset Hol(\mathbb{Z}/p) \int \Sigma_{p^m}.$$

induces a restriction map

$$\Delta^* : H^*(Hol(\mathbb{Z}/p) \int \Sigma_{p^m}; \mathbb{Z}) \longrightarrow H^*(Hol(\mathbb{Z}/p); \mathbb{Z}) \subset H^*(\mathbb{Z}/p; \mathbb{Z})$$

mapping into $\mathbb{Z}\left[w^{p^m}\right]/pw^{p^m}$. This can be seen by looking at the diagonal restriction map

$$\rho^* : H^*(Hol(\mathbb{Z}/p) \times \cdots \times Hol(\mathbb{Z}/p); \mathbb{Z}) \longrightarrow H^*(Hol(\mathbb{Z}/p); \mathbb{Z}).$$

Because elements of odd degree are mapped to zero, ρ^* factors through

$$H^*(Hol(\mathbb{Z}/p); \mathbb{Z}) \otimes \cdots \otimes H^*(Hol(\mathbb{Z}/p); \mathbb{Z}) = \mathbb{Z}[w_1, \cdots w_{p^m}]/p(w_1, \cdots, w_{p^m})$$

with each w_i of degree $2(p-1)$ and $(w_1, \cdots, w_{p^m})$ denoting the ideal generated by these elements. But all elementary symmetric functions in the variables $w_1, \cdots, w_{p^m}$ map via ρ^* to 0 so that the image of Δ^* is contained in the subalgebra generated by the image of Πw_i, which is w^{p^m}. It follows that $p(\Sigma_p \int \Sigma_{p^m}) \geq 2(p-1)p^m$, completing the proof.

Remark. Note that our proof shows that $p(\Sigma_p \int \Sigma_{p^m}) = 2(p-1)p^m$ and

$$p(Gl_n(\mathbb{Z})) = 2(p-1)\left[\frac{n}{p-1}\right]_p.$$

As already mentioned, the Yagita invariant coincides with the p-period in the p-periodic case, and a group of finite virtual cohomological dimension is p-periodic if and only if its elementary Abelian p-subgroups are cyclic (cf. Brown's book [B]).

Corollary 2.2. *Let p be an arbitrary prime and $n \geq 1$. Then the following holds:*

(i) Aut(F_n) is p-periodic if and only if Out(F_n) is.

(ii) Aut(F_n) and Out(F_n) are p-periodic if and only if $p - 1 \leq n < 2(p-1)$.

(iii) If Aut(F_n) and Out(F_n) are p-periodic, their p-period equals $2(p-1)$.

3. The p-primary cohomology

The complexity of the p-primary part of the cohomology of $Aut(F_n)$ and $Out(F_n)$ is simplest in the p-periodic case. We will completely determine the Farrell cohomology in the first case, that is, in case $n = p - 1$. The next two cases, $n = p$ and $n = p + 1$ respectively, will be considered in a forthcoming paper (cf. [GM]). Since the Krull dimension of $Aut(F_{p-1})$, $\kappa_p(Aut(F_{p-1}))$, equals one, the p-part of the Farrell cohomology of $Aut(F_{p-1})$ is given by (cf. Brown's book [B])

$$\hat{H}^\bullet(Aut(F_{p-1}); \mathbb{Z})_p = \prod_{C \in K} \hat{H}^\bullet(N_A(C); \mathbb{Z})_p$$

where K denotes a set of representatives of conjugacy classes of subgroups C of order p of $Aut(F_{p-1})$, and $N_A(C)$ the normalizer of $C \subset Aut(F_{p-1})$. Elements of order p in the automorphism group of a free group have been classified in [DS]. In particular, if $\alpha \in Aut(F_{p-1})$ has order p, then according to [DS] there exists a basis $(x_1, \cdots, x_{p-1})$ of F_{p-1} such that

$$\alpha(x_i) = \begin{cases} x_{i+1}, & 1 \leq i < p - 1 \\ (x_1 x_2 \cdots x_{p-1})^{-1}, & i = p - 1. \end{cases}$$

Theorem 3.1. *Let p be an odd prime. Then all elements of order p in $Aut(F_{p-1})$ are conjugate. Moreover, if $\alpha \in Aut(F_{p-1})$ has order p then $N_A(\langle \alpha \rangle)$, the normalizer of the subgroup $\langle \alpha \rangle$ generated by α in $Aut(F_{p-1})$, injects into $N_O(\langle \bar{\alpha} \rangle)$, the normalizer of the subgroup $\langle \bar{\alpha} \rangle$ generated by the image $\bar{\alpha}$ of α in $Out(F_{p-1})$.*

Proof That the elements of order p in $Aut(F_{p-1})$ are conjugate follows from the classification theorem cited above. If ϕ lies in the kernel of $N_A(\langle \alpha \rangle) \longrightarrow N_O(\langle \bar{\alpha} \rangle)$ then ϕ is an inner automorphism of F_{p-1}, say $\phi(x) = txt^{-1}, t \in F_{p-1}$. But, assuming that α has order p, ϕ^{p-1} will centralize α so that $\phi^{p-1} \circ \alpha = \alpha \circ \phi^{p-1}$ and therefore $t^{p-1}\alpha(x)t^{1-p} = \alpha(t^{p-1})\alpha(x)\alpha(t^{1-p})$ for all $x \in F_{p-1}$. This shows that $\alpha(t^{p-1}) = t^{p-1}$, because the center of F_{p-1} is trivial for $p > 2$. Hence $t^{p-1} \in F_{p-1}^{\langle \alpha \rangle}$, the group fixed by α, which is known to be a trivial group ([DS]). As a consequence, $t = e \in F_{p-1}$, since F_{p-1} is torsion-free, and we conclude that $N_A(\langle \alpha \rangle) \longrightarrow N_O(\langle \bar{\alpha} \rangle)$ is injective.

To establish the finiteness of the normalizers $N_A(\langle\alpha\rangle)$ for α as in Theorem 3.1, it suffices therefore to consider the normalizer $N_O(\langle\alpha\rangle) \subset Out(F_{p-1})$. For this we consider the spine $\mathcal{K}_n$ of the space introduced by Culler and Vogtmann ([CV]), which was also used in [SV1]. It is a contractible $(2n-3)$-dimensional simplicial complex with vertices certain equivalence classes of *marked admissible graphs*, and $Out(F_n)$ acts on $\mathcal{K}_n$ simplicially with finite stabilizers, such that if $\gamma \in Out(F_n)$ fixes a point $x \in \mathcal{K}_n$, then it fixes the carrier of x pointwise. If $C \subset Out(F_n)$ is any subgroup, the normalizer $N_O(C)$ acts on the fixed-point space $\mathcal{K}_n^C$ and thus $\mathcal{K}_n^C = \{pt\}$ implies that $N_O(C)$ is a finite group.

Theorem 3.2. *Let p be an odd prime and $C \subset Out(F_{p-1})$ a subgroup of order p. Then $N_O(C)$, the normalizer of C in $Out(F_{p-1})$, is a finite group.*

Proof The fixed point space $\mathcal{K}_n^G$ is known to be contractible for any finite subgroup G of $Out(F_n)$, see Krstić and Vogtmann [KV], or White [W]. In particular, $\mathcal{K}_{p-1}^C$ is non-empty and connected. We show that $\mathcal{K}_{p-1}^C = \{pt\}$ by showing that $\mathcal{K}_{p-1}^C$ does not contain any 1-simplex of $\mathcal{K}_{p-1}$. A vertex of $\mathcal{K}_{p-1}$ is represented by a pair (g, G), G an admissible graph of rank $p-1$ and $g : R \longrightarrow G$ a homotopy equivalence, where R is a *rose* (i.e., a graph with a single vertex; see [SV1] for the notion and discussion of admissible graphs). The admissible graph G associated with a vertex of $\mathcal{K}_{p-1}$ is unique up to homeomorphism and is called the *graph type* of the vertex; the stabilizer of the vertex is isomorphic to the automorphism group of the graph G (which is the same as the group of components of the group of homeomorphisms of G, since G is admissible). According to [SV1], there is a unique graph of rank $p-1$ admitting an automorphism of order p (namely, the graph $G(p)$ considered in the previous section). On the other hand, the two vertices of a 1-simplex of $\mathcal{K}_{p-1}$ have always distinct graph type (one is obtained from the other by a *forest collapse*). Thus $\mathcal{K}_{p-1}^C$ contains no 1-simplex, which proves that $N_O(C)$ is finite.

We can now prove the first half of the Theorem of the introduction.

Theorem 3.3. *If p is an odd prime then $\hat{H}^*(Aut(F_{p-1}); \mathbb{Z})_p \cong \mathbb{F}_p[w, w^{-1}]$, with w of degree $2(p-1)$.*

Proof We know from our earlier discussion that $Aut(F_{p-1})$ contains elements of order p. By Theorem 3.1 we conclude that there is a unique conjugacy class of subgroups of order p in $Aut(F_{p-1})$ and, if C is such a subgroup, the natural map $N_A(C) \longrightarrow Aut(C) \cong \mathbb{Z}/p-1$ is surjective. Since $N_A(C)$ is contained in the finite group $N_O(C)$, it injects via the natural map $Aut(F_{p-1}) \longrightarrow Gl_{p-1}(\mathbb{Z})$ into $Gl_{p-1}(\mathbb{Z})$, which by Minkowski's Theorem [M] does not contain any p-subgroups of order larger than p. It follows that the p-Sylow subgroup of the (finite) group $N_A(C)$ is cyclic of order p. Hence, $N_A(C)$ is a split extension of C by a finite group of order prime to p and one has therefore a natural surjection onto the holomorph of C, $N_A(C) \longrightarrow C \rtimes Aut(C) =: Hol(C)$ with finite kernel Q of order prime to p. In particular, this projection will induce an isomorphism on the p-primary part of cohomology. Thus

$$H^*(N_A(C); \mathbb{Z})_p \cong H^*(Hol(C); \mathbb{Z})_p \cong \mathbb{Z}[w]/pw$$

with w of degree $2(p-1)$. Moreover, we infer that

$$\hat{H}^*(Aut(F_{p-1}); \mathbb{Z})_p \cong \hat{H}^*(N_A(C); \mathbb{Z})_p \cong \mathbb{F}_p[w, w^{-1}]$$

with w of degree $2(p-1)$.

To obtain the corresponding result for $Out(F_{p-1})$, we first need to verify that $Out(F_{p-1})$ contains just one conjugacy class of subgroups of order p. This follows from the following Lemma.

Lemma 3.4. *Let p be an odd prime. Every subgroup $C \subset Out(F_{p-1})$ of order p lifts to a subgroup $\tilde{C} \subset Aut(F_{p-1})$ of order p.*

Proof We make use of Zimmermann's Realization Theorem [Z], (see also Culler [C]), which implies the following. Given $C \subset Out(F_{p-1})$ of order p, there exists a finite, connected graph H containing a subgroup $D \subset Aut(H)$ of order p such that D maps onto C via an isomorphism $\pi_1(H) \longrightarrow F_{p-1}$. To see that C lifts to a subgroup $\tilde{C} \subset Aut(F_{p-1})$ of order p, it suffices therefore to check that the D-action on H has a fixed-point. For this, let $f \in D$ denote a generator and consider $f_* : H_1(H; \mathbb{Q}) \longrightarrow H_1(H; \mathbb{Q})$. Since f maps to an element of order p in $Gl_{p-1}(\mathbb{Q})$ via the maps

$$Aut(H) \longrightarrow Out(F_{p-1}) \longrightarrow Gl_{p-1}(\mathbb{Z}) \longrightarrow Gl_{p-1}(\mathbb{Q}),$$

we conclude that the trace $\text{Tr}(f_*) = -1$. We use here the well-known fact that every matrix in $Gl_{p-1}(\mathbb{Q})$ of order p has trace -1. It follows that the Lefschetz number of f satisfies $\Lambda(f) = 1 - (-1) = 2$, and we conclude that $f : H \longrightarrow H$ has a fixed-point.

Remark. The result that $Out(F_{p-1})$ contains only one conjugacy class of subgroups of order p follows also immediately from the fact that there is only one admissible graph of rank $p - 1$ on which the cyclic group of order p can act ([SV2]), together with the Culler–Zimmermann realization theorem (loc. cit.).

Theorem 3.5. *Let p be an odd prime. Then $\hat{H}^*(Out(F_{p-1}); \mathbb{Z})_p \cong \mathbb{F}_p[w, w^{-1}]$ with w of degree $2(p-1)$.*

Indeed, $Out(F_{p-1})$ contains a unique conjugacy class of subgroups of order p and the normalizer $N_O(C)$ of such a subgroup is a finite group, mapping onto

$Hol(C)$ with kernel of order prime to p. The rest of the argument is then as in the proof for Theorem 3.3. Note, because $vcd(Out(F_n)) = 2n - 3$ (cf. [CV]), Theorem 3.5 determines also the ordinary cohomology of $Out(F_{p-1})$ in degrees $> 2p - 5$.

References

[A]   A. Ash, *Farrell Cohomology of GL(n, $\mathbb{Z}$)*, Israel. J. Math. (1989), **67** pp. 327–336.

[B]   K. S. Brown, *Cohomology of groups*, Springer-Verlag New York, Heidelberg, Berlin (1982).

[BE]   B. Bürgisser and B. Eckmann, *The p-torsion of the Farrell-Tate cohomology of GL(n, $O_S(K)$) and SL(n, $O_S(K)$)*, Mathematika (1984), **31** pp. 89–97.

[C]   M. Culler, *Finite groups of outer automorphisms of a free group*, Contemporary Mathematics (1984), **33** pp. 197–207.

[CV]   M. Culler and K. Vogtmann, *Moduli of automorphisms of free groups*, Invent. math. (1986), **84** pp. 91–119.

[DS]   J. Dyer and G. P. Scott, *Periodic automorphisms of free groups*, Communications in Algebra (1975), **3** pp. 195–201.

[GLT]   H. H. Glover, I. J. Leary and C. B. Thomas, *The Yagita invariant of general linear groups*, in: Algebraic Topology: New Trends in Localization and Periodicity Progress in Mathematics, **136** Birkhäuser, Basel (1996), pp. 185–192.

[GM]   H. H. Glover and G. Mislin, *On the p-primary cohomology of Out(F_n) in the p-rank one case*, submitted for publication.

[GMX]   H. H. Glover, G. Mislin and Y. Xia, *On the Yagita invariant of mapping class groups*, Topology (1994), **33** pp. 557–574.

[KV]   S. Krstić and K. Vogtmann, *Equivariant outer space and automorphisms of free-by-finite groups*, Comment. Math. Helv. (1993), **68** pp. 216–262.

[M]   H. Minkowski, *Zur Theorie der positiven quadratischen Formen*, J. Reine Angew. Math. **101** (1887), pp. 196–202.

[SV1]   J. Smillie and K. Vogtmann, *A generating function for the Euler characteristic of Out(F_n)*, J. Pure and Appl. Algebra (1987), **44** pp. 329–348.

[SV2]   J. Smillie and K. Vogtmann, *Automorphisms of graphs, p-subgroups of Out(F_n) and the Euler characteristic of Out(F_n)*, J. Pure and Appl. Algebra (1987), **49** pp. 187–200.

[W]   T. White, *Fixed points of finite groups of free group automorphisms*, Proc. Amer. Math. Soc. (1993), **3** pp. 681–688.

[X]   Y. Xia , *The p-torsion of the Farrell cohomology of* $\Gamma_{(p-1)/2}$, in: Topology 90 (B. Apanasov, W.D. Neumann, A.W. Reid and L. Siebenmann, eds.), Walter de Gruyter, (1992) pp. 391–398.

[Y]   N. Yagita, *On the dimension of spheres whose product admits a free action by a non-abelian group,* Quart. J. Math. Oxford (1985), **36** pp. 117–127.

[Z]   B. Zimmermann, *Über Homöomorphismen n-dimensionaler Henkelkörper und endliche Erweiterungen von Schottky-Gruppen,* Comment. Math. Helv. **56** (1981), pp. 424–486.

On Tychonoff Groups

R.I. Grigorchuk

1. Introduction

The class AG of amenable groups can be characterized by the property that the group has a fixed point for any action by affine transformations on a convex compact subset of a locally convex topological vector space.

Now let us suppose that instead of a compact set we have a nonzero cone. What kind of fixed-point theorems may hold in this situation? There is a number of conditions under which a selftransformation of a cone has a nonzero fixed point. We will consider the situation when a group acts by affine transformations on a convex cone with compact base. The groups for which any such action has an invariant ray are called Tychonoff groups and were defined by H. Furstenberg [5].

The Tychonoff property was also considered in [3,6] in the greater generality of locally compact groups and used for the description of positive harmonic functions on groups which contain closed nilpotent subgroups with compact quotient.

In this paper we try to attract the attention of the reader to Tychonoff groups again, and begin the systematic investigation of abstract Tychonoff groups (with the discrete topology).

Among other observations we show that the class of Tychonoff groups is closed under the operations of taking factor groups and direct limits, but it is not closed under taking subgroups. It is also not closed with respect to extensions but some sufficient conditions for a semidirect product of Tychonoff groups to be Tychonoff are given.

Groups with the Margulis property (every positive μ-harmonic function is constant on the cosets of the commutator subgroup) are Tychonoff but not every Tychonoff group has this property.

The results presented here were obtained under the financial support of the Russian Fund for Fundamental Research Grants 93-01-00239, 94-01-00820.

Using arguments of Margulis we extend the Margulis theorem about positive harmonic functions from the class of nilpotent groups to the class of ZA-groups and from finitary probability distributions to arbitrary such distributions. We show that for nilpotent groups the Tychonoff property is equivalent to the Margulis property.

Infinite, finitely generated Tychonoff groups are indicable (i.e. they can be mapped onto the infinite cyclic group) and have some other interesting properties.

At the end of the paper we consider bounded actions of groups of subexponential growth on convex cones with compact base and prove a fixed-point theorem for such actions.

2. The definition and some properties of Tychonoff groups

Let us recall some notions. A selfmap $A : E \longrightarrow E$ of a topological vector space E is called affine on a convex subset $V \subset E$ if for any $x, y \in V$ and $p, q \geq 0$, with $p + q = 1$, $A(px + qy) = pAx + qAy$.

A set $K \subset E$ is called a cone if
1. $K + K \subset K$
2. $\lambda K \subset K$ for any number $\lambda \geq 0$
3. $K \cap (-K) = \{0\}$.

The ray in a cone K is any halfline $L_x = \{\lambda x : \lambda \geq 0\}$, where $x \in K$, $x \neq 0$. A cone K has a compact base if there is a continuous linear functional Φ on E such that $\Phi(x) > 0$, if $x \in K$, $x \neq 0$ and such that the set $B = \{x \in K : \Phi(x) = 1\}$ is compact. Any such set B is called the base of the cone K.

Definition 2.1. A group G is called Tychonoff if for any action of G by continuous affine transformations on a convex cone K with compact base in a locally convex topological vector space there is a G-invariant ray.

Let TG be the class of Tychonoff groups. We will see later that $TG \subset AG$. Throughout this paper K denotes a cone with compact base B determined by a functional Φ.

Examples

2.2. *Any finite group is Tychonoff.*

If a finite group G acts on a cone K then for any $x \in K$, $x \neq 0$ the nonzero point

$$\xi = \frac{1}{|G|} \sum_{g \in G} gx$$

is G-invariant and so the ray L_ξ is G-invariant as well.

2.3. *The infinite cyclic group $\mathbb{Z}$ is Tychonoff.*

If $A : K \longrightarrow K$ is the affine transformation determined by the generator element of a cyclic group, then the transformation

$$\tilde{A} : B \longrightarrow B$$
$$\tilde{A}(x) = \frac{A(x)}{\Phi(A(x))}$$

is continuous and by Tychonoff's theorem it has a fixed point $\xi \in B$. The ray L_ξ is $\mathbb{Z}$-invariant.

Later we will see that nilpotent and in particular commutative groups are Tychonoff. The following example shows that a virtually commutative group is not necessarily Tychonoff.

2.4. *The infinite dihedral group is not Tychonoff.*

This group is given by one of the following presentations (by means of generators and relations):

$$G = < a, b | a^2 = b^2 = 1 >$$
$$= < a, c | a^2 = (ac)^2 = 1 > \qquad (1)$$

where $b = ac$. Hence G is isomorphic to the group generated by matrices

$$A = \begin{pmatrix} 0 & 1 \\ 1 & 0 \end{pmatrix}, \qquad C = \begin{pmatrix} \frac{1}{2} & 0 \\ 0 & 2 \end{pmatrix}$$

which acts by linear transformations on 2-dimensional vector space. The first quarter

$$K = \{(x, y) \in \mathbb{R}^2 : x, y \geq 0\}$$

is G-invariant and has compact base. It is easy to see that there are no invariant rays for this action.

Proposition 2.5. *A factor group of a Tychonoff group is Tychonoff.*

This is obvious.

Proposition 2.6. *Let G be a directed (by inclusion) union of Tychonoff groups H_i, $i \in I$ (that is $G = \cup_i H_i$ and for any H_i and H_j there is $H_k \supset (H_i \cup H_j)$). Then $G \in TG$.*

Proof. Let G act by affine transformations on a cone K with compact base B and let B_i be the compact nonempty set of traces of G_i-invariant rays on the base B so that $B_i = \{x \in B : gx = \lambda_g x, \quad g \in G_i, \quad \lambda_g > 0\}$. The system $\{B_i\}_{i \in I}$ satisfies the finite intersection property and so the intersection $B_\infty = \cap_{i \in I} B_i$ is nonempty. Any $x \in B_\infty$ determines a G-invariant ray L_x.

The next statement was remarked in [5].

Proposition 2.7. *The strict inclusion $TG \subset AG$ holds.*

Proof. Let us prove that TG is a subset of AG. Let $l_\infty(G)$ be the space of bounded functions on G with uniform norm, $l_\infty^*(G)$ be the space of continuous functionals equipped by the weak-$*$ topology and let $B \subset l_\infty^*(G)$ be the set of means on G that is the set of linear positive functionals $m \in l_\infty^*(G)$ such that $m(\mathbf{1_G}) = 1$, where $\mathbf{1_G}$ is constant on G with value 1.
Now let K be the cone generated by B:

$$K = \{0\} \cup \{x \in l_\infty^*(G) : \lambda x \in B \quad \text{for some} \quad \lambda > 0\}.$$

Then B is the base of cone K, determined by the functional Φ given by $\Phi(m) = m(\mathbf{1_G}), m \in l_\infty^*(G)$.
By the Alaoglu theorem this base is compact in the weak-$*$ topology. The group G acts on $l_\infty(G)$ by left shifts: $(L_g f)(x) = f(g^{-1}x)$ and this action in a canonical way induces the action on the conjugated space: $(gm)(f) = m(L_g f), \quad m \in l_\infty^*(G)$.
The cone K is G-invariant, so there is a G-invariant ray L_x, $x \in K$, $x \neq 0$. But the base B is G-invariant as well. Thus $m = B \cap L_x$ is an invariant point for the action of G and m is a left invariant mean on G. So $G \in AG$. The inclusion $TG \subset AG$ is strict, because the infinite dihedral group is amenable but not Tychonoff.

An extension of one Tychonoff group by another need not be a Tychonoff group, as the example of the infinite dihedral group shows.
A subgroup of a Tychonoff group also need not be Tychonoff. The corresponding example will be constructed later.
However, a subgroup of finite index in a Tychonoff group is Tychonoff [3], [6]. Now we are going to consider some types of extensions preserving the Tychonoff property.

Proposition 2.8. *Let $G = M \times N$, where $M, N \in TG$. Then $G \in TG$.*

Proof. Let G act on K with base B and let $B_0 \subset B$ be the nonempty subset determined by the traces of M-invariant rays on B. Let $x_0 \in B_0$ and $mx_0 = \lambda(m)x_0$, $m \in M$, where $\lambda : M \to \mathbb{R}_+$ is some homomorphism. We define K_0 as a (nonzero convex closed) subcone of K consisting of vectors x with the property $mx = \lambda(m)x$. The cone K_0 is N-invariant. Really, if $x \in K_0$ then $mnx = nmx = \lambda(m)nx$. Because $N \in TG$ there is N-invariant ray L_ξ which is G-invariant as well.

Corollary 2.9. *Every commutative group is Tychonoff.*

Let us agree that in this paper by a character of a group G we mean any homomorphism $G \longrightarrow \mathbb{R}_+$ where $\mathbb{R}_+$ is the multiplicative group of positive real numbers.

Proposition 2.10. *Let $G = \mathbb{Z} \ltimes_A \mathbb{Z}^d$ be a semidirect product of the infinite cyclic group $\mathbb{Z}$ and a free abelian group of rank $d \geq 2$, where a generator of $\mathbb{Z}$ acts on $\mathbb{Z}^d$ as the automorphism determined by a matrix $A \in GL_n(\mathbb{Z})$ with the following condition: A has no eigenvalues on the unit circle other than 1. Then $G \in TG$.*

Proof. Let G act on K with compact base B determined by a functional Φ. Because $\mathbb{Z}^d \in TG$ there is a vector $\xi \in B$ such that for some character $\varphi : \mathbb{Z}^d \longrightarrow \mathbb{R}_+$ and any $g \in \mathbb{Z}^d$ the equality $g\xi = \varphi(g)\xi$ holds.

Let $K_\varphi = \{x \in K : gx = \varphi(g)x, \quad g \in \mathbb{Z}^d\}$. Then K_φ is convex subcone of K determined by some nonempty compact base $B_\varphi \subset B$.

If a is a generator of the infinite cyclic group $\mathbb{Z}$ then $aK_\varphi = K_{\varphi^a}$, where the action of a on φ is determined by the relation $\varphi^a(g) = \varphi(g^{-1}ag)$.

For any $b \in \mathbb{Z}^d$ and $x \in K_\varphi$ we have $bgx = gg^{-1}bgx = \varphi(g^{-1}bg)gx$ and so we have uniform on $n \in \mathbb{Z}$ upper bound:

$$\varphi(a^{-n}ba^n) = \frac{\Phi(ba^n x)}{\Phi(a^n x)} \leq \sup_{y \in B} \frac{\Phi(by)}{\Phi(y)} < \infty.$$

We see that for any $b \in \mathbb{Z}^d$ the sequence $\varphi^{a^n}(b)$ when n ranges over $\mathbb{Z}$ is bounded. Let us prove that φ is invariant under A.

Let $a_1, \cdots, a_d$ be the basis of the group $\mathbb{Z}^d$. Any character on $\mathbb{Z}^d$ is determined by the vector $\overline{\chi} = (\chi_1, \cdots, \chi_d)$ of positive numbers: if $g = a_1^{m_1}, \ldots, a_d^{m_d}$ then $\chi(g) = \chi_1^{m_1}, \cdots, \chi_d^{m_d}$.

Denote by μ the additive character $\mu = \log \chi$, $\mu(g) = m_1 \log \chi_1 + \cdots + m_d \log \chi_d$. The action of an automorphism a on $\mathbb{Z}^d$ corresponds to the mapping $\overline{m} \longrightarrow \overline{m}A$ of integer vectors $\overline{m} = (m_1, \cdots, m_d)$ which determines elements of the group $\mathbb{Z}^d$. Thus $\mu^{a^n}(g) = m_1^{(n)} \log \chi_1 + \cdots + m_d^{(n)} \log \chi_d$, where $(m_1^{(n)}, \cdots, m_d^{(n)}) = (m_1, \cdots, m_d)A^n$ and

$$\mu^{a^n}(g) = \langle \overline{\log \chi}, \overline{m}A^n \rangle = \langle \overline{\log \chi}(A')^n, \overline{m} \rangle$$

where $\langle , \rangle$ is scalar product and A' is the matrix transpose to A.

Lemma 2.11. *Let A be a linear transformation of $\mathbb{R}^d$ which has no eigenvalues on the unit circle not equal to 1. If for some x the set of vectors $\{A^n x\}_{n=-\infty}^{+\infty}$ is bounded then $Ax = x$.*

The proof is identical to the proof of lemma 4.1 from [5] and is omitted.

If the sequence of vectors $\{\overline{\log \varphi}(A')^n\}_{n=-\infty}^{+\infty}$ (φ is the character defined above) is bounded then due to Lemma 2.11 and the condition of Proposition 2.10 the vector $\overline{\log \varphi}$ is invariant with respect to A'. Thus φ is a-invariant and so the cone K_φ is a-invariant. An arbitrary a-invariant ray in K_φ will be G-invariant as well.

The following three statements are similar to the one given above.

Proposition 2.12. *Let $G = N \ltimes H$ be a semidirect product, where N and H are Tychonoff and let N act trivially on the set of characters of the group H. Then $G \in TG$.*

Proof. The cone $K_\varphi = \{x \in k : gx = \varphi(g)x, \quad g \in H\}$, where φ is a character for which there is a vector $\xi \in K$, $\xi \neq 0$, with $g\xi = \varphi(g)\xi$ for any $g \in H$ is N-invariant and therefore any N-invariant ray in K_φ will be G-invariant.

There is a bijective correspondence (given by the function log) between multiplicative characters $G \longrightarrow \mathbb{R}_+$ and additive characters $G \longrightarrow \mathbb{R}$. We say that the set of multiplicative characters $G \longrightarrow \mathbb{R}_+$ is finite dimensional if the space Char (G) of additive characters $G \longrightarrow \mathbb{R}$ is finite dimensional.

Proposition 2.13. *Let $G = \mathbb{Z} \ltimes_A H$, where $H \in TG$. Suppose that the space Char (H) is finite dimensional and the matrix A determining the action of generator of $\mathbb{Z}$ on the space Char (H) has no eigenvalues on the unit circle other than 1. Then $G \in TG$.*

The arguments are similar to those given in the proof of Proposition 2.12.

Examples

2.14. *The Metabelian group $G = \langle a, b | a^{-1}ba = b^2 \rangle$ is Tychonoff.*

Proof. The group G is isomorphic to a semi-direct product $\mathbb{Z} \ltimes H$, where $\mathbb{Z}$ is the infinite cyclic group generated by the element a and H is the group of rational numbers of the form $\frac{k}{2^n}$, $k, n \in \mathbb{Z}$ with the operation of addition. The element a acts on H as multiplication by 2.

Any character φ on H is determined by the value $\varphi(1)$ so the space of characters is 1-dimensional. We have $\varphi^{a^n}(g) = \varphi(2^n g) = [\varphi(g)]^{2^n}$ and thus the orbit $\{\varphi^{a^n}(g)\}_{n=-\infty}^{+\infty}$ is bounded if and only if $\varphi(g) = 1$ and if this holds for any g then φ is trivial. Thus $G \in TG$.

2.15. *Let $\mathbb{Z}_k = \mathbb{Z}/k\mathbb{Z}$, $G = \mathbb{Z}^d wr \mathbb{Z}_k$ (wr means the wreath product). The group G can be defined as $G = \mathbb{Z}^d \ltimes (\mathbb{Z}_k)^{\mathbb{Z}^d}$, where the group $\mathbb{Z}^d$ acts on the space $(\mathbb{Z}_k)^{\mathbb{Z}^d}$ of $\mathbb{Z}_k$-configurations on $\mathbb{Z}^d$ by shifts.*

The group $\mathbb{Z}_k^{\mathbb{Z}^d}$ has only the trivial character. Thus $G \in TG$.

2.16. *The group $\mathbb{Z} wr \mathbb{Z}$ is not Tychonoff, because it can be mapped onto the infinite dihedral group.*

2.17. *The Tychonoff property may not be preserved when passing to a subgroup.*

Proof. Let $H = < a, b | [a, b] = c, \quad [a, c] = [b, c] = 1 >$ be the nilpotent Heisenberg group and let the automorphism $\varphi \in \operatorname{Aut} H$ be defined by

$$\varphi \begin{cases} a \longrightarrow ab \\ b \longrightarrow a. \end{cases}$$

Then φ induces an automorphism of the group $\mathbb{Z}^2 = H/[H, H]$ determined by the matrix

$$A = \begin{pmatrix} 1 & 1 \\ 1 & 0 \end{pmatrix},$$

the eigenvalues of which $\lambda_{1,2} = (1 \pm \sqrt{5})/2$ do not belong to the unit circle. The automorphism φ acts on the generator c of the center $Z(H)$, as $c = [a, b] \xrightarrow{\varphi} [b, a] = c^{-1}$. Let

$G = \mathbb{Z} \ltimes_\varphi H$

$\quad = < a, b, c, d \, | [a, b] = c, [a, c] = [b, c] = 1, \quad d^{-1}ad = ab, \quad d^{-1}bd = a >$

By Proposition 2.13 the group G is Tychonoff. At the same time G contains the subgroup $< c, d \, | d^{-1}cd = c^{-1} >$ which can be mapped onto the infinite dihedral group $< x, y \, | x^2 = y^2 = 1 >$ by the map $c \mapsto xy$, $d \mapsto x$, and so is not Tychonoff.

3. Harmonic functions and the Tychonoff property

Let G be a countable group and $p(g)$ be a probability distribution on $G : p(g) \geq 0, \quad \sum_{g \in G} p(g) = 1$.
A function $f : G \longrightarrow \mathbb{R}$ is called μ-harmonic (μ is a real number) if $Pf = \mu f$, where P is the Markovian operator determined by the relation:

$$(Pf)(g) = \sum_{h \in G} p(h) f(gh).$$

The left shift of a μ-harmonic function is again a μ-harmonic function. If $\mu = 1$ then we get the standard notion of a harmonic function.
A distribution $p(g)$ is called generating if its support $\operatorname{supp} p(g) = \{g \in G : p(g) \neq 0\}$ generates G.

Proposition 3.1. *If given G there is a generating probability distribution with finite support such that for any $\mu > 0$ every positive μ-harmonic function is constant on cosets of the commutator subgroup then G is Tychonoff.*

Proof. It follows from the assumption that every bounded harmonic function on G is a constant and this implies the amenability of G by the theorem of Azencoff and Guivarc'h [2].

Let $p(g)$ be a distribution on G for which every μ-harmonic function is constant on the cosets of the commutator subgroup when $\mu > 0$ and let G act by affine transformations on a cone K with compact base B determined by the functional Φ.

We can define affine continuous mapping $T : K \longrightarrow K$,

$$Tx = \sum_{g \in G} p(g)gx$$

for which there is an invariant ray L_ξ, $\xi \in B$. Assume that $T\xi = \mu\xi$. The function $f(g) = \Phi(g\xi)$ is positive and μ-harmonic:

$$(Pf)(g) = \sum_{h \in G} p(h)f(gh) = \sum_{h \in G} \Phi(gh\xi)p(h) =$$

$$= \Phi(g \sum_{h \in G} p(h)h\xi) = \Phi(gT\xi) = \mu\Phi(g\xi) = \mu f(g).$$

By our assumption, this function is constant on cosets of the commutator subgroup. In particular, $f(g) = 1$ if $g \in G' = [G, G]$. Thus the G'-orbit of the point ξ belongs to the base B.

We can consider the action of G' on the convex closure of the orbit $\{g\xi\}_{g \in G'}$ and use the amenability of G' to claim the existence of a G'-invariant point $\eta \in B$.

Now let $K' \subset K$ be the nonempty convex closed cone of G'-fixed points. The cone K' is G-invariant. Thus the action of G' on K' induces the action of $G_{ab} = G/G'$ on K' by affine transformations. This action has an invariant ray which is G-invariant as well.

Definition 3.2. A group G is a ZA-group if G has an increasing transfinite central chain of normal subgroups

$$1 = G_1 < \cdots < G_\alpha < \cdots < G_\gamma = G \tag{2}$$

where $G_\lambda = \cup_{\alpha < \lambda} G_\alpha$ if λ is a limit ordinal and for any α, $G_{\alpha+1}/G_\alpha < Z(G/G_\alpha)$ (as usual $Z(H)$ denotes the center of a group H).

The following statement is similar to the main result of [9] and our proof follows the one given in [9]. We observe only that lemma 2 of [9] must be slightly modified, either in part of the formulation or in part of the proof, and that the condition on the probability distribution to have a finite support that was included in the formulation of the theorem in [9] can be omitted.

Let us say that G has the Margulis property if for any generating probability distribution $p(g)$ and any $\mu > 0$, every positive μ-harmonic function is constant on cosets of the commutator subgroup.

Theorem 3.3. *Any countable ZA-group has the Margulis property.*

Corollary 3.4. *Any nilpotent group is Tychonoff and so any locally nilpotent group is Tychonoff as well.*

Remark 3.5. By a theorem of A. Malcev [5] any finitely generated ZA-group is nilpotent. Thus ZA is a proper sub-class of the class of locally nilpotent groups. It is not known if every locally nilpotent group has the Margulis property.

There is a number of direct proofs that nilpotent groups are Tychonoff. The first one was given in [3,6]. The arguments used in [6] can be applied for proving Theorem 3.3 but under the assumption that $p(g)$ has finite support. using the Tychonoff property of nilpotent groups. On the other hand the adapted Margulis proof given below uses only the Tychonoff fixed-point theorem, and works in the case of any distribution $p(g)$.

Proof of Theorem 3.3 Let p(g) be the generating distribution on the ZA-group G and for some $\mu > 0$ let the set of μ-harmonic functions be nonempty. We fix this μ and denote by K the convex cone of positive μ-subharmonic functions that is functions with the property $Pf \leq \mu f$. The first step of our considerations is to get the analogue of the Martin representation for μ-harmonic functions.

Let $V(G)$ be the space of real valued functions on G endowed with the topology of pointwise convergence. If $f_n \in K$, $n \in N$ is a net and $f_n \longrightarrow f$ then

$$Pf \leq \lim_n Pf_n \leq \mu \lim_n f_n = \mu f.$$

Thus the cone $K \subset V(G)$ is closed.

Let Φ be the functional on $V(G)$ determined by the relation $\Phi(f) = f(1)$. Then the base $B = \{f \in K : \Phi(f) = 1\}$ is a compact set because from $Pf \leq \mu f$ it is easy to deduce that

$$f(g) \leq \frac{\mu^n}{p(g_1)\cdots p(g_n)} \tag{3}$$

where the elements g_i, $i = 1, \cdots, n$ are selected in such a manner that $g = g_1 \cdots g_n$ and $p(g_i) > 0$, $i = 1, \cdots, n$.

From (3) it follows that all functions from B are majorized by the function from the right-hand side of (3), which gives the compactness of B.

We can introduce the partial ordering on the cone K : $x < y$, $x, y \in K$ if $y - x \in K$. Then the cone K is a lattice: for any $x, y \in K$ there is an infimum $z = \inf(x, y)$ that is the element such that $x - z, y - z \in K$ and if $x - z', y - z' \in K$ for some other $z' \in K$ then $z - z' \in K$.

In our case z is determined by the relation $z(g) = \min\{x(g), y(g)\}$. The following statement follows from theorem of Choquet and Deny and is a part of a more general statement from [5] (theorem 6.2).

Proposition 3.6. *The set E of extremal points of B is a Borel set and any point $b \in B$ is a resultant of some unique probabilistic measure $d\nu$ defined on E; i.e. b can be presented in the form*

$$b = \int_E x d\nu(x).$$

If f is a μ-harmonic function then the corresponding measure $d\nu$ is concentrated on μ-harmonic functions. Indeed, let f be μ-harmonic and

$$f = \int_E x d\nu_f(x).$$

Then

$$Pf \le \int_E P x d\nu_f(x)$$

and

$$0 = Pf - \mu f \le \int_E (Px - \mu x) d\nu_f(x).$$

But $Px - \mu x \le 0$, so for any $g \in G$ the set

$$F_g = \{x \in E : (Px - \mu x)(g) = 0\}$$

has ν_f-measure 1 and thus

$$\nu_f\left(\bigcap_{g \in G} F_g\right) = 1$$

because G is countable.

Now we are going to characterize extremal μ-harmonic functions as characters of the group G. Any such function will be denoted by $k(x)$.

Lemma 3.7. *If $z \in Z(G)$ then*

$$k(xz) = k(x)k(z). \tag{4}$$

Proof. Since the left shift of μ-harmonic function is again μ-harmonic and z is an element of the center we get the relation

$$k(x) = p\frac{k(xz)}{k(z)} + q\frac{c(xz)}{c(z)},$$

where the function c is defined as $c(xz) = k(x) - bk(xz)$, the number b is selected to satisfy the inequality

$$0 < b < \frac{k(x)}{k(xz)}$$

and $p = bk(z)$, $q = 1 - p$.

Now we observe that the functions

$$\frac{k(xz)}{k(z)}, \quad \frac{c(xz)}{c(z)}$$

belong to B and as k is an extremal point, k coincides with each of these functions and this leads to (4).

Lemma 3.8. *Let $x, y \in G$ and $z = [x, y] \in Z(G)$. Then $k(z) = 1$.*

Proof. If $[x_1, y], [x_2, y] \in Z(G)$ then $[x_1 x_2, y] = [x_1, y][x_2, y] \in Z(G)$. This shows that if $[x, y] \in Z(G)$ then $[x^n, y] = [x, y]^n \in Z(G)$.

Let K be the closed cone generated by functions $k(y^n x)$, $n \in \mathbb{Z}$, let the base B be defined as $B = \{f \in K : f(1) = 1\}$ and let T_y be the continuous map from K to K, which preserves B and is defined by

$$(T_y f)(x) = \frac{f(yx)}{f(y)}.$$

Every function $h \in K$ satisfies the relation $h(xz) = h(x)k(z)$, when $z \in Z(G)$. By a theorem of Tychonoff there is $h \in K$ such that $Th = h$, that is $h(x)h(y) = h(yx)$ for any $x \in G$.

Besides this there is a constant $b > 0$ such that for any $n \in \mathbb{Z}$, the inequality

$$0 < b < \frac{h(x^n y)}{h(x^n)}$$

holds. Since $x^n y = yx^n[x^n, y]$, $h(x^n y) = h(yx^n)k([x, y]^n) = h(yx^n)k^n([x, y])$, and we get for any $n \in \mathbb{Z}$

$$k^n([x, y]) = \frac{h(x^n y)}{h(yx^n)} = \frac{h(x^n y)}{h(y)h(x^n)} > \frac{b}{h(y)}.$$

This leads to the inequality $k([x, y]) \geq 1$, which holds for any $x, y \in G$. Therefore $k([x, y]) = 1$.

Now let us finish the proof of the theorem. For this purpose we will prove by transfinite induction on α that $k(x)$ is constant on cosets of the subgroup $[G, G_\alpha] < G$.

Let (2) be a central series of a group G. If $x \in G, y \in G_3$ then $[x, y] \in G_2 < Z(G)$ and by Lemma 3.8 $k([x, y]) = 1$. So k is equal to 1 on the subgroup $[G, G_3]$ and by Lemma 3.7 k is constant on cosets of $[G, G_3]$.

Let us pass from (1) to the central series of "smaller" length:

$$1 < G_3/[G, G_3] < \cdots < G_\alpha/[G, G_3] < \cdots < G/[G, G_3]$$

After such factorization the distribution $p(g)$ on G will be projected on some distribution $p^{(3)}(g)$ on the group $G^{(3)} = G/[G, G_3]$ and the function k will be projected on positive μ-harmonic with respect to the distribution $p^{(3)}(g)$ function $k^{(3)}$ on the group $G^{(3)}$. Moreover, $k^{(3)}$ will be an extremal point in the base of the corresponding cone of μ-harmonic function on $G^{(3)}$.

Let us suppose now, that for some ordinal λ every positive μ-harmonic function on G is constant on cosets of any subgroup $[G, G_\alpha]$ $\alpha < \lambda$. In case λ is a limit ordinal this property can be extended on cosets of the subgroup $[G, G_\lambda]$ as well.

If λ is not a limit ordinal and $\lambda = \mu + 1$ then let us consider the central series

$$1 < G_\mu/[G, G_\mu] < G_\lambda/[G, G_\mu] < \cdots < G/[G, G_\mu]. \tag{5}$$

Let $p^{(\mu)}(g)$ and $k^{(\mu)}$ be a distribution and μ-harmonic function on $G^{(\mu)} = G/[G, G_\mu]$ that are projections of μ and k respectively.
If $x \in G^{(\mu)}$, $\quad y \in G_\lambda/[G, G_\mu]$ then

$$[x, y] \in G_\mu/[G, G_\mu] < Z(G/[G, G_\mu])$$

and so the function $k^{(\mu)}$ is constant on cosets of the subgroup

$$[G/[G, G_\mu], \, G_\lambda/[G, G_\mu]] < G/[G, G_\mu].$$

Thus we can pass from (5) to the series $1 < G_\lambda/[G, G_\lambda] < \cdots < G/[G, G_\lambda]$ and define on the group $G^{(\lambda)} = G/[G, G_\lambda]$ projections $p^{(\lambda)}(g)$ and $k^{(\lambda)}$ of $p(g)$, k respectively.
This gives the possibility to apply the inductive assumption and to prove the theorem.

Remark 3.9. The space of μ-harmonic functions on a group G with the Margulis property can be identified with the space of μ-harmonic functions on the abelianization G_{ab} and extremal μ-harmonic functions are characters in this case. In particular, every bounded harmonic function is constant.

Remark 3.10 There are examples showing that a group containing a nilpotent subgroup of finite index can have extremal μ-harmonic functions that are not characters. Here is the simplest one.

Let G be the infinite dihedral group, given by the presentation (1) and the distribution p be uniform on the set $\{a, b\}$ of generators: $p(a) = p(b) = \frac{1}{2}$. The elements of G can be identified with words over the alphabet $\{a, b\}$ which do not contain the subwords aa, bb.
The Cayley graph of G looks like the Cayley graph of the infinite cyclic group

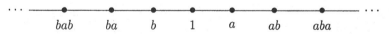

$$\cdots \quad bab \quad ba \quad b \quad 1 \quad a \quad ab \quad aba \quad \cdots$$

and the Markovian operator T acts on functions on G analogously to the operator $\tilde{T}$ on the group $\mathbb{Z}$, where $(\tilde{T}f)(n) = 1/2(f(n-1) + f(n+1))$.
Extremal rays of the cone of positive solutions of the equation $\tilde{T}f = \mu f$, $\mu \geq 1$, have the form $f_\xi(n) = \xi^n$ where ξ is some positive number satisfying the equation

$$\mu \xi^n = \frac{1}{2}(\xi^{n-1} + \xi^{n+1}),$$

i.e.

$$\xi = \frac{\mu \pm \sqrt{\mu^2 - 1}}{2}.$$

Respectively the function $f_\xi(g) = \xi^{\sigma(g)}$, where $\sigma(g)$ is the length of an element g taken with the sign $+$ if irreducible form of g starts on a and taken with the sign $-$ otherwise, is an extremal μ-harmonic function on G but is not a character.

Remark 3.11 There are groups with the Tychonoff property having nonconstant bounded harmonic functions. For instance, any group $G = \mathbb{Z}^d wr \mathbb{Z}_k$ is Tychonoff (see Example 2.15) and has nonconstant bounded harmonic functions when $d \geq 3$ [7]. Thus the class TG does not coincide with the class of groups for which every bounded harmonic function is constant.

4. The Tychonoff property and indicability

A group is called indicable if it can be mapped onto the infinite cyclic group.

Theorem 4.1. *Any infinite, finitely generated Tychonoff group is indicable.*

Proof. Let G be such a group. The theorem will be proven if we construct an action of G without fixed points by affine transformations on convex cone K with compact base B. Indeed, then the action of G on any fixed ray L_ξ, $\xi \in B$ given by $g\xi = \varphi(g)\xi$ determines the desired homomorphism $\varphi : G \longrightarrow \mathbb{R}_+$ with infinite cyclic image.

We need to show that the cone K of positive μ-harmonic functions satisfies the required property where μ is any fixed number greater than 1 and the distribution $p(g)$ has a finite support which generates G. This follows, for instance, from general results on Markov chains [10]. For the convenience of the reader we include the proof of this fact in this special case.

The group G acts on K by left shifts. This action is affine and has no nonzero fixed points. The same arguments which were given in the proof of Theorem 3.3 show that the base $B = \{f \in K : f(1) = 1\}$ is compact in the topology of pointwise convergence. Thus the only point remaining is to prove that the cone K is nonzero.

Let $p(g)$ be a distribution on G with finite support A which generates G. Let $p(n, x, y)$ be the probability of transmission from x to y in n steps in a right random walk on G, determined by distribution $p(g)$: starting from x we can reach xa in one step with probability $p(a)$. The Markovian operator P corresponding to this random walk is determined by the relations:

$$(Pf)(x) = \sum_{y \in G} p(y)f(xy) = \sum_{y \in G} p(1, x, y)f(y).$$

For any λ, $|\lambda| < 1$ the series

$$g^\lambda(x, y) = \sum_{n=0}^{\infty} \lambda^n p(n, x, y)$$

converges and we can define the generalized Martin's kernels

$$k_y^\lambda(x) = \frac{g^\lambda(x,y)}{g^\lambda(1,y)}$$

and functions

$$\Pi_y^\lambda(x) = \sum_{i=0}^\infty \lambda^i f(i,x,y),$$

where $f(i,x,y)$ is the probability of first getting from x into y on the ith step. It is clear that $p(n,x,y) = \sum_{i=0}^n f(i,x,y)p(n-i,y,y)$ and so $g^\lambda(x,y) = \Pi_y^\lambda(x)g^\lambda(y,y)$ because

$$\sum_{n=0}^\infty \lambda^n p(n,x,y) = \sum_{n=0}^\infty \lambda^n \sum_{i=0}^n f(i,x,y)p(n-i,y,y)$$

$$= \sum_{n=0}^\infty \sum_{i=0}^n \lambda^i f(i,x,y)\lambda^{n-i}p(n-i,y,y)$$

$$= \sum_{i=0}^\infty \lambda^i f(i,x,y) \sum_{n=i}^\infty \lambda^{n-i}p(n-i,y,y) = \Pi_y^\lambda(x)g^\lambda(y,y).$$

Thus

$$k_y^\lambda(x) = \frac{\Pi_y^\lambda(x)}{\Pi_y^\lambda(1)}.$$

Lemma 4.2. *The following equality*

$$k_y^\lambda(x) - \lambda P k_y^\lambda(x) = \begin{cases} 0 & \text{if } x \neq y \\ \frac{1}{g^\lambda(1,y)} & \text{if } x = y \end{cases}$$

holds.

Proof. First of all observe that

$$Pk_y^\lambda(x) = \frac{1}{g^\lambda(1,y)} \sum_{h \in G} p(1,x,h)g^\lambda(h,g)$$

$$= \frac{1}{g^\lambda(1,y)} \sum_{h \in G} p(1,x,h) \sum_{n=0}^\infty \lambda^n p(n,h,y)$$

$$= \frac{1}{g^\lambda(1,y)} \sum_{n=0}^\infty \lambda^n \sum_{h \in g} p(1,x,h)p(n,h,y)$$

$$= \frac{1}{g^\lambda(1,y)} \sum_{n=0}^\infty \lambda^n p(n+1,x,y)$$

$$= \frac{1}{g^\lambda(1,y)} \left[\frac{1}{\lambda}g^\lambda(x,y) - \frac{1}{\lambda}p(0,x,y) \right].$$

We claim that when x and λ are fixed the set of numbers $\{k_y^\lambda(x), \quad y \in G\}$ is bounded.

Indeed let $A = \text{supp}(g)$ be the set of generators of G, and C_G be the Cayley graph of the group G constructed with respect to the generating set A.

For every element $x \in G$, fix a path l_x in C_G that joins 1 and x. Let $p(l_x)$ be the probability of the path l_x (the product of probabilities of transmission along links of this path) and let t_x be the length of l_x.

If $y \notin l_x$, then $f(i, 1, y) > p(l_x) \cdot f(i - t_x, x, y)$. Therefore

$$\frac{\lambda^{i-t_x} f(i - t_x, x, y)}{\lambda^i f(i, 1, y)} \leq \frac{1}{\lambda^{t_x} p(l_x)}$$

and

$$k_y^\lambda(x) = \frac{\Pi_y^\lambda(x)}{\Pi_y^\lambda(1)} = \frac{\sum_{i=0}^\infty \lambda^i f(i, x, y)}{\sum_{i=0}^\infty \lambda^i f(i, 1, y)}$$

$$\leq \frac{\sum_{i=0}^\infty \lambda^i f(i, x, y)}{\sum_{i=0}^\infty \lambda^{i+t_x} f(i + t_x, 1, y)} \leq \frac{1}{\lambda^{t_x} p(l_x)}$$

Thus the set of functions $\{k_y^\lambda(x)\}_{y \in G}$ is majorized by the function $(\lambda^{t_x} p(l_x))^{-1}$. Now we take any sequence $y_n \in G$, $y_n \to \infty$ and extract a subsequence y_{n_k} such that the sequence $k_{y_{n_k}}^\lambda$ converges to some positive function $k^\lambda(x)$ which is $\frac{1}{\lambda}$-harmonic. Since the distribution $p(g)$ has finite support, $Pk^\lambda(x) = P \lim_{k\to\infty} k_{y_{n_k}}^\lambda(x) = \lim_{k\to\infty} Pk_{y_{n_k}}(x)$, and passing to the limit in the relation

$$\frac{1}{\lambda} k_{y_{n_k}}^\lambda(x) - Pk_{y_{n_k}}^\lambda(x) = \begin{cases} 0, & \text{if } x \neq y_{n_k} \\ \frac{1}{\lambda g^\lambda(1, y_{n_k})}, & \text{if } x = y_{n_k} \end{cases}$$

we get the relation

$$\frac{1}{\lambda} k^\lambda(x) = Pk^\lambda(x).$$

We have proven that on any infinite finitely generated group for any $\mu > 1$ there is a positive μ-harmonic function. The cone of such functions is nonzero and satisfies all necessary conditions. The theorem is proven.

5. A fixed-point theorem for actions on cones

Let a group G act by affine transformations on a cone K. We shall call such an action bounded if the orbit of any point $\xi \in K$ is bounded. Thus the orbits can accumulate to zero, but not to infinity.

A finitely generated group G is called a group of subexponential growth if

$$\gamma = \lim_{n\to\infty} \sqrt[n]{\gamma(n)} = 1,$$

where $\gamma(n)$ is the growth function of the group G with respect to some finite system of generators ($\gamma(n)$ is equal to the number of elements of G that can be presented as a product of $\leq n$ of generators and its inverses).

Theorem 5.1. *Let G be a group of subexponential growth acting by affine transformations on a nonzero convex cone K with compact base in a locally convex topological vector space, and suppose that this action is bounded. Then there is a G-fixed point $\xi \in K$, $\xi \neq 0$.*

Proof. Let $p(g)$ be a symmetric (that is $p(g) = p(g^{-1})$ for any $g \in G$) probability distribution the support of which is finite and generates G, and let P be the corresponding Markovian operator

$$(Pf)(g) = \sum_{h \in G} p(h) f(gh).$$

We can define a continuous map $T : K \to K$ by

$$Tx = \sum_{h \in G} p(h) hx.$$

By the Tychonoff theorem there is a T-invariant ray L_ξ, $\xi \in B$, that is $T\xi = \lambda \xi$ for some $\lambda > 0$.

We are going to prove that $\lambda = 1$. Let Φ be a functional determining the base B of the cone K and let f be a function on G determined by the relation $f(g) = \Phi(g\xi)$. Then f is λ-harmonic as shown in the proof of Proposition 3.1. From the relation $(P^n f)(1) = \lambda^n f(1)$ we get the inequality $p(n, 1, 1) \leq \lambda^n$ where $p(n, 1, 1)$ is the probability of returning to the unit after n-steps in the right random walk on a group G.

It is well-known that any group of subexponential growth is amenable and by a theorem of H. Kesten the spectral radius

$$r = \lim_{n \to \infty} \sup \sqrt[n]{p(n, 1, 1)}$$

is equal to 1 for a symmetric random walk on any amenable group. Thus $\lambda \geq 1$ in our case. But because the action of G on K is bounded, one can find a number $d > 0$ such that $\Phi(f) \cdot \lambda^n = \Phi(P^n f) \leq d$ when $n \geq 1$, which leads to the equality $\lambda = 1$. Thus the function $f(g)$ is a bounded harmonic function on the group G.

By the theorem of Avez [1] the function f is constant and so the orbit O_ξ of the point ξ is a subset of the base B. The closed convex hull $\overline{O}_\xi$ of the orbit O_ξ will be a compact set on which G acts by affine transformations, and this action has a fixed point because G is amenable. This finishes the proof of the theorem.

Remark 5.2. The statement of Theorem 5.1 holds for any group for which there is a symmetric probability distribution with finite support having the property that every positive bounded harmonic function is constant. There are some solvable groups of exponential growth having this property. The simplest example would be the group of the form $\mathbb{Z} \ltimes_A \mathbb{Z}^d$ where not all eigenvalues of the matrix A lie on the unit circle.

6. Final remarks and open questions

The notion of Tychonoff group can be naturally generalized as follows.

Definition 6.1. A group G is called k-Tychonoff if for any action of G by continuous affine transformations on a convex cone with compact base in a locally convex topological vector space there is a convex G-invariant cone with $\leq k$ extremal rays. Hence 1-Tychonoff is the same as Tychonoff. A group has the generalized Tychonoff property if it has the k-Tychonoff property for some $k \geq 1$.

A group containing a Tychonoff subgroup of finite index has the generalized Tychonoff property.

Some of the statements given above for Tychonoff groups can be proven for generalized Tychonoff groups as well.

There are many open questions about Tychonoff groups. Here are two of them.

Problem 6.2. Is it true that every Tychonoff group belongs to the class EG of elementary amenable groups (that is groups that can be obtained from finite and commutative groups by operations of extension and direct limit)?

Remark 6.3. Problem 7 in the Problem Session of [4] is stated incorrectly. The correct question is given in Problem 6.2.

Problem 6.4. To describe solvable Tychonoff groups.

Remark 6.5. The description of polycyclic Tychonoff groups was given by A. Starkov [11].

Acknowledgements

This paper was written during my stay at the Max Planck Mathematical Institute in Bonn and the Institute of Mathematics of the Hebrew University in Jerusalem. I would like to thank F. Hirzebruch and A. Lubotzky for giving me the opportunity for these fruitful visits.

I also thank B. Weiss, L. Saloff-Coste, Y. Shalom and A. Starkov for their helpful remarks.

References

[1]   A. Avez, *Theorem de Choquet-Deny pour les groupes a croissance non exponentielle*, C.R. Acad. Sci. Paris Ser. A **279** (1974), pp. 25–28.

[2] A. Avez, *Croissance des groupes de type fini et fonctions harmoniques,* in: Theory Ergodique, Rennes 1973/1974 Lecture Notes in Mathematics, **532** Springer, Berlin (1976), pp. 35–49.

[3] J.P. Conze and Y. Guivarc'h, *Proprieté de droite fixe et fonctions propres des operateurs de convolution,* Collection: Séminaire de Probabilités, I, Universite de Rennes.

[4] Anon., *Problem Session,* in: Combinatorial and Geometric Group Theory, Edinburgh 1993 (A.J. Duncan, N.D. Gilbert and J. Howie, eds.), London Mathematical Society Lecture Notes Series, **284** Cambridge University Press (1994).

[5] H. Furstenberg, *Invariant cones of functions on semi-simple Lie groups,* Translation in Bull. Amer. Math. Soc. **71** (1965), pp. 271–326.

[6] Y. Guivarc'h and J.P. Conze, *Propriete de droit fixe et functions harmoniques positive,* Lecture Notes in Mathematics, **404** Springer, Berlin (1974), pp. 126–132.

[7] V.A. Kaimanovich and A.M. Vershik, *Random walks on discrete groups: boundary and entropy,* Ann. Probab. **11** (1983), pp. 457–490.

[8] A.I. Malcev, *Nilpotent torsion-free groups,* Izvestiya Akad. Nauk SSSR, Ser. Mat. **13** (1949), pp. 201–212.

[9] G. Margulis, *Positive harmonic functions on nilpotent groups,* Dokl. AN SSSR **166 no. 5** (1966), pp. 1054–1057.

[10] W.E. Pruitt, *Eigenvalues of nonnegative matrices,* Ann. Math. Statistics **35** (1964), pp. 1797–1800.

[11] A.N. Starkov, *Tychonoff property for linear groups,* to appear in Mathematical Notes.

Word Growth of Coxeter Groups

D. L. Johnson

Mathematics Department, University of Nottingham, University Park, Nottingham.

Let (W, S) be a Coxeter system, $W_X(t)$ the growth series of the subgroup $W_X = \langle X \rangle$ of W for $X \subseteq S$, and $\ell(S)$ the length of the unique longest word in W when the latter is finite. Then the famous formula of Bourbaki ([2], Exercise 26, p.45; see also [4]) asserts that

$$\sum_{X \subseteq S} \frac{(-1)^{|X|}}{W_X(t)} = \frac{t^{\ell(S)}}{W_S(t)}$$

when $W = \langle S \rangle$ is finite, and is zero otherwise. A proof of this formula is to be found in [3].

Since the list of finite Coxeter groups is short and well known (see [1], which also lists the corresponding values of $\ell(S)$), it would be advantageous to have a formula for computing $W_S(t)$ in terms of only the *finite* subgroups $\langle X \rangle$ of W, $X \subseteq S$. Such a formula can be obtained by a simple purely combinatorial device as follows.

We evaluate the double sum (over Y and X)

$$\Sigma(t) - \sum_{X \subseteq Y \subseteq S} \frac{(-1)^{|X|}(-1)^{|Y|}}{W_X(t)}$$

in two different ways.

(i) $\qquad \Sigma(t) = \sum_{Y \subseteq S} (-1)^{|Y|} \sum_{X \subseteq Y} \frac{(-1)^{|X|}}{W_X(t)} = \sum_{Y \subseteq S, |W_Y| < \infty} (-1)^{|Y|} \frac{t^{\ell(Y)}}{W_Y(t)}.$

(ii) $\qquad \Sigma(t) = \sum_{X \subseteq S} \frac{(-1)^{|X|}}{W_X(t)} \sum_{X \subseteq Y \subseteq S} (-1)^{|Y|} = \frac{1}{W_S(t)}$

since the inner sum (over Y, $X \subseteq Y \subseteq S$) is $\pm$ the alternating sum of the $|S - X|$-row of Pascal's triangle, which is 0 unless $X = S$, and $(-1)^{|S|}$ in this case.

Proposition. *Let (W, S) be a Coxeter system and let $\ell(Y)$ denote the length of the unique longest word in W_Y for all $Y \subseteq S$ with $|W_Y| < \infty$. Then*

$$\frac{1}{W_S(t)} = \sum_{Y \subseteq S, |W_Y| < \infty} \frac{(-1)^{|Y|} t^{\ell(Y)}}{W_Y(t)}.$$

References

[1] C. T. Benson and L. C. Grove, *Finite reflection groups,* Bogden and Quigley, Tarrytown (1971).

[2] N. Bourbaki, *Groupes et algèbres de Lie,* Chapters 4, 5 and 6 Hermann, Paris (1968).

[3] L. Paris, *Growth series of Coxeter groups,* in: Group theory from a geometrical viewpoint (E. Ghys, A. Haefliger and A. Verjovsky, eds.), World Scientific (1991), pp. 302–310.

[4] R. Steinberg, *Endomorphisms of linear algebraic groups,* Mem. Amer. Math. Soc., **80** American Mathematical Society, Providence (1968).

Poly-surface Groups

Francis Johnson

Department of Mathematics, University College London, Gower Street, London, WC1E 6BT.

Abstract. Cocompact lattices in linear semisimple Lie groups have a number of desirable properties; in particular, they are (I) residually finite; (II) satisfy the Tits' alternative; (III) are virtually Poincaré Duality groups of type FL; (IV) admit, up to commensurability, a unique decomposition into irreducibles, of which there are infinitely many; (V) act properly discontinuously with compact quotient on a Euclidean space endowed with some non-euclidean geometry; (VI) satisfy a *finite rigidity* property.

In this paper we compare this behaviour with that of a collection of Poincaré Duality groups which are in general not cocompact lattices, namely the class of poly-Surface groups. We show that properties (I)–(IV) are satisfied without change, and that (V) is virtually satisfied. For poly-Surface groups, the finite rigidity property seems rather difficult to analyse because of our ignorance of the various geometries which arise; nevertheless, for each $n \geq 2$, there are infinitely many distinct commensurability classes of poly-Surface groups which act as discrete cocompact isometry groups on bounded domains in $\mathbf{C}^n$.

1. Preliminaries on Surface groups

By a Surface group (or $\mathcal{S}$-group, for short) we shall mean the fundamental group of a closed orientable surface of genus $g \geq 2$; that is, a group having a presentation of the form

$$\Sigma_g = \langle X_1, \ldots, X_{2g} : \prod_{i=1}^{g}[X_{2i-1}, X_{2i}]\rangle$$

Alternatively, the uniformisation theory of complex curves allows us to regard $\mathcal{S}$-groups as torsion free discrete cocompact subgroups in $PSL_2(\mathbf{R})$. A subgroup of finite index in an $\mathcal{S}$-group is also an $\mathcal{S}$-group with genus determined by the following:

Theorem 1.1. *(Riemann-Hurwitz) If G is an S-group of genus g, and H is a subgroup of finite index d in G, then H is an S-group of genus $1 + d(g-1)$.*

We summarise the algebraic properties of S-groups thus:

(1.2) An S-group has no nontrivial finite normal subgroup.

(1.3) The centre of an S-group is trivial.

(1.4) An abelian subgroup of an S-group is cyclic.

(1.5) A nontrivial normal subgroup of infinite index in an S-group is a free group of infinite rank.

(1.6) Let H be a subgroup of an S-group G; then H is itself an S-group if and only it has finite index in G.

Generalising (1.5), there is a sort of Noetherian property:

(1.7) Let $H \in S$ and let $H_0 \subset H_1 \subset \ldots \subset H_n = H$ be a sequence of finitely generated subgroups such that $H_r \triangleleft H_{r+1}$ for each r; then there exists m , $1 \le m \le n$, such that H_r has finite index in H for m $\le$ r, and $H_r = \{1\}$ for $r < m$.

The above statements are all well known, and although it is difficult to give a single reference, they can be easily obtained from a combination of references, for example [5] and [17].

2. Poly-S-groups and their filtrations

Let C be a class of abstract groups; a group G is called a poly-C group when it possesses a subnormal filtration $\mathcal{G} = (G_r)_{0 \le r \le n}$ with $G_0 = \{1\}$, and $G_n = G$, $G_{r-1} \triangleleft G_r$ and $G_r/G_{r-1} \in C$ for each r. Such a filtration is called *strongly poly-C* when, in addition, $G_r \triangleleft G$ for each r, and *characteristically poly-C* when G_{r-1} is a characteristic subgroup of G_r for each r. The following are easy to verify:

(2.1) A characteristic poly-C filtration is strong.

(2.2) Let H be a subgroup of finite index in a strongly poly-S group G; then H is also a strongly poly-S group.

If all C-groups have trivial centre, the same is true of poly-C groups. In particular, the centre of any poly-S group is trivial, so enabling us to describe, in principle at least, the construction of all poly-S groups; if $\mathcal{G} = (G_r)_{0 \le r \le n}$ is a poly-S filtration of length n , the extension $1 \to G_{n-1} \to G_n \to G_n/G_{n-1} \to 1$ is determined, up to congruence, by an operator homomorphism $h_{n-1} : G_n/G_{n-1} \to Out(G_{n-1})$; we can regard G_n as a fibre product $G_n = Aut(G_{n-1}) \underset{\lambda, h_{n-1}}{\times} G_n/G_{n-1}$, where $\lambda : Aut(G_{n-1}) \to Out(G_{n-1})$ is the canonical mapping. Inductively, the study of poly-S groups of length n may be reduced to that of the outer automorphism groups of poly-S groups of length (n-1). By a *stable* poly-S filtration $\mathcal{G} = (G_r)_{0 \le r \le n}$ on a group $G = G_n$ we shall mean one for which $\operatorname{rank}(G_r/G_{r-1}) < \operatorname{rank}(G_{r+1}/G_r)$ for

$1 \leq r \leq n-1$; a poly-$\mathcal{S}$ group is called stable when it admits a stable poly-$\mathcal{S}$ filtration. We shall prove:

Theorem 2.3. *A stable poly-$\mathcal{S}$ filtration is characteristic.*

Proof Using (1.7), the general case reduces to showing that if

$$1 \to K \to G \xrightarrow{p} Q \to 1$$

is an exact sequence of groups in which $Q \in \mathcal{S}$ and K is finitely generated with rank$(K) <$ rank(Q), then K is characteristic in G. For this, if α is an automorphism of G for which $\alpha(K) \neq K$, then $p\alpha(K)$ is a nontrivial finitely generated normal subgroup of Q in which rank$(p\alpha(K)) <$ rank(Q); however, this violates the Riemann-Hurwitz Theorem. Hence $\alpha(K) = K$, and K is characteristic.

Proposition 2.4. *A poly-$\mathcal{S}$ group contains a subgroup of finite index which is stably (and hence characteristically) poly-$\mathcal{S}$.*

Proof By (2.3), it suffices to prove the stability statement. Let n denote the length of a poly-$\mathcal{S}$ filtration. The case n = 1 is trivial, so suppose we have proved the proposition for n-1 and let $\mathcal{G} = (G_r)_{0 \leq r \leq n}$ be a poly-$\mathcal{S}$ filtration of length n. Inductively, choose a subgroup H_{n-1} of finite index in G_{n-1} admitting a poly-$\mathcal{S}$ filtration $(H_r)_{0 \leq r \leq n-1}$ such that rank$(H_r/H_{r-1}) <$ rank(H_{r+1}/H_r) for all $r \in \{1, \ldots, n-2\}$. A finitely generated group has only a finite number of subgroups of a given finite index; thus$\{\alpha(H_{n-1}) : \alpha \in Aut(G_{n-1})\}$ is a finite set. Put $S(H_{n-1}) = \{\alpha \in Aut(G_{n-1}) : \alpha(H_{n-1}) = H_{n-1}\}$. Then $S(H_{n-1})$ is a subgroup of finite index in $Aut(G_{n-1})$. Let $c : G_n \to Aut(G_{n-1})$ denote the conjugation map, and put $\widetilde{H_n} = c^{-1}(S(H_{n-1}))$. Observe that H_{n-1} is normal in $\widetilde{H_n}$, and we have a surjective homomorphism $\widetilde{H_n}/H_{n-1} \to \widetilde{H_n}/(G_{n-1} \cap \widetilde{H_n})$ in which $\widetilde{H_n}/(G_{n-1} \cap \widetilde{H_n})$, being a subgroup of finite index in G_n/G_{n-1}, is itself in $\mathcal{S}$. In particular, $\widetilde{H_n}/(G_{n-1} \cap \widetilde{H_n})$ contains subgroups of arbitrary finite index. It follows easily from the Riemann Hurwitz Theorem that we may choose a subgroup Q of finite index in $\widetilde{H_n}/(G_{n-1} \cap \widetilde{H_n})$ such that rank$(H_{n-1}/H_{n-2}) <$ rank(Q). Let $\phi : \widetilde{H_n} \to \widetilde{H_n}/H_{n-1}$ denote the identification mapping; then $H_n = \phi^{-1}(Q)$ is a subgroup of finite index in G_n, and the poly-$\mathcal{S}$ filtration $(H_r)_{0 \leq r \leq n-1}$ satisfies the condition rank$(H_r/H_{r-1}) <$ rank(H_{r+1}/H_r) for all $r \in \{1, \ldots, n-1\}$. This completes the proof.

3. Properties (I)–(III)

(I) Residual finiteness

A group G is residually finite when, for each nontrivial element $g \in G$, there exists a homomorphism to a finite group $\phi : G \to \Phi$ such that $\phi(g) \neq 1$. It is straightforward to see that if H is a subgroup of finite index in a finitely generated group G then H is residually finite if and only if G is. Mal'cev showed that any finitely generated linear group is residually finite([28]). Here we establish this for poly-$\mathcal{S}$ groups.

Proposition 3.1. *Let* $1 \to \Phi \to G \xrightarrow{p} Q \to 1$ *be an extension with* Φ *finite and* $Q \in \mathcal{S}$; *then* G *contains an* $\mathcal{S}$-*subgroup* G' *of finite index.*

Proof For each positive integer n, Q contains a subgroup Q' of index n; Q' is necessarily in $\mathcal{S}$; in fact, Q' can be taken to be the fundamental group of a n-fold cyclic covering of the surface corresponding to Q.

First consider the case where Φ is finite and central in G; then the extension $\mathcal{E} = (1 \to \Phi \to G \xrightarrow{p} Q \to 1)$ is completely determined by a cohomology class $c(\mathcal{E}) \in H^2(Q; \Phi) \cong \Phi$. Let Q' be a subgroup of Q with index n = exponent(Φ), and let $\mathcal{E}'$ be the extension $\mathcal{E}' = (1 \to \Phi \to G' \xrightarrow{p} Q' \to 1)$ where $G' = p^{-1}(Q')$. It is easy to see that $c(\mathcal{E}') = 0$ so that G' splits as a direct product $G' \cong \Phi \times Q'$. The result follows since G' has finite index in G.In the general case, let $c : G \to Aut(\Phi)$ be the homomorphism induced by conjugation, and put $G' = Ker(c), \Phi' = \Phi \cap G'$ and $Q' = p(G')$. Then the extension $\mathcal{E}'$ is in the case considered above. The result follows since G' has finite index in G.

As an immediate consequence we obtain:

Corollary 3.2. *Let* $1 \to \Phi \to G \to Q \to 1$ *be an extension with* Φ *finite and* $Q \in \mathcal{S}$; *then* G *contains an* $\mathcal{S}$-*subgroup of finite index.*

Let $1 \to K \to G \to Q \to 1$ be an exact sequence of groups in which K is a finitely generated residually finite group, and $Q \in \mathcal{S}$. Let g be a nontrivial element of G; there exists a subgroup K_1 of finite index in K such that $g \notin K_1$. Since K is also finitely generated, K_1 may be assumed to be characteristic in K, so that G/K_1 occurs in an extension $1 \to K/K_1 \to G/K_1 \to Q \to 1$. Observe that $\pi(g) \neq 1$ where $\pi : G \to G/K_1$ is the canonical epimorphism; since $Q \in \mathcal{S}$, we may, by (3.1), choose an $\mathcal{S}$ subgroup H of finite index in G/K_1. H is residually finite, as it admits a faithful finite dimensional real linear representation [28], so we can ensure that $\pi(g) \notin H$; $G' = \pi^{-1}(H)$ is a subgroup of finite index in G such that $g \notin G'$. Since G is finitely generated, we may choose a normal subgroup G'' of finite index in G such that $G'' \subset G' \subset G$. Then $\psi(g) \neq 1$ where ψ is the canonical epimorphism of G onto G/G''. We have established:

Theorem 3.3. *Let* $1 \to K \to G \to Q \to 1$ *be an exact sequence in which* K *is a finitely generated residually finite group and* $Q \in \mathcal{S}$; *then* G *is residually finite.*

Following (3.3), an induction on the length of a filtration shows that:

Theorem 3.4. *A poly-$\mathcal{S}$ group is residually finite.*

(II) The Tits' alternative

A class $\mathcal{C}$ of groups is said to satisfy the "Tits' alternative" when, given a group Γ in $\mathcal{C}$ and a subgroup Δ of Γ, then either Δ is polycyclic or else contains a non-abelian free group. In [26] Tits showed that the class of finitely generated linear groups has this property. We will verify it for the class of poly-$\mathcal{S}$ groups.

Proposition 3.5. *Let* $1 \to K \to G \overset{p}{\to} Q \to 1$ *be an extension where* Q *is a nonabelian subgroup of an $\mathcal{S}$-group. Then* G *contains a nonabelian free group.*

Proof Let $Q' \subset Q$ be a nonabelian free group and put $G' = p^{-1}(Q')$. Then the extension $1 \to K \to G' \overset{p}{\to} Q' \to 1$ splits, so that G', and hence also G, contains a subgroup isomorphic to Q'.

Proposition 3.6. *Let* Γ *be a poly-$\mathcal{S}$ group and let* Δ *be a nontrivial subgroup of* Γ. *If* Δ *is soluble, then* Δ *is poly-{infinite cyclic}.*

Proof If $\Gamma \in \mathcal{S}$ and Δ is a nontrivial soluble subgroup of Γ, then Δ is infinite cyclic. The result follows by induction on the length of a poly-$\mathcal{S}$ filtration on Γ.

It follows easily by induction that:

Corollary 3.7. *The class of poly-$\mathcal{S}$ groups satisfies the Tits' alternative.*

(III) Poincaré Duality

$\mathcal{S}$-groups are Poincaré Duality groups in the sense of [16]. Since the class of Poincaré Duality groups is closed under extension, it follows that a poly-$\mathcal{S}$ group is a Poincaré Duality group. Moreover, by a result of Serre [23], finitely presented groups of type FL are also closed under extension, we see that poly-$\mathcal{S}$ groups are of type FL.

4. Decomposition into irreducibles

Two abstract groups G_1, G_2 are said to be *commensurable*, written $G_1 \sim G_2$, when there exists a group H, and injections $\iota_r : H \to G_r (r = 1, 2)$, such that $\iota_r(H)$ has finite index in G_r. An infinite group G is reducible when it is commensurable to a direct product $G \sim H_1 \times H_2$ where H_1, H_2 are infinite groups; otherwise, G is irreducible. It is straightforward to see that:

Proposition 4.1. *A finitely generated infinite group G is irreducible if and only if it contains no subgroup of finite index which is isomorphic to a direct product of infinite groups.*

As a consequence of the Borel Density Theorem, a lattice in a connected linear semisimple Lie group admits, up to commensurability, an essentially unique decomposition into a product of irreducible semisimple lattices ([22], p.86). Here we establish an analogous decomposition for poly-$\mathcal{S}$ groups into products of irreducible poly-$\mathcal{S}$ groups. If $\mathcal{C}$ is a class of groups, we say that a group G admits a $\mathcal{C}$-product structure when it is the internal direct product, written $G = G_1 \circ \cdots \circ G_m$, of a finite sequence $(G_1, \ldots, G_m)$ of normal subgroups of G such that each $G_i \in \mathcal{C}$. Two product structures $\mathcal{P} = (G_\lambda)_{\lambda \in \Lambda}, \mathcal{Q} = (H_\omega)_{\omega \in \Omega}$ on a group G are said to be *equivalent* when there exists a bijection $\sigma : \Lambda \to \Omega$ such that for all $\lambda \in \Lambda, G_\lambda \cong H_{\sigma(\lambda)}$, and *strongly equivalent* when, in addition, we have equality $G_\lambda = H_{\sigma(\lambda)}$ for all $\lambda \in \Lambda$.

It is easier to work within a wider context. Let $\mathcal{L}$ denote the class of finitely generated infinite groups of finite cohomological dimension with the property that every subgroup of finite index has trivial centre; $\mathcal{L}$ contains all poly-$\mathcal{S}$ groups. Let $\mathcal{L}_0$ denote the subclass of $\mathcal{L}$ consisting of irreducible groups; we show that an $\mathcal{L}_0$-product structure on a group is unique up to commensurability.

Proposition 4.2. *If H and K are commensurable groups having $\mathcal{L}_0$-product structures $H = H_1 \circ \cdots \circ H_m$ and $K = K_1 \circ \cdots \circ K_n$ respectively, then $m = n$, and for some permutation τ of the indices, $H_i \cap K_j$ has finite index in both H_i and $K_{\tau(i)}$ if $j = \tau(i)$, and is trivial if $j \neq \tau(i)$.*

Proof Without loss we may suppose that $m \leq n$. Put $G = H \cap K$, which has finite index in both H and K, and define $L_i = G \cap K_i (= H \cap K_i)$, and $L = L_1 \circ \cdots \circ L_n$. Each L_i has finite index in K_i, so that L has finite index in K. Since $L \subset G$, L also has finite index in H. Fix $\mu \in \{1, \ldots, m\}$ and let $\pi_\mu : H \to H_\mu$ denote the projection map. Since H_μ is irreducible and $\pi_\mu(L_i)$ centralises $\pi_\mu(L_j)$ for $i \neq j$, there exists a unique element $\tau(\mu) \in \{1, \ldots, n\}$ such that $\pi_\mu(L_i) = \{1\}$ for $i \neq \tau(\mu)$ and $\pi_\mu(L_\tau(\mu))$ has finite index in H_μ. τ defines a function $\tau : \{1, \ldots, m\} \to \{1, \ldots, n\}$; since each L_i is nontrivial, for each $i \in \{1, \ldots, n\}$ there exists $\mu \in \{1, \ldots, m\}$ such that $\pi_\mu(L_i) \neq \{1\}$; that is, τ is surjective. Thus $m = n$ and hence τ is also bijective. It is easy to check that, for all μ, $L_{\tau(\mu)}$ is contained in H_μ with finite index. If $H_i \cap K_j \neq \{1\}$, then, since K_j is torsion free, $H_i \cap K_j$ must be infinite; thus $j = \tau(i)$, otherwise our previous claim that $L_{\tau(i)}$ is contained in H_i with finite index is contradicted. Thus $H_i \cap K_j$ is trivial for $j \neq \tau(i)$. This completes the proof.

A straightforward argument now yields:

Corollary 4.3. *Any two $\mathcal{L}_0$-product structures on a group are strongly equivalent.*

An easy induction on cohomological dimension shows that any $\mathcal{L}$-group G has a finite index subgroup H of the form $H \cong H_1 \times \cdots \times H_m$, with each H_i an irreducible $\mathcal{L}$-group. If α is an automorphism of G, then $H \sim \alpha(H) = \alpha(H_1) \circ \cdots \circ \alpha(H_m)$, and by (4.2), there is a unique permutation $\alpha_* : \{1,\ldots,m\} \to \{1,\ldots,m\}$ with the property that $H_i \cap \alpha(H_j)$ is trivial if $j \neq \alpha_*(i)$, and $H_i \cap \alpha(H_{\alpha_*}(i))$ has finite index in both H_i and $\alpha(H_{\alpha_*}(i))$.

Put $G_0 = G_1 \circ \cdots \circ G_m$, where, for each i, $G_i = \bigcap_{\alpha \in Aut(G)} \alpha(H_{\alpha_*}(i))$. It is easy to see that G_0 is a characteristic subgroup of finite index in G. We have established:

Theorem 4.4. *An $\mathcal{L}$ group G contains a characteristic subgroup G_0 of finite index such that $G_0 \cong H_1 \times \cdots \times H_m$, where $H_1,\ldots,H_m$ are $\mathcal{L}_0$-groups.*

When G is a poly-$\mathcal{S}$ group, (4.4) provides a characteristic subgroup G_0 of finite index which is a direct product of irreducible $\mathcal{L}$-groups. We strengthen this to show that G_0 may be chosen to be a direct product of irreducible strongly poly-$\mathcal{S}$ groups. Say that group G has property $\mathcal{N}$ when every nontrivial normal subgroup of G is nonabelian. Clearly $\mathcal{S}$-groups have property $\mathcal{N}$; moreover, in an extension $1 \to H_1 \to G \to H_2 \to 1$, if both H_1 and H_2 have property $\mathcal{N}$ then so also does G. We obtain:

Proposition 4.5. *Each poly-$\mathcal{S}$ group has property $\mathcal{N}$.*

We write $G = G_1 \circ G_2$ when the group G is the internal direct product of its normal subgroups G_1, G_2. If $G_1,\ldots,G_k$ are groups then a (normal) subgroup H of $\prod_{i=1}^k G_i$ is called a (normal) subdirect product when for each i, $\pi_i(H) = G_i$ where π_i is projection onto the ith factor. If H is a normal subdirect product in $G = G_1 \times \cdots \times G_k$, then $[G_1,G_1] \times \cdots \times [G_k,G_k] \subset H$ (see, for example Proposition (1.2) of [9]). If, in addition, $G_1,\ldots,G_k$ are torsion free nonabelian groups, it follows that H contains a free abelian group of rank n. Suppose that $G = G_1 \circ G_2$ where G_1, G_2 both have property $\mathcal{N}$, and let H be a torsion free normal subgroup of G, with the property that every abelian subgroup of H is cyclic. Then H is a normal subdirect product of $H_1 \circ H_2$, where H_i is the image of H under the projection of $G_1 \circ G_2$ onto G_i. As noted above, $[H_1,H_1] \circ [H_2,H_2] \subset H$. For i = 1, 2, H_i is a normal subgroup of G_i;

if H_i is nontrivial, then since G_i has property $\mathcal{N}$, $[H_i, H_i] \neq \{1\}$, and since H is torsion free, $[H_i, H_i]$ contains an infinite cyclic group. If both H_1, H_2 are nontrivial, then, H contains a free abelian subgroup of rank 2, contradicting our assumption that every abelian subgroup of H is cyclic. Thus at least one projection $H_i = \pi_i(H)$ must be trivial, from which we see that:

Proposition 4.6. *Let H be a torsion free normal subgroup of $G_1 \circ G_2$, where G_1, G_2 both have property $\mathcal{N}$. If H has the property that every abelian subgroup is cyclic, then either $H \subset G_1$ or $H \subset G_2$.*

Proposition 4.7. *Let $G = K_1 \circ K_2$ be the (internal) direct product of nontrivial normal subgroups K_1, K_2; if G is a strongly poly-$\mathcal{S}$ group, then K_1, K_2 are also strongly poly-$\mathcal{S}$ groups.*

Proof Let $(G_r)_{0 \leq r \leq n}$ be a strong poly-$\mathcal{S}$ filtration on a group $G = G_n = K_1 \circ K_2$. By (4.5), G has property $\mathcal{N}$, so that K_1, K_2 also have property $\mathcal{N}$. From (4.6), it follows that either $G_1 \subset K_1$ or $G_1 \subset K_2$. Without loss of generality, we may suppose that $G_1 \subset K_1$. Then $G/G_1 \cong (K_1/G_1) \times K_2$ and G/G_1 admits a strong poly-$\mathcal{S}$ filtration of length n-1. The result now follows by an easy induction, beginning with the case n = 1, which is empty since $\mathcal{S}$ are indecomposable as direct products.

Combined with (4.4) and (2.2), we have proved

Theorem 4.8. *An $\mathcal{S}$-group G contains a characteristic subgroup G_0 of finite index such that $G_0 \cong H_1 \times \cdots \times H_m$ where $H_1, \ldots, H_m$ are irreducible strongly poly-$\mathcal{S}$-groups.*

5. The number of commensurability classes

The class $\mathcal{S}$ represents a single commensurability classes of abstract groups. For poly-$\mathcal{S}$ groups, the situation is quite different. In this section, we shall prove:

Theorem 5.1. *For each $n \geq 2$, the irreducible poly-$\mathcal{S}$ groups of dimension $2n$ represent infinitely many distinct commensurability classes of abstract groups.*

This result is consistent with the situation for irreducible semisimple lattices; for each $n \geq 2$, there are infinitely many commensurability classes of irreducible lattices in the n-fold product $PSL_2(\mathbf{R})^{(n)}$. In fact, it can be shown using results of [8],[13], [26] that if G is any connected linear semisimple Lie group which is C-isotypic and non-simple, then G contains infinitely many commensurability classes of irreducible lattices. Since $\mathcal{S} \subset \mathcal{L}_0$, it follows easily from (4.3) and (4.7) that any two $\mathcal{S}$-product structures on a group G are strongly equivalent. The following rigidity theorem for subgroups is a relative version of this.

Proposition 5.2.   *Let $L = L_1 \times \cdots \times L_n$ and $K = K_1 \times \cdots \times K_m$ be groups admitting S-product structures, and suppose that L has a finite index subgroup H which contains K as a normal subgroup. Then $m \le n$, and there is a (unique) injective mapping $\sigma : \{1, \ldots, m\} \to \{1, \ldots, n\}$ with the property that for all i, K_i is a subgroup of finite index in $L_{\sigma(i)}$.*

Proof Dimension considerations show immediately that m $\le$ n. Let $\pi_j : L \to L_j$ denote the projection for $j \in \{1, \ldots, n\}$, and let J denote the set of indices j for which $\pi_j(K) \ne \{1\}$. Obviously J $\ne \emptyset$, since K is nontrivial. For $j \in J$, $\pi_j(H)$ is a S-group as it has finite index in L_j, and so $\pi_j(K)$ is also a S-group, since it is a finitely generated nontrivial normal subgroup of $\pi_j(H)$. However, $\pi_j(K_1), \ldots, \pi_j(K_n)$ are mutually centralising normal subgroups of $\pi_j(K)$, and as $\pi_j(K) = \pi_j(K_1) \ldots \pi_j(K_n)$, then a straightforward argument using (1.4), (1.5), (1.6) gives the existence of a mapping $\tau : J \to \{1, \ldots, n\}$ with the property that $\pi_j(K_i) = \{1\}$ if $i \ne \tau(j)$, and $\pi_j(K_i) = \pi_j(K)$ if $i = \tau(j)$; τ is surjective since $\bigcap_{j=1}^n Ker(\pi_j) = \{1\}$. There are only two possibilities.

  (I) Card(J) = m and $\tau : J \to \{1, \ldots, m\}$ is bijective;

(II) Card(J) > m.

Suppose that (II) holds. After a permutation of the indices, we may suppose that for some $k \ge 2$, $\tau(r) = 1$ for all r such that $r \le k$, and that $\tau(\text{r}) > 1$ for all r such that $r > k$. In particular, $K_1 \subset L_1 \times \cdots \times L_k$. Moreover, $\pi_j(H) \ne \{1\}$ for $j = 1, \ldots, k$, and as $\pi_j(K_1) = \pi_j(K)$ is a nontrivial finitely generated normal subgroup of the S-group $\pi_j(H)$, it follows also that $\pi_j(K_1)$ is an S-group. However, K_1 is a normal subdirect product of $\pi_1(K_1) \times \cdots \times \pi_p(K_1)$, so that K_1 contains a free abelian group of rank p ≥ 2 by (4.6). This is a contradiction as K_1 is a S-group. Thus (II) is false and (I) is true. Let $\sigma : \{1, \ldots, m\} \to J \subseteq \{1, \ldots, n\}$ be the inverse mapping of τ; then for all i, $K_i \subset L_{\sigma(i)}$ and the index $[L_{\sigma(i)} : K_i]$ is finite by (1.6).

Proposition 5.3.   *Let Σ be a S-group, and let $N_1, \ldots, N_k$ be distinct mutually centralising subgroups such that $N = N_1 \ldots N_k$ is a nonabelian subgroup of Σ. Then there exists a unique index j such that $N_j = N$ and $N_i = \{1\}$ for $i \ne j$.*

Proof If Λ is a nonabelian subgroup of a S-group G, then the centraliser of Λ in G is trivial. Thus if N_1, N_2 are mutually centralising subgroups of G and both N_1 and N_2 are nontrivial, then $N_1.N_2$ is abelian. The rest now follows by induction.

The following observation is needed at several points. For a proof see, for example, [25].

Proposition 5.4.   *Let G be a group satisfying n-dimensional Poincaré Duality, and let H be a subgroup of G. If H also satisfies n-dimensional Poincaré Duality then the index $[G : H]$ is finite.*

By a pre-$\mathcal{M}_n$ structure we mean a pair (G, K) where G is a group, and K is a normal subgroup of G which admits a product structure $K = K_1 \times \cdots \times K_n$ in which $K_1, \ldots, K_n$ and G/K are all $\mathcal{S}$-groups; if (G, K) is a pre-$\mathcal{M}_n$ structure then G has cohomological dimension equal to $2n + 2$. Up to congruence, the canonical extension $(1 \to K \to G \to G/K \to 1)$ is classified by the operator homomorphism $\phi : G/K \to Out(K_1 \times \cdots \times K_n)$. We say that (G, K) is an $\mathcal{M}_n$-structure when, in addition, the operator homomorphism assumes the form

$$\phi = (\phi_1, \ldots, \phi_n) : G/K \to Out(K_1) \times \cdots \times Out(K_n) \subset Out(K_1 \times \cdots \times K_n),$$

in which for each i, $Im(\phi_i)$ is an infinite subgroup of $Out(K_i)$. As we shall see, the notion of $\mathcal{M}_n$-structure is an intrinsic one; the defining extension is uniquely determined by the isomorphism class of the group G. First observe that:

Proposition 5.5. *Let (G, K) be an $\mathcal{M}_n$ structure; then $K_{\alpha_1} \times \cdots \times K_{\alpha_m}$ is normal in G for any sequence $1 \le \alpha_1 < \ldots < \alpha_m \le n$.*

Theorem 5.6. *Let (G, K) be a pre-$\mathcal{M}_n$ structure and (G, L) an $\mathcal{M}_n$-structure on the same group G with $n \ge 2$. Then for some (unique) permutation τ of $\{1, \ldots, n\}$, $L_i = K_{\tau(i)}$ for all i; in particular, $K = L$.*

Proof Let $p : G \to G/K$ and $\pi : G \to G/L$ denote the canonical surjections, and suppose that $\pi(K) \ne \{1\}$; then $\pi(K)$ is a nontrivial finitely generated normal subgroup of the $\mathcal{S}$-group G/L. In particular, $\pi(K)$ has finite index in G/L. Since $\pi(K) = \pi(K_1) \ldots \pi(K_n)$, some $\pi(K_i)$ must be nontrivial. After permuting the indices we may suppose that $\pi(K_n) \ne \{1\}$. For $j \ne n$, K_j centralises K_n, so that $\pi(K_j)$ centralises $\pi(K_n)$. Thus $\pi(K_j) = \{1\}$ for $j \ne n$, and so also $\pi(K_n)$ has finite index in G/L, so that, the restriction of ϕ_1' to $\pi(K_n)$ has infinite image. Hence $K_1 \times \cdots \times K_{n-1} \subset Ker(\pi) = L$. Since $K_1 \times \cdots \times K_{n-1}$ is normal in G it is also normal in L. By (5.2), we can find indices r, s such that $K_r \subset L_s$. Again after permuting indices, if necessary, we may suppose that $K_1 \subset L_1$. Since K_1, L_1 are both $\mathcal{S}$-groups, K_1 has finite index in L_1. Now K_n centralises K_1, so that as K_1 has finite index in L_1, the action of K_n on L_1 by conjugation induces only a finite group of automorphisms of L_1. This contradicts our previous deduction that the image of the restriction of ϕ_1' to $\pi(K_n)$ is infinite, and refutes our initial assumption that $\pi(K) \ne \{1\}$. We have established that $K \subset Ker(\pi) = L$. The quotient group L/K must now be finite, by (1.6) above, and as it is contained in the torsion free group G/K, it follows that $L = K$ as claimed. Finally, from (4.3) we deduce the existence of the permutation τ with the property that for all i, $L_i = K_{\tau(i)}$.

Theorem 5.7. *Let (G, L) and (H, K) be $\mathcal{M}_n$-structures with $n \ge 2$, where H is a subgroup of G. Then for some (unique) permutation σ of $\{1, \ldots, n\}$, $K_i \subset L_{\sigma}(i)$; in particular, $K \subset L$.*

Proof Let $p : G \to G/L$ and $\pi : H \to G/K$ denote the canonical surjections. By (5.4), H necessarily has finite index in G. Thus $p(H)$ is a subgroup of finite index in G/L, and hence is a $\mathcal{S}$-group. Suppose $p(K) \neq \{1\}$. By (5.2), and after a permutation of indices, we can suppose that $p(K_i) = \{1\}$ for $i < n$ and that $p(K_n)$ is a subgroup of finite index in G/L. Hence $K_1 \circ \cdots \circ K_{n-1} \subset L = Ker(p)$. Moreover, $K_1 \circ \cdots \circ K_{n-1}$ is normal in $H \cap L$ which has finite index in L. Applying (5.2) again, we find a pair of indices (i, j) such that $K_i \subset L_j$. Let $\phi = (\phi_1, \ldots, \phi_n) : G/L \to Out(L_1) \times \cdots \times Out(L_n)$ denote the operator homomorphism for the extension $1 \to L \to G \overset{p}{\to} G/L \to 1$. Since $p(K_n)$ has finite index in G/L, and $Im(\phi_j)$ is infinite, K_n acts by conjugation on L_j as an infinite group of automorphisms. However, K_n centralises K_i, and K_i is a subgroup of finite index in L_j, so that K_n acts by conjugation on L_j as an infinite group of automorphisms. This is a contradiction, refuting our initial assumption; thus $p(K) = \{1\}$, or equivalently, $K \subset L$. The conclusion follows immediately from (4.3).

Theorem 5.8. *If G admits an $\mathcal{M}_n$-structure then it is irreducible.*

Proof Let (G, K) be an $\mathcal{M}_n$-structure and suppose that G is reducible; then G contains a subgroup of finite index $H = H_1 \circ H_2$ where H_1 and H_2 are infinite. Let $p : G \to G/K$ denote the canonical map; then $p(H_1).p(H_2)$ has finite index in Q, so that $p(H_1).p(H_2)$ is a $\mathcal{S}$-group, and $p(H_1), p(H_2)$ are mutually centralising subgroups of $p(H_1).p(H_2)$. It follows that either $p(H_1) = \{1\}$ and $p(H_2)$ has finite index in Q or $p(H_1)$ has finite index in Q and $p(H_2) = \{1\}$. Without loss of generality, assume the former, so that $H_1 \subset K$ and $(H_1 \circ H_2) \cap K = H_1 \circ (H_2 \cap K)$. Moreover, $(H_1 \circ H_2) \cap K = H_1 \circ (H_2 \cap K)$ has finite index in K. Let $\pi_r : K \to K_r$ denote the projection to the rth factor, where $K = K_1 \times \cdots \times K_n$ is the $\mathcal{S}$-product structure on K. For each r, $\pi_r(H_1)$ and $\pi_r(H_2 \cap K)$ are mutually centralising subgroups, and $\pi_r(H_1).\pi_r(H_2 \cap K)$ has finite index in K_r. Now decompose $\{1, \ldots, n\}$ as a union $I \cup J$ where

(i) $\pi_i(H_1)$ has finite index in K_i and $\pi_i(H_2 \cap K) = \{1\}$ for $i \in I$;

(ii) $\pi_j(H_1) = \{1\}$ and $\pi_j(H_2 \cap K)$ has finite index in K_j for $j \in J$.

Clearly $I \cap J = \emptyset$. Define $K^I = \underset{j \in J}{\cap} Ker(\pi_j)$ and $K^J = \underset{i \in I}{\cap} Ker(\pi_i)$. Then $K = K^I \circ K^J$ and $H_1 \subset K^I$ and $H_2 \cap K \subset K^J$. Moreover, H_1 has finite index in K^I. Now H_2 centralises H_1 and so the group of automorphisms of K^I induced by conjugation of H_2 on K_I is finite. As $p(H_2)$ has finite index in Q, this contradicts our hypothesis that $Im(\phi_i)$ is infinite for each $i \in I$. Hence our initial assumption is false, and G is irreducible.

In [12] we pointed out that for the class of fundamental groups of smooth closed aspherical 4n-manifolds, the invariant $Sign/\chi$, when defined, is an invariant of commensurability class. Of course, this fails in dimensions congruent to 2 mod 4. Here we show how to modify this invariant to detect commensurability classes amongst groups in $\mathcal{M}_n$, which can be of any even dimension. Thus suppose that G admits an $\mathcal{M}_n$-structure (G,K). In this case the Euler characteristic $\chi(G) \neq 0$.

Define $K_i' = K_1 \circ \cdots \circ K_{i-1} \circ K_{i+1} \circ \cdots \circ K_n$. Then K_i' is normal in G, so we may define $G_i = G/K_i'$. Observe that G/K and $K/K_i' \cong K_i$ are $\mathcal{S}$-groups, and there is an extension $1 \to K/K_i' \to G_i \to G/K \to 1$. We have shown elsewhere, (e.g. [11]) that any such extension is smoothable; that is:

Proposition 5.9. *For each i, G_i is the fundamental group of a smooth closed aspherical 4-manifold.*

In general, the Poincaré Duality group G_i will not be orientable. For any surface Σ of genus ≥ 1, the group we denote by $Out_+(\pi_1(\Sigma))$ the subgroup of index 2 in $Out(\pi_1(\Sigma))$ consisting of those elements which act trivially upon $H_2(\Sigma; \mathbf{Z})$. If $\phi = (\phi_1, \ldots, \phi_n) : G/K \to Out(K_1) \times \cdots \times Out(K_m)$ is the operator homomorphism for the defining extension of G, then the extension $1 \to K/\widehat{K_i} \to G/\widehat{K_i} \to G/K \to 1$ is classified by $\phi_i : G/K \to Out(K/\widehat{K_i}) \cong Out(K_i)$. It is straightforward to check that:

Proposition 5.10. *G_i is an orientable Poincaré Duality group if and only if $Im(\phi_i) \subset Out_+(K_i)$.*

We shall say that (G, K) is an orientable $\mathcal{M}_n$-structure when each G_i is orientable. It is clear that:

Proposition 5.11. *If (G, K) is an $\mathcal{M}_n$-structure then there exists an orientable $\mathcal{M}_n$-structure $(\widetilde{G}, K)$ such that $\widetilde{G}$ has finite index index in G.*

If (G, K) is an orientable $\mathcal{M}_n$-structure, we define

$$\nu(G) = \frac{2^n}{\chi(G/K)^n} \prod_{i=1}^n \frac{Sign(G_i)}{\chi(K_i)}.$$

The factor 2^n is not strictly necessary, but prevents the unnecessary proliferation of powers of 2 arising in the denominator from the fact that the Euler characteristic of G/K is even.The expression $\nu(G)$ is obviously an invariant of the equivalence class of the defining extension of G. By (5.6), a group G admits at most one $\mathcal{M}_n$-structure, so that:

Theorem 5.12. *$\nu(G)$ is an invariant of isomorphism class for groups which admit an orientable $\mathcal{M}_n$ structure.*

In view of the subgroup rigidity property (5.2) above, it is straightforward to see that this can be strengthened as follows:

Theorem 5.13. *$\nu(G)$ is an invariant of commensurability class for groups which admit an orientable $\mathcal{M}_n$-structure.*

We can now extend the invariant ν to all $\mathcal{M}_n$-groups. If (G, K) is an $\mathcal{M}_n$-structure, then by (4.4), there exists an orientable $\mathcal{M}_n$-structure $(\widetilde{G}, K)$ for which $\tilde{G}$ is a subgroup of finite index in G. In view of (4.7), the value of $\nu(\tilde{G})$ is independent of the particular orientable $\mathcal{M}_n$-structure chosen, so that we may define $\nu(G) = \nu(\tilde{G})$.

It remains to show that, for each n $\geq$ 2, the invariant ν assumes infinitely many values. To do this, we recall that Kodaira [19] constructs for each h $\geq$ 3, a fibration $\Sigma^{\alpha(h)} \to Z(h) \to \Sigma^{\beta(h)}$ in which fibre and base are real surfaces with genera $\alpha(h), \beta(h)$ respectively. The fibration is holomorphic and locally trivial in the C^∞ category, but is not holomorphically locally trivial. Moreover, the signature of the complex algebraic surface Z(h) is nonzero. In Atiyah's formulation [1], the particular values of $\alpha(h), \beta(h), Sign(Z(h))$ are as follows;

$$\alpha(h) = 2h; \beta(h) = 1 + 2^{2h}(h-1); Sign(Z(h)) = -\chi(\Sigma^{\beta(h)}) = 2^{2h+1}(h-1).$$

Applying π_1 to the above fibration gives a group extension $1 \to \Sigma_{\alpha(h)} \to G(h) \to \Sigma_{\beta(h)} \to 1$ where we write $\Sigma_g = \pi_1(\Sigma^g)$ and $G(h) = \pi_1(Z(h))$. Let $\rho_h : \Sigma_{\beta(h)} \to Out(\Sigma_{\alpha(h)})$ denote the operator homomorphism of this extension. If $Im(\rho_h)$ is finite, it is easy to see that $Z(h)$ has a finite covering $\widehat{Z(h)}$ which is diffeomorphic to a direct product $\Sigma^{\alpha(h)} \times \Sigma^{\gamma(h)}$ where $\Sigma^{\gamma(h)}$ is a finite covering of $\Sigma^{\beta(h)}$. This implies that $Sign(\widehat{Z(h)}) = 0$, and so also $Sign(Z(h)) = 0$, which is a contradiction. We obtain:

Proposition 5.14. *$Im(\rho_h)$ is infinite for all h $\geq$ 3.*

Now fix an integer n $\geq$ 2, let

$$\phi_h : \Sigma_{\beta(h)} \to \underbrace{Out(\Sigma_{\alpha(h)}) \times \cdots \times Out(\Sigma_{\alpha(h)})}_{\text{(n copies)}}$$

be the homomorphism $\phi_h = \rho_h \times \cdots \times \rho_h$, and let $G(h)$ be the group defined by the extension

$$1 \to \Sigma_{\alpha(h)} \times \cdots \times \Sigma_{\alpha(h)} \to G(h) \to \Sigma_{\beta(h)} \to 1$$

with ϕ_h as operator homomorphism. In this case, each $G(h)_i \cong \pi_1(Z(h))$ so that the value of $\nu(G(h))$ is calculated as follows:

$$\nu(G(h)) = \frac{2^n Sign(Z(h))^n}{\chi(\Sigma_{\alpha(h)})^n \chi(\Sigma_{\alpha(h)})^n}.$$

Since $Sign(Z(h)) = -\chi(\Sigma_{\beta(h)})$, we obtain

$$\nu\frac{(G(h) = (-1)^n 2^n}{\chi(\Sigma_{\alpha(h)})^n} = \frac{1}{(\alpha(h)-1)^n}.$$

However, $\alpha(h) = 2h$ so that

$$\nu(G(h)) = \frac{1}{(\alpha(h) - 1)^n} = \frac{1}{(2h - 1)^n}.$$

Evidently $\nu(G(h))$ assumes infinitely many values, and thus, for $n \geq 2$, the class $\mathcal{M}_n$ represents infinitely many distinct commensurability classes of abstract groups. In fact, the same computation works also for $n = 1$, giving the same conclusion as [12]. We obtain:

Theorem 5.15. *For each $m \geq 2$, the irreducible poly-$\mathcal{S}$ groups of dimension $2m$ represent infinitely many distinct commensurability classes of abstract groups.*

The construction of $G(h)$ can be varied, for example, by choosing $h_1, \ldots, h_n$ independently, and finding a real surface Σ^β which covers each of the surfaces $\Sigma^{\beta(h_1)}, \ldots, \Sigma^{\beta(h_n)}$. Let ρ'_i be the restriction of ρ_{h_i} to Σ_β , and put $\phi'_h = \rho'_1 \times \cdots \times \rho'_n$. Finally, let $G'(h)$ be the extension determined by the operator homomorphism ϕ'_h. It is straightforward to calculate that, in this case, the invariant ν assumes the value

$$\nu = \frac{1}{(2h_1 - 1) \ldots (2h_n - 1)}.$$

6. Automorphisms

Let G be a group given as a group extension $\mathcal{E} = (1 \to K \to G \xrightarrow{p} Q \to 1)$. In general, there is no easy relationship between the automorphism groups of G, K and Q. However, in the case which interests us, when K and Q are poly-$\mathcal{S}$ groups, it is possible to say something. We start by considering the group $Aut(\mathcal{E})$ of automorphisms which preserve $\mathcal{E}$; to be precise, $Aut(\mathcal{E})$ consists of those automorphisms $\alpha : G \to G$ such that $\alpha(K) = K$; α then induces an automorphism α_Q on the quotient Q. We obtain a "restriction" homomorphism $\rho : Aut(\mathcal{E}) \to Aut(K) \times Aut(Q)$ given by $\rho(\alpha) = (\alpha_K, \alpha_Q)$, where α_K denotes the restriction of α to K. The kernel of ρ is the group $C(\mathcal{E})$ of self-congruences of $\mathcal{E}$, so that we have an exact sequence

$$1 \to C(\mathcal{E}) \to Aut(\mathcal{E}) \xrightarrow{\rho} Aut(K) \times Aut(Q).$$

The homomorphism $c : G \to Aut(K)$ obtained from conjugation, $c(g)(k) = gkg^{-1}$, induces the so-called "operator homomorphism" $\phi : Q \to Out(K) = Aut(K)/Inn(K)$; the centre $\mathcal{Z}(K)$ of K is naturally a module over $Out(K)$, and becomes a module over Q via the operator homomorphism. It is easy to check that for $\alpha \in C(\mathcal{E})$, the assignment $x \mapsto \alpha(x)x^{-1}$ is a function on G taking values in $\mathcal{Z}(K)$. Moreover, the function $z_\alpha : Q \to \mathcal{Z}(K)$ defined by $z_\alpha(p(y)) = z_\alpha(y)$ is an element of $Z^1(Q, \mathcal{Z}(K))$, the (abelian) group of 1-cocycles of Q with values in $\mathcal{Z}(K)$. Moreover, the mapping $C(\mathcal{E}) \to Z^1(Q, \mathcal{Z}(K)), \alpha \mapsto z_\alpha$, is an isomorphism of groups. When K has trivial centre, matters simplify to give:

(6.1) If K has trivial centre, the group of congruences $C(\mathcal{E})$ is trivial, so that the exact sequence $1 \to C(\mathcal{E}) \to Aut(\mathcal{E}) \overset{\rho}{\to} Aut(K) \times Aut(Q)$ reduces to an injection $Aut(\mathcal{E}) \overset{\rho}{\hookrightarrow} Aut(K) \times Aut(Q)$.

If $\mathcal{G} = (G_r)_{0 \le r \le n}$ is a poly-$\mathcal{S}$ filtration of length n on G_n, we define

$$Aut(\mathcal{G}) = \{\alpha \in Aut(G_n) : \alpha(G_r) = G_r \text{for all} r, 1 \le r \le n\}.$$

Since groups in $\mathcal{S}$ all have trivial centre, we see inductively that:

Theorem 6.2. *Let $\mathcal{G}$ be a poly-$\mathcal{S}$ filtration on $G = G_n$; then $Aut(\mathcal{G})$ imbeds as a subgroup of $Aut(Q_1) \times \cdots \times Aut(Q_m)$, where $Q_1, \ldots, Q_m$ are the successive quotients, $Q_r = G_r/G_{r-1}$.*

Although $\mathcal{S}$-groups themselves fail to be rigid in the sense of Mostow, poly-$\mathcal{S}$ groups nevertheless exhibit a very strong form of rigidity; the author has recently shown (February 1996) the following, which completely answers a question raised in [14].

Theorem 6.3. *Any group G admits at most a finite number of poly-$\mathcal{S}$ filtrations.*

The details will appear in [15]. As an immediate corollary, one obtains:

Theorem 6.4. *If $\mathcal{G}$ is a poly-$\mathcal{S}$ filtration on G, then $Aut(\mathcal{G})$ is a subgroup of finite index in $Aut(G)$.*

Corollary 6.5. *If $G$is a group with poly-$\mathcal{S}$ filtration $\mathcal{G} = (G_r)_{0 \le r \le n}$, then $Aut(G)$ is commensurable with a subgroup of $Aut(Q_1) \times \cdots \times Aut(Q_m)$ where $Q_1, \ldots, Q_m$ are the successive quotients, $Q_r = G_r/G_{r-1}$.*

7. Smoothing

A discrete group Γ is said to be smoothable with model the smooth closed manifold X_Γ when X_Γ is aspherical with $\pi_1(X_\Gamma) \cong \Gamma$. There is a relative condition which is a priori stronger; we say that Γ has the extension smoothing property when, for any extension $\mathcal{E} = (1 \to \Gamma \to G \to Q \to 1)$ in which Q is smoothable with model X_Q, the canonical fibration

$$\mathcal{K}(\mathcal{E}, 1) = (K(\Gamma, 1) \to K(G, 1) \to K(Q, 1))$$

is fibre homotopy equivalent to a smooth fibre bundle $X_\Gamma \to E \to X_Q$ in which the fibre X_Γ is a smooth finite dimensional manifold of homotopy type

$K(\Gamma, 1)$. If $\mathcal{E} = (1 \to K \to G \to Q \to 1)$ is a group extension in which Γ has the fibre smoothing property and Q is smoothable it is tautological that G is also smoothable. More important for our purposes is the following *devissage* result, whose proof is a straightforward exercise in exact sequences and fibre bundles.

Proposition 7.1. *Suppose that in the extension* $1 \to \Gamma_1 \to \Gamma \to \Gamma_2 \to 1$, *both* Γ_1 *and* Γ_2 *have the extension smoothing property and that* Γ_1 *is a characteristic subgroup of* Γ; *then* Γ *has the extension smoothing property.*

Next we show that $\mathcal{S}$-groups possess the extension smoothing property. If Σ is a closed surface of genus ≥ 2, and $\mathcal{D}\mathrm{iff}(\Sigma)$ is the group of diffeomorphisms of Σ, topologised with the C^∞-topology, then by a theorem of Baer [2] , the natural homomorphism $\phi : \mathcal{D}\mathrm{iff}(\Sigma) \to Out(\pi_1(\Sigma))$ is surjective and has as its kernel the identity component $\mathcal{D}\mathrm{iff}_0(\Sigma)$. However, $\mathcal{D}\mathrm{iff}_0(\Sigma)$ is contractible, by a result of Earle and Eells [4]. The classifying space functor $G \mapsto BG$ preserves homotopy equivalences [3] , so that $B\mathcal{D}\mathrm{iff}_0(\Sigma)$ is also contractible, and the induced map $B\phi : B\mathcal{D}\mathrm{iff}(\Sigma) \to BOut(\pi_1(\Sigma))$ is a homotopy equivalence. If $\mathcal{B}_\Sigma(X)$ denotes the set of smooth equivalence classes of smooth bundles with fibre Σ over a smooth connected manifold X, then standard approximation arguments show that $\mathcal{B}_\Sigma(X)$ is naturally equivalent to the set of based homotopy classes $[X, B\mathcal{D}\mathrm{iff}(\Sigma)] \cong [X, BOut(\pi_1(\Sigma))]$. However, fibrations with fibre Σ over a CW complex X are classified by $[X, BG(\Sigma)]$ where $G(\Sigma)$ is the monoid of homotopy equivalences of Σ. Since Σ is aspherical and $\pi_1(\Sigma)$ has trivial centre, the identity component $G_0(\Sigma)$ is contractible, and the natural mapping $G(\Sigma) \to Out(\pi_1(\Sigma))$ also induces a homotopy equivalence $BG(\Sigma) \to BOut(\pi_1(\Sigma))$, so that the classification of smooth fibre bundles with fibre Σ coincides with that of fibrations with fibre Σ. In particular, any fibration $K(\pi_1(\Sigma), 1) \to X_G \to X_Q$ over a smooth base X_Q is fibre homotopy equivalent to a smooth fibre bundle $\Sigma \to E \to X_Q$. That is, we have proved:

Theorem 7.2. $\mathcal{S}$-*groups possess the extension smoothing property.*

From (2.3), (7.1) and (7.2), it follows immediately that

Corollary 7.3. *Any stably poly-$\mathcal{S}$ group has the extension smoothing property.*

Since, by (2.4), an arbitrary poly-$\mathcal{S}$ group contains a stably poly-$\mathcal{S}$ group with finite index, we see that:

Corollary 7.4. *Any poly-$\mathcal{S}$ group contains a smoothable subgroup of finite index.*

Let Γ be a discrete group which is smoothable with model X_Γ. We say that X_Γ is finitely rigid when each finite subgroup $\Psi \subset Out(\Gamma)$ can be realised as a group of diffeomorphisms on X_Γ. It is a consequence of the Mostow Rigidity Theorem [20] and Kerckhoff's solution to the Nielsen Conjecture [18] that any torsion free discrete cocompact subgroup of a semisimple Lie group admits a finitely rigid model [10]. This is not known for poly-$\mathcal{S}$ groups, and it is here that the analogy between poly-$\mathcal{S}$ groups and discrete cocompact subgroups is farthest from being complete; it is precisely this which stands in the way of showing that all poly-$\mathcal{S}$ groups are smoothable. One difficulty is that it is by no means clear how to endow even the smooth models that we can construct with geometries tractable enough to approach this problem. For example, a typical poly-$\mathcal{S}$ group need not be the fundamental group of any compact Kähler manifold [7]. Indeed, in real dimension four, there is a "rigidity theorem" due to A.N. Parshin which shows that, if the diffeomorphism types of base and fibre are fixed, there are only finitely many operator homomorphisms $\pi_1(Base) \to Out(\pi_1(Fibre))$ for which the associated fibration admits the structure of a Kodaira fibration. See [21], [14]. In some cases, however, the manifold we construct does admit the structure of a complex projective variety. In particular, this is true for the $\mathcal{M}_n$ groups of §5. The approach of Griffiths [6] and Shabat [24] using the Bers Simultaneous Uniformisation Theorem can be generalised to show

Theorem 7.5. *Each $\mathcal{M}_n$ group acts as a discrete cocompact group of isometries on a bounded domain in $\mathbf{C}^{n+1}$.*

The details will appear elsewhere.

References

[1] M.F. Atiyah, *The signature of fibre bundles,* in: Collected Papers in Honour of K. Kodaira, Tokyo University Press (1969).

[2] R. Baer, *Isotopie von Kurven auf orientierbaren, geschlossenen Flachen und ihr Zusammenhang mit der topologischen Deformation der Flachen,* J. Reine angew. Math. **159** (1928), pp. 101–111.

[3] J. M. Boardman and R. M. Vogt, *Homotopy invariant algebraic structures on topological spaces,* Lecture Notes in Mathematics, **347** Springer, Berlin (1973).

[4] C.J. Earle and J.Eells, *A fibre bundle description of Teichmuller theory,* J. Diff Geom. **3** (1969), pp. 19–43.

[5] L. Greenberg, *Discrete groups of motions,* Canad. J. Math. **12** (1960), pp. 416–425.

[6]   P.A. Griffiths, *Complex-analytic properties of certain Zariski open sets on algebraic varieties,* Annals of Math. **94** (1971), pp. 21–51.

[7]   F.E.A. Johnson, *A class of non-Kählerian manifolds,* Math. Proc. Camb. Phil. Soc. **100** (1986), pp. 519–521.

[8]   F.E.A. Johnson, *On the existence of irreducible discrete subgroups in isotypic Lie groups of classical type,* Proc. London Math. Soc. **56** (1988), pp. 51–77.

[9]   F.E.A. Johnson, *On normal subgroups of direct products,* Proc. Edinburgh Math. Soc. **33** (1990), pp. 309–319.

[10] F.E.A. Johnson, *Extending group actions by finite groups,* Topology **31** (1992), pp. 407–420.

[11] F.E.A. Johnson, *Surface fibrations and automorphisms of non-abelian extensions,* Quart. Jour. Math. **44** (1993), pp. 199–214.

[12] F.E.A. Johnson, *A rational invariant for infinite discrete groups,* Math. Proc. Camb.Phil. Soc. **113** (1993), pp. 473–478.

[13] F.E.A. Johnson, *On the uniqueness of arithmetic structures,* Proc. Roy. Soc. Edinburgh **124A** (1994), pp. 1037–1044.

[14] F.E.A. Johnson, *A group theoretic analogue of the Parshin-Arakelov rigidity theorem,* Archiv der Math. **63** (1994), pp. 354–361.

[15] F.E.A. Johnson, *A rigidity theorem for group extensions,* preprint University College London (February, 1996).

[16] F.E.A. Johnson and C.T.C. Wall, *On groups satisfying Poincaré Duality,* Ann. of Math. **96** (1972), pp. 592–598.

[17] S. Katok, *Fuchsian groups,* University of Chicago Press (1992).

[18] S.P. Kerckhoff, *The Nielsen realisation problem,* Ann. of Math. **117** (1983), pp. 235–265.

[19] K. Kodaira, *A certain type of irregular algebraic surface,* Jour. Anal. Math. **19** (1967), pp. 207–215.

[20] G.D. Mostow, *Strong rigidity of locally symmetric spaces,* Annals of Math. Studies, **78** Princeton Univ. Press (1973).

[21] D. Mumford, *Curves and their Jacobians,* University of Michigan Press, Ann Arbor (1971).

[22] M.S. Raghunathan, *Discrete subgroups of Lie groups,* Ergebnisse der Math., **68** Springer (1972).

[23] J-P. Serre, *Cohomologie des groupes discrets,* in: Prospects in Mathematics Annals of Math. Studies, **70** Princeton Univ. Press (1971), pp. 77–169.

[24] G.B. Shabat, *The complex structure of domains covering algebraic surfaces,* Functional Analysis and Applications **11** (1977), pp. 135–142.

[25] R. Strebel, *A note on subgroups of infinite index in Poincaré Duality groups,* Comment. Math. Helv. **52** (1977), pp. 317–324.

[26] J. Tits, *Free subgroups in linear groups,* J. Algebra **20** (1972), pp. 250–270.

[27] D.J.St.H. Webber, *On the existence of irreducible discrete subgroups in isotypic Lie groups of exceptional type,* Ph.D. Thesis University of London (1985).

[28] B.A.F. Wehrfritz, *Infinite linear groups,* Ergebnisse der Math., **76** Springer (1973).

ANALYTIC VERSIONS OF THE ZERO DIVISOR CONJECTURE

PETER A. LINNELL

Math, VPI, Blacksburg, VA 24061–0123, USA
email: linnell@math.vt.edu

1. INTRODUCTION

This is an expanded version of the three lectures I gave at the Durham conference. The material is mainly expository, though there are a few new results, and for those I have given complete proofs. While the subject matter involves analysis, it is written from an algebraic point of view. Thus hopefully algebraists will find the subject matter comprehensible, though analysts may find the analytic part rather elementary.

The topic considered here can be considered as an analytic version of the zero divisor conjecture over $\mathbb{C}$: recall that this states that if G is a torsion free group and $0 \neq \alpha, \beta \in \mathbb{C}G$, then $\alpha\beta \neq 0$. Here we will study the conjecture that if $0 \neq \alpha \in \mathbb{C}G$ and $0 \neq \beta \in L^p(G)$, then $\alpha\beta \neq 0$ (precise definitions of some of the terminology used in this paragraph can be found in later sections). We shall also discuss applications to L^p-cohomology.

Since these notes were written, the work of Rosenblatt and Edgar [19, 54] has come to my attention. This is closely related to the work of Section 6.

2. NOTATION AND TERMINOLOGY

All rings will have a 1, and to say that R is a field will imply that R is commutative (because we use the terminology division ring for not necessarily commutative "fields"). A nonzero divisor in a ring R will be an $a \in R$ such that $ab \neq 0 \neq ba$ for all $b \in R \backslash 0$. To say that the ring R is a domain will mean that if $a, b \in R \backslash 0$, then $ab \neq 0$; equivalently $R \backslash 0$ is the set of nonzero divisors of R. We shall use the notation $\mathbb{C}$, $\mathbb{R}$, $\mathbb{Z}$, $\mathbb{N}$ and $\mathbb{P}$ for the complex numbers, real numbers, integers, nonnegative integers and positive integers respectively. Ring homomorphisms will preserve the 1, and unless otherwise stated, mappings will be on the left and modules will be right modules. If $n \in \mathbb{N}$, then M^n will indicate the direct sum of n copies of the R-module M. As usual, $\ker \theta$ and $\operatorname{im} \theta$ will denote the kernel and image of the map θ. The closure of a subset X in a Banach space will be denoted by $\overline{X}$; in particular if θ is a continuous map between Banach spaces, then $\overline{\operatorname{im} \theta}$ denotes the closure of the image of θ. If $\mathcal{H}$ is a Hilbert space and $\mathcal{K}$ is a subspace of $\mathcal{H}$, we shall let $\mathcal{L}(\mathcal{H})$ denote the set of bounded linear operators on $\mathcal{H}$, and $\mathcal{K}^\perp$ denote the orthogonal complement of $\mathcal{K}$ in $\mathcal{H}$. We shall let $\mathrm{M}_n(R)$ indicate the set of $n \times n$ matrices over a ring R, $\mathrm{GL}_n(R)$ the set of invertible elements of $\mathrm{M}_n(R)$,

1_n the identity matrix of $M_n(R)$, and 0_n the zero matrix of $M_n(R)$. If $t \in \mathbb{P}$ and $A_i \in M_{n_i}(R)$ $(1 \le i \le t)$, then $\mathrm{diag}(A_1, \ldots, A_t)$ denotes the matrix in $M_{n_1 + \cdots + n_t}(R)$

$$\begin{pmatrix} A_1 & 0 & \cdots & 0 \\ 0 & A_2 & \cdots & 0 \\ \vdots & \vdots & \ddots & \vdots \\ 0 & 0 & \cdots & A_t \end{pmatrix}.$$

For any ring R, we let $K_0(R)$ denote the Grothendieck group associated with the category of all finitely generated projective R-modules: thus $K_0(R)$ has generators $[P]$ where P runs through the class of finitely generated projective R-modules, and relations $[P] = [Q] \oplus [U]$ whenever P, Q and U are finitely generated projective R-modules and $P \cong Q \oplus U$.

When R is a right Noetherian ring, the Grothendieck group associated with the category of all finitely generated R-modules will be denoted by $G_0(R)$: thus $G_0(R)$ has generators $[M]$ where M runs through the class of finitely generated R-modules, and relations $[L] = [M] \oplus [N]$ whenever L, M and N are finitely generated R-modules and there is a short exact sequence $0 \to M \to L \to N \to 0$. There is then a natural map $K_0(R) \to G_0(R)$ given by $[P] \to [P]$, and in the case R is semisimple Artinian, this map is an isomorphism.

We shall use the notation $G *_A H$ for the free product of the groups G and H amalgamating the subgroup A, $[G : A]$ for the index of A in G, G' for the commutator subgroup of G, and $\mathcal{F}(G)$ for the set of finite subgroups of G. If the orders of the subgroups in $\mathcal{F}(G)$ are bounded, we shall let $\mathrm{lcm}(G)$ stand for the lcm (lowest common multiple) of the orders of the subgroups in $\mathcal{F}(G)$. The characteristic subgroup of G generated by its finite normal subgroups will be indicated by $\Delta^+(G)$. If S is a subset or an element of G, then $\langle S \rangle$ will denote the subgroup generated by S. For $g \in G$, we shall let $C_G(g)$ indicate the centralizer of $g \in G$. If $\mathcal{X}$ and $\mathcal{Y}$ are classes of groups, then $G \in \mathcal{X}\mathcal{Y}$ will mean that G has a normal subgroup $X \in \mathcal{X}$ such that $G/X \in \mathcal{Y}$.

3. DEFINITIONS AND $L^p(G)$

Here we will define the Banach spaces $L^p(G)$ and discuss some elementary results from functional analysis. Throughout this section G will be a group. As usual, we define the complex group ring

$$\mathbb{C}G = \{\sum_{g \in G} \alpha_g g \mid \alpha_g \in \mathbb{C} \text{ and } \alpha_g = 0 \text{ for all but finitely many } g\}.$$

For $\alpha = \sum_{g \in G} \alpha_g g$, $\beta = \sum_{g \in G} \beta_g g \in \mathbb{C}G$, the multiplication is defined by

$$\alpha\beta = \sum_{g,h \in G} \alpha_g \beta_h gh = \sum_{g \in G} \Big(\sum_{x \in G} \alpha_{gx^{-1}} \beta_x\Big) g.$$

Then for $1 \leq p \in \mathbb{R}$, we define

$$L^p(G) = \{\alpha = \sum_{g \in G} \alpha_g g \mid \alpha_g \in \mathbb{C} \text{ and } \sum_{g \in G} |\alpha_g|^p < \infty\},$$

$$\|\alpha\|_p = \left(\sum_{g \in G} |\alpha_g|^p\right)^{1/p}.$$

Thus $L^p(G)$ is a Banach space under the norm $\|.\|_p$ (of course $L^p(G)$ can also be defined for $p < 1$, but then it would no longer satisfy the triangle inequality $\|\alpha + \beta\|_p \leq \|\alpha\|_p + \|\beta\|_p$ and so would not be a Banach space). Also we define

$$L^\infty(G) = \{\alpha = \sum_{g \in G} \alpha_g g \mid \alpha_g \in \mathbb{C} \text{ and } \sup_{g \in G} |\alpha_g| < \infty\},$$

$$C_0(G) = \{\alpha = \sum_{g \in G} \alpha_g g \mid \alpha_g \in \mathbb{C} \text{ and given } \epsilon > 0,$$

$$\text{there exist only finitely many } g \text{ such that } |\alpha_g| > \epsilon\},$$

$$\|\alpha\|_\infty = \sup_{g \in G} |\alpha_g|.$$

Then $L^\infty(G)$ and $C_0(G)$ are Banach spaces under the norm $\|.\|_\infty$. If $\alpha \in L^\infty(G)$, then $\alpha_g \in \mathbb{C}$ is determined by the formula $\alpha = \sum_{g \in G} \alpha_g g$. For $p < q$,

$$\mathbb{C}G \subseteq L^p(G) \subseteq L^q(G) \subseteq C_0(G) \subseteq L^\infty(G),$$

and there is equality everywhere if and only if $|G| < \infty$ and strict inequality everywhere if and only if $|G| = \infty$. The multiplication in $\mathbb{C}G$ extends to a multiplication

$$L^1(G) \times L^\infty(G) \to L^\infty(G)$$

according to the formula

(3.1) $$\sum_{g \in G} \alpha_g g \sum_{g \in G} \beta_g g = \sum_{g,h \in G} \alpha_g \beta_h gh = \sum_{g \in G} \left(\sum_{x \in G} \alpha_{gx^{-1}} \beta_x\right) g,$$

and this also induces a multiplication $L^1(G) \times L^p(G) \to L^p(G)$ for all $p \geq 1$; in the case $p = 1$, this makes $L^1(G)$ into a ring. Another multiplication is $L^2(G) \times \mathbb{C}G \to L^2(G)$; this is useful because it means that $L^2(G)$ can be viewed as a right $\mathbb{C}G$-module, as we do in Section 11.

The central topic of these notes is the following:

Problem 3.1. *Let G be a torsion free group and let $1 \leq p \leq \infty$. Does $0 \neq \alpha \in \mathbb{C}G$ and $0 \neq \beta \in L^p(G)$ imply $\alpha\beta \neq 0$?*

We shall also consider generalizations of this to groups with torsion and to matrix rings. Since this can be considered as an extension of the classical zero divisor conjecture, let us consider the current status of that problem.

4. THE CLASSICAL ZERO DIVISOR CONJECTURE

We shall briefly review the status of the classical zero divisor conjecture. Recall that the group G is right ordered means that there exists a total order $\leq$ on G such that $x \leq y$ implies that $xz \leq yz$ for all $x, y, z \in G$. The class of right ordered groups includes all torsion free abelian groups, all free groups, and is closed under taking subgroups, directed unions, free products, and group extension (i.e. H and G/H are right ordered implies that G is right ordered). It also includes the class of locally indicable groups, where G is locally indicable means that if $H \neq 1$ is a finitely generated subgroup of G, then there exists $H_0 \lhd H$ such that $H/H_0 \cong \mathbb{Z}$. Furthermore if G has a family of normal subgroups $\{H_i \mid i \in \mathcal{I}\}$ for some indexing set $\mathcal{I}$ such that G/H_i is right orderable for all $i \in \mathcal{I}$ and $\bigcap_{i \in \mathcal{I}} H_i = 1$, then G is right orderable. These results can be found in [44, §7.3]. Then the usual argument which shows that a polynomial ring is a domain can be extended to show

Theorem 4.1. *Let k be a field and let G be a right ordered group. Then kG is a domain.*

Variants of this result have been around in the literature for a long time. For instance back in 1940, Higman [29] proved the above result in the case G is locally indicable.

Little further progress was made until the 1970's, though in 1959 Cohn proved that the free product of two domains amalgamating a common division ring is also a domain [13, theorem 2.5]. The significance of this result was not realized for group rings until Lewin applied it to show that under fairly mild restrictions, the group ring of a free product with amalgamation is a domain. To describe his results, we need to recall the definition of the Ore condition. Let R be a ring, let S be the set of nonzero divisors in R, and let S_0 be a subset of R which is closed under multiplication and contains 1. Then R satisfies the right Ore condition with respect to S_0 means that for each $r \in R$ and $s \in S_0$, there exists $r_1 \in R$ and $s_1 \in S_0$ such that $rs_1 = sr_1$, and then we can form the ring RS_0^{-1} which consists of elements $\{rs^{-1} \mid r \in R, s \in S_0\}$. Normally S_0 will be contained in S, but this is not essential. We say that R satisfies the right Ore condition if it satisfies the right Ore condition with respect S. Also a classical right quotient ring for R is a ring Q which contains R such that every element of S is invertible in Q, and every element of Q can be written in the form rs^{-1} with $r \in R$ and $s \in S$. If such a ring Q exists, then R satisfies the right Ore condition and $RS^{-1} \cong Q$. In the case that R is also domain, this is equivalent to saying that R can be embedded as a right order in a division ring D; in other words, each element of D can be written in the form rs^{-1} where $r, s \in R$ and $s \neq 0$. It is well known that a semiprime right Noetherian ring satisfies the right Ore condition.

A right Ore domain will mean a domain which satisfies the right Ore condition; thus by the above, a right Noetherian domain is a right Ore domain. Of course one can replace "right" with "left" in all of the above, and then an Ore domain will mean a domain which satisfies the Ore condition; i.e. both the right and left Ore condition. If G is a solvable group and k is a field such that kG is a domain, then the proposition of [36] shows that kG satisfies the Ore condition. Then one of the consequences of Lewin's results for example, is (see [36, theorem 1])

Theorem 4.2. *Let k be a field and let $G = G_1 *_H G_2$ be groups such that $H \lhd G$. Suppose kG_1 and kG_2 are domains, and kH satisfies the right Ore condition. Then kG is a domain.*

This result was applied by Formanek [25] to prove that if k is a field and G is a torsion free supersolvable group, then kG is a domain.

The next step was made by Brown, Farkas and Snider [6, 24] who realized that a combination of ring and K-theoretic techniques could be applied to the problem, especially solvable groups. Their techniques established that if k is a field of characteristic zero and G is a torsion free polycyclic-by-finite group, then kG is a domain. Building on these ideas, Cliff [8] established the zero divisor conjecture for group rings of polycyclic-by-finite groups over fields of arbitrary characteristic.

At this time it was already folklore that a suitable generalization of some well known K-theoretic theorems on polynomial rings, in particular on the Grothendieck group G_0, would yield stronger results for the zero divisor conjecture, especially for solvable groups. Let G be a group, let R be a ring, and let $R * G$ be a crossed product (see [47]). Thus $R * G$ is an associative ring with a 1, and it may be viewed as a free R-module with basis $\{\bar{g} \mid g \in G\}$, where each $\bar{g}$ is a unit in $R * G$. Another way of describing $R * G$ is that it is a G-graded ring with a unit in each degree (see [47, chapter 1, §2]). Of course $R * G$ is not uniquely determined by R and G in general, but this never seems to cause any confusion. Also it is clear that if $H \leqslant G$, then $R * H$ (the free R-submodule of $R * G$ with R-basis the elements of H) is also a crossed product and is a subring of $R * G$. Many theorems for group rings go over immediately to the crossed product situation. Thus for example, Theorem 4.1 becomes

Theorem 4.3. *Let k be a domain, let G be a right ordered group, and let $k * G$ be a crossed product. Then $k * G$ is a domain.*

To make induction arguments work, we would prefer to work with $R * G$ rather than the group ring RG. Indeed if $H \lhd G$, then a crossed product $R * G$ can be expressed as the crossed product $RH * [G/H]$, whereas the corresponding result for group rings, that if k is a field then kG can be expressed as the group ring $kH[G/H]$, is decidedly false.

The importance of G_0 for the zero divisor conjecture is as follows. If G is a torsion free group and $k * G$ is a crossed product, then one can often prove that $k * G$ can be embedded in a matrix ring $M_n(D)$ over a division ring D for some $n \in \mathbb{P}$ in a "nice way". Clearly what we need is that $n = 1$. If I is a minimal right ideal of $M_n(D)$, then $G_0(M_n(D)) = \langle [I] \rangle$, so we would like to prove that $G_0(M_n(D)) = \langle [M_n(D)] \rangle$. With the right setup, the inclusion of

$k * G$ in $M_n(D)$ induces an epimorphism of $G_0([k * G])$ onto $G_0(M_n(D))$, so it will be sufficient to prove that $G_0(k * G) = \langle[k * G]\rangle$.

If G is a finitely generated free abelian group, k a right Noetherian ring, and $k * G$ a crossed product, then by exploiting the fact that G can be ordered it has been known for a long time that the natural map $G_0(k) \to G_0(k * G)$ is an epimorphism; in particular if k is a field, then $G_0(k * G) = \langle[k * G]\rangle$. However for a long time better K-theoretic results (at least for applications to the zero divisor conjecture) seemed hard to come by. Then in 1986, John Moody came up with the following remarkable theorem (proved in [43, theorem 1]).

Theorem 4.4. *Let G be a finitely generated abelian-by-finite group, let R be a right Noetherian ring, and let $R * G$ be a crossed product. Then the induced map*

$$\bigoplus_{H \in \mathcal{F}(G)} G_0(R * H) \to G_0(R * G)$$

is surjective.

For an exposition of this result, see [9, 23] and [47, chapter 8]. Thus in the special case R is a division ring and G is torsion free finitely generated abelian-by-finite, we have that $G_0(R * G) = \langle[R * G]\rangle$, and using earlier remarks of this section, it is not difficult to prove that $R * G$ is a domain. Also an easy induction argument shows that Theorem 4.4 remains valid if G is replaced by an arbitrary polycyclic-by-finite group (this is in fact how Theorem 4.4 is stated in [43, theorem 1]). Another consequence of Theorem 4.4 is the following result, well known from when Theorem 4.4 was established.

Corollary 4.5. *Let G be an abelian-by-finite group, let k be a division ring, and let $k * G$ be a crossed product. If $k * H$ is a domain whenever H is a finite subgroup of G, then $k * G$ is an Ore domain.*

Proof (sketch). We may assume that G is finitely generated and $\Delta^+(G) = 1$. Let $A \lhd G$ with A free abelian and $[G : A] < \infty$. If $S = k * A \backslash 0$, then we can form the ring $k * GS^{-1}$, which will be an $n \times n$ matrix ring over a division ring for some $n \in \mathbb{P}$. Note that $k * H$ is a division ring whenever H is a finite subgroup of G. By Theorem 4.4 $G_0(k * G) = \langle[k * G]\rangle$, and by [34, lemma 2.2] the inclusion $k * G \to k * GS^{-1}$ induces an epimorphism $G_0(k * G) \to G_0(k * GS^{-1})$. Therefore $G_0(k * GS^{-1}) = \langle[k * GS^{-1}]\rangle$ and we deduce that $n = 1$, i.e. $k * GS^{-1}$ is a division ring. The result follows. $\square$

Another induction argument now gives the zero divisor conjecture for crossed products of torsion free solvable groups over right Noetherian domains; in fact it shows that if G is a torsion free solvable group, R is a right Ore domain and $R * G$ is a crossed product, then $R * G$ is also a right Ore domain. Roughly the argument goes as follows. To prove that $R * G$ is a right Ore domain, we may assume that G is finitely generated. Then there exists $H \lhd G$ such that G/H is finitely generated abelian-by-finite and H is "smaller" than G, so by induction we may assume that $R * F$ is a right Ore domain whenever F/H is a finite subgroup of G/H; let us say that $R * H$ is a right order in the division ring D. We now form the crossed product $D * [G/H]$, and since $D * [F/H]$ is a domain for all finite subgroups F/H of G/H, we deduce from Corollary 4.5 that $D * [G/H]$ is an Ore domain. It now follows easily that $R * G$ is a right Ore domain.

These arguments also apply to the case when G is an elementary amenable group. Recall that the class of elementary amenable groups, which we shall denote by $\mathcal{C}$, is the smallest class of groups which

(i) Contains all cyclic and all finite groups,
(ii) Is closed under taking group extension,
(iii) Is closed under directed unions.

Then $\mathcal{C}$ contains all solvable groups, and every elementary amenable group is amenable (see [48, 49] for much information on amenable groups). Then the arguments of above establish the following result.

Theorem 4.6. *Let $G \in \mathcal{C}$ and let R be a right Noetherian domain. If G is torsion free, then $R * G$ is a domain. In fact, $R * G$ is a right order in a division ring.*

More results along these lines can be found in [34].

Theorem 4.4 is very useful for Problem 3.1 and related problems. Whenever you can prove a conjecture related to zero divisors for a class of groups $\mathcal{D}$, then with the aid of Theorem 4.4, it is usually easy to prove it also for the class of groups $\mathcal{DC}$; an exception to this is Theorem 4.1.

Finally results of Lazard [35] imply that if p is an odd prime and G is the kernel of the natural epimorphism $\mathrm{GL}_n(\mathbb{Z}) \to \mathrm{GL}_n(\mathbb{Z}/p\mathbb{Z})$ (i.e. G is a congruence subgroup), then $\mathbb{Z}_p G$ is a domain (where $\mathbb{Z}_p$ denotes the p-adic integers; a similar result holds for $p = 2$). This is described in [23]; see also [46].

When proving the zero divisor conjecture and related problems, it seems in nearly all cases that one needs to not only show that the group ring is domain, but that it embeds in a division ring in some nice way. This is the case, for example, in Theorem 4.6.

We shall see that for the case $p = 2$, many of the above techniques are still relevant for Problem 3.1, but in the case $p > 2$, at least at the moment, they do not seem to be helpful and methods from Fourier analysis appear to be more useful.

5. ELEMENTARY RESULTS AND L^p-COHOMOLOGY

If G is a group with torsion, say $g \neq 1 = g^n$ for some $g \in G$ and $n \in \mathbb{P}$, then $(1 + g + \cdots + g^{n-1})(1 - g) = 0$, so there are zero divisors. Thus the simplest nontrivial case to consider is when G is infinite cyclic, say $G = \langle x \rangle$ where x has infinite order. If $L = L^p(G)$, $C_0(G)$ or $\mathbb{C}G$, and $\alpha \in \mathbb{C}G$, let us say that α is a zero divisor in L if there exists $\beta \in L \backslash 0$ such that $\alpha\beta = 0$, and that α is a nonzero divisor in L if no such β exists.

Theorem 5.1. *Let $G = \langle x \rangle$ where x has infinite order, and let $\xi \in \mathbb{C}$ where $|\xi| = 1$. Then*

(i) $x - \xi$ is a zero divisor in $L^\infty(G)$.

(ii) If $0 \neq \alpha \in \mathbb{C}G$, then α is a nonzero divisor in $C_0(G)$.

Proof. (i) $(x - \xi) \sum_{n=-\infty}^\infty \xi^{-n} x^n = 0$.

(ii) Write $\alpha = cx^m(x - a_1)\ldots(x - a_n)$ where $c, a_i \in \mathbb{C}$, $m \in \mathbb{Z}$, and $c \neq 0$. Then by induction on n, we may assume that $n = 1$, $m = 0$ and $c = 1$; in other words we may assume that $\alpha = x - a$ where $a \in \mathbb{C}$. Suppose $\alpha\beta = 0$ where $\beta \in C_0(G)$. Write $\beta = \sum_{n=-\infty}^\infty b_n x^n$ where $b_n \in \mathbb{C}$ for all n. Equating coefficients of x^{n+1}, we obtain $b_n = ab_{n+1}$ for all $n \in \mathbb{Z}$. Without loss of generality, we may assume that $|a| \leq 1$ and $b_1 \neq 0$. But then our equation on the coefficients yields $|b_n| \geq |b_1|$ for all $n \in \mathbb{P}$, which contradicts the hypothesis that $\beta \in C_0(G)$. $\square$

Thus though we cannot expect Problem 3.1 to have an affirmative answer in the case $p = \infty$, it seems plausible that it may have an affirmative answer in all other cases (and also in the case when $L^p(G)$ is replaced by $C_0(G)$).

Let us give some motivation for the problem from L^p-cohomology. For more detailed information we refer the reader to [7, 10, 11] and [26, §8]. Let X be a simplicial complex on which G acts freely, let X_r denote the set of r-simplices of X, let $C_r(X)$ denote the free abelian group with basis X_r, and let $\partial_r \colon C_r(X) \to C_{r-1}(X)$ denote the boundary map. For simplicity, we shall assume that X_r has only finitely many orbits for each $r \in \mathbb{N}$. Now define

$$L^p(X_r) = \{f \colon X_r \to \mathbb{C} \mid \sum_{\sigma \in X_r} |f(\sigma)|^p < \infty\}.$$

Then $L^p(X_r)$ is a Banach space under the norm $\|f\| = \left(\sum_{\sigma \in X_r} |f(\sigma)|^p\right)^{1/p}$; in fact it is isomorphic to $L^p(G)^{d_r}$ where d_r is the number of orbits of X_r. The coboundary map $\delta_r \colon L^p(X_r) \to L^p(X_{r+1})$ which obeys the rule $(\delta_r f)\sigma = f(\partial_{r+1}\sigma)$ for all $\sigma \in X_{r+1}$, is clearly a well defined bounded linear operator on $L^p(X_r)$. Thus $\ker \delta_r$ is a closed subspace of $L^p(X_r)$, but $\operatorname{im} \delta_r$ need not be closed. We now define the L^p-cohomology groups by

$$l_p \bar{\mathrm{H}}^r(X) = \frac{\ker \delta_r}{\operatorname{im} \delta_{r-1}}.$$

Since ∂_r commutes with the action of G, it follows that ∂_{r+1} is described by a $d_r \times d_{r+1}$ matrix all of whose entries are in $\mathbb{Z}G$, and δ_r is described by the transpose of this matrix. Therefore δ_r is described by a matrix all of whose entries are in $\mathbb{Z}G$. To determine $l_p \bar{\mathrm{H}}^r(X)$, we need to know about $\ker \delta_r$ and in particular when it is nonzero. The simplest case is when δ_r is 1×1 matrix. Thus we have come up against the problem stated in Problem 3.1.

In the case of L^2-cohomology, we can exploit the fact that $L^2(G)$ is a Hilbert space (see Section 8). Let M_r denote the orthogonal complement of $\operatorname{im} \delta_{r-1}$ in $\ker \delta_r$. Then M_r is a closed subspace and also a $\mathbb{C}G$-submodule of $L^2(X_r)$. It follows that M has a well defined von Neumann dimension $\dim_G(M)$ (which will be described precisely in Section 11). Then for $r \in \mathbb{N}$, the L_2-Betti numbers are defined by $b^r_{(2)}(X : G) = \dim_G M_r$. In the case G is a group whose finite subgroups have bounded order, results from studying Problem 3.1 show for example, that if G has a normal subgroup F such that F is a direct product of free groups and G/F is elementary amenable, then $\operatorname{lcm}(G)\, b^r_{(2)}(X:G) \in \mathbb{N}$ for all $r \in \mathbb{N}$ and for all X.

6. The case $p > 2$ and G abelian

In view of Theorem 5.1, it seems surprising that the answer to Problem 3.1 is negative if G is a noncyclic abelian group and $p > 2$. The work of this section describes work of my research student Mike Puls.

Throughout this section $d \in \mathbb{P}$, and G is a finitely generated free abelian group of rank d. Let $\mathbb{T}$ denote the torus, which we will think of as $[-\pi, \pi]/\{-\pi \sim \pi\}$, and let $\mathbb{T}^d = \mathbb{T} \times \cdots \times \mathbb{T}$, the d-torus. We can view $\mathbb{T}$ as the abelian group $\mathbb{R}/2\pi\mathbb{Z}$, and then $\mathbb{T}^d$ is also a group. This means that we can talk about cosets in $\mathbb{T}^d$; a coset of $\mathbb{T}^d$ will mean a coset of the form Ht where $H \leqslant \mathbb{T}^d$ and $t \in \mathbb{T}^d$, and the coset will be proper if $H \neq \mathbb{T}^d$. Let $\{x_1, \ldots, x_d\}$ be a $\mathbb{Z}$-basis for G. If $g = x_1^{n_1} \ldots x_d^{n_d} \in G$ (where $n_i \in \mathbb{Z}$), then we can define the Fourier transform $\hat{g} \colon \mathbb{T}^d \to \mathbb{C}$ by

$$\hat{g}(t_1, \ldots, t_d) = e^{i(n_1 t_1 + \cdots + n_d t_d)}$$

(where $t_i \in \mathbb{T}$). If $\alpha = \sum_{g \in G} \alpha_g g \in L^1(G)$, then we set

$$\hat{\alpha} = \sum_{g \in G} \alpha_g \hat{g} \colon \mathbb{T}^d \longrightarrow \mathbb{C},$$

and this extends the Fourier transform to $L^1(G)$. Set $Z(\alpha) = \{t \in \mathbb{T}^d \mid \hat{\alpha}(t) = 0\}$. Then Puls [52] proved the following result.

Theorem 6.1. *Suppose $\alpha \in L^1(G)$ and $Z(\alpha)$ is contained in a finite union of proper closed cosets. Then α is a nonzero divisor in $C_0(G)$.*

Let us indicate how this theorem is proved. If E is a closed subset of $\mathbb{T}^d$, then we define $I(E) = \{\beta \in L^1(G) \mid E \subseteq Z(\beta)\}$, $j(E)$ to be the set of all $\beta \in L^1(G)$ such that there exists an open subset O in $\mathbb{T}^d$ such that $E \subseteq O \subseteq Z(\beta)$, and $J(E)$ to be the closure of $j(E)$ in $L^1(G)$. Then $j(E) \subseteq J(E) \subseteq I(E)$, $J(E)$ and $I(E)$ are closed ideals in $L^1(G)$, and $\alpha \in I(Z(\alpha))$. We say that E is an *S-set* (or *set of spectral synthesis*) if $J(E) = I(E)$. We require the next result on the existence of S-sets, which follows from [55, Theorem 7.5.2], the remark just preceding that theorem, namely that C-sets are S-sets, and the remark immediately following that theorem, namely that C-sets are invariant under translation.

Proposition 6.2. *A finite union of closed cosets is an S-set.*

Define

$$\Phi(E) = \{h \in L^\infty(G) \mid \beta h = 0 \text{ for all } \beta \in I(E)\},$$
$$\Psi(E) = \{h \in L^\infty(G) \mid \beta h = 0 \text{ for all } \beta \in J(E)\}.$$

Then $\Phi(E) \subseteq \Psi(E)$ because $J(E) \subseteq I(E)$, and if F is a closed subset of E, then $\Psi(F) \subseteq \Psi(E)$. Now for $\alpha \in L^1(G)$, it follows from [55, Corollary 7.2.5a] that $J(Z(\alpha)) \subseteq \overline{\alpha L^1(G)}$, where $^-$ denotes the closure in $L^1(G)$. Therefore if $h \in L^\infty(G)$, then

(6.1) $\alpha h = 0$ implies $h \in \Psi(Z(\alpha))$.

We say that E is a set of *uniqueness* if $\Psi(E) \cap C_0(G) = 0$; clearly if E is a set of uniqueness and F is a closed subset of E, then F is also a set of uniqueness. It follows from (6.1) that if $Z(\alpha)$ is contained in a set of uniqueness, then α is a nonzero divisor in $C_0(G)$. Conversely if α is a nonzero divisor in $C_0(G)$ and $Z(\alpha)$ is an S-set, then $\Phi(Z(\alpha)) = \Psi(Z(\alpha))$ and we deduce that $Z(\alpha)$ is a set of uniqueness. Thus we have

Lemma 6.3. *Let $\alpha \in L^1(G)$.*

(i) *If $Z(\alpha)$ is contained in a set of uniqueness, then α is a nonzero divisor in $C_0(G)$.*

(ii) *If α is a nonzero divisor in $C_0(G)$ and $Z(\alpha)$ is an S-set, then $Z(\alpha)$ is a set of uniqueness.*

Proof of Theorem 6.1. For this proof, let us say that a hypercoset in $\mathbb{T}^d$ is a set of the form $Z(g - \xi)$ where $g \in G \backslash 1$, $\xi \in \mathbb{C}$ and $|\xi| = 1$. From [55, section 2.1], it is not difficult to see that every proper closed coset of $\mathbb{T}^d$ is contained in a hypercoset. Since $Z(\beta\gamma) = Z(\beta) \cup Z(\gamma)$ for $\beta, \gamma \in L^1(G)$, we see that any finite union of hypercosets in $\mathbb{T}^d$ is of the form $Z\big(\prod_i(g_i - \xi_i)\big)$ where $g_i \in G \backslash 1$, $\xi_i \in \mathbb{C}$ and $|\xi_i| = 1$.

If $1 \neq g \in G$ and $\xi \in \mathbb{C}$, then the same argument as in Theorem 5.1(ii) shows that $g - \xi$ is a nonzero divisor in $C_0(G)$. It follows that $\prod_i(g_i - \xi_i)$ is a nonzero divisor in $C_0(G)$ whenever $g_i \in G \backslash 1$ and $\xi_i \in \mathbb{C}$; the relevant case here is when $|\xi_i| = 1$ for all i. Using Proposition 6.2 and Lemma 6.3(ii), we see that any finite union of hypercosets is a set of uniqueness. Therefore α is a nonzero divisor in $C_0(G)$ by Lemma 6.3(i).   □

Let us now describe Puls's proof that if $G \cong \mathbb{Z}^2$, then there exists $\alpha \in \mathbb{C}G \backslash 0$ which is a zero divisor in $L^q(G)$ for some $q < \infty$ (we shall consider the case $G \cong \mathbb{Z}^d$ where $d > 2$ later, where it will be seen that better values of q can be obtained). Let $\{x, y\}$ be a basis for G. For $i, j \in \mathbb{Z}$ and $\beta \in L^\infty(G)$, we shall write β_{ij} or $\beta_{i,j}$ for $\beta_{x^i y^j}$. Given a bounded measure μ on $\mathbb{T}^2$, we can define its Fourier transform $\tilde{\mu} \in L^\infty(G)$ by

$$\tilde{\mu} = \sum_{m,n \in \mathbb{Z}} \tilde{\mu}_{mn} x^m y^n \quad \text{where} \quad \tilde{\mu}_{mn} = \int_{\mathbb{T}^2} e^{-i(ms+nt)}\, d\mu(s,t).$$

Then we can state

Proposition 6.4. *Let $\alpha \in L^1(G)$ and let μ be a bounded measure on $\mathbb{T}^2$. If μ is concentrated on $Z(\alpha)$, then $\alpha\tilde{\mu} = 0$.*

Proof. We need to prove that $(\alpha\tilde{\mu})_{ij} = 0$ for all $i, j \in \mathbb{Z}$. Replacing α with $\alpha x^{-i} y^{-j}$, we see that it is sufficient to prove that $(\alpha\tilde{\mu})_1 = 0$. Now

$$(\alpha\mu)_1 = \sum_{m,n} \alpha_{mn} \tilde{\mu}_{-m,-n} = \sum_{m,n} \alpha_{mn} \int_{\mathbb{T}^2} e^{i(ms+nt)} d\mu(s,t)$$

$$= \int_{\mathbb{T}^2} \sum_{m,n} \alpha_{mn} e^{ims} e^{int} d\mu(s,t) = \int_{Z(\alpha)} \hat{\alpha}(s,t) d\mu(s,t) = 0,$$

as required. $\qquad\square$

Thus it is easy to construct zero divisors α in $L^p(G)$ by choosing a nonzero μ; all that we need to verify is that $\tilde{\mu} \in L^p(G)$. To make this verification, we require theorems from Fourier analysis. Let $a, b \in \mathbb{R}$ such that $-\pi \leq a < b \leq \pi$, and let $\alpha \in L^1(G)$. Suppose $Z(\alpha)$ contains $\{(t, \theta(t)) \mid a \leq t \leq b\}$ where $\theta \colon [a, b] \to [-\pi, \pi]$ is smooth (i.e. infinitely differentiable). Define a measure μ on $\mathbb{T}^2$ by $\int_{\mathbb{T}^2} f d\mu = \int_a^b f(t, \theta(t)) \, dt$ for all measurable f. Then

$$\tilde{\mu}_{mn} = \int_a^b e^{-i(mt+n\theta(t))} \, dt$$

and $\tilde{\mu} \neq 0$ because $\tilde{\mu}_{00} = b - a$. What we need is that $\sum_{m,n\in\mathbb{Z}} |\tilde{\mu}_{mn}|^p < \infty$ for p large enough. This certainly will not be true in general, for example take $\theta = 0$. In fact if $\frac{d^2\theta}{dt^2}(t) = 0$ for all $t \in (a, b)$, then it is not difficult to see that $\tilde{\mu} \notin C_0(G)$. This is not surprising in view of Theorem 6.1, which in this case says that if $Z(\alpha)$ is contained in a finite union of lines with rational slope, then α is a nonzero divisor in $C_0(G)$. Let us assume that there exists $k \in \mathbb{P}$ such that for each $t \in [a, b]$, there exists $l \in \mathbb{P}$ such that $l \leq k$ and $\frac{d^l\theta}{dt^l}(t) \neq 0$ (where l depends on t). We need the following result from Fourier analysis, for which we refer to [57, §8.3].

Proposition 6.5. *In the above situation, there exists $A \in \mathbb{R}$ such that $|\tilde{\mu}_{mn}| \leq A(m^2 + n^2)^{-1/(2k)}$ for all $m, n \in \mathbb{Z}$.*

It now follows easily that if $p > 2k$, then $\sum_{m,n\in\mathbb{Z}} |\tilde{\mu}_{mn}|^p < \infty$ and hence $\tilde{\mu} \in L^p(G)$ for all $p > 2k$.

Example 6.6. *Let $\alpha = 2xy - x + y - 2 \in \mathbb{C}G$. Then α is a zero divisor in $L^p(G)$ for all $p > 4$.*

Proof. For $(s, t) \in \mathbb{T}^2$ (where $-\pi \leq s, t \leq \pi$), we have $\hat{\alpha}(s, t) = 2e^{is}e^{it} - e^{is} + e^{it} - 2$, thus $\hat{\alpha}(s, t) = 0$ when $e^{it} = \frac{e^{is}+2}{2e^{is}+1}$. Therefore $Z(\alpha) = \{(t, \theta(t)) \mid -\pi \leq t \leq \pi\}$ where $e^{i\theta(t)} = \frac{e^{it}+2}{2e^{it}+1}$ and we may write $\theta(t) = -i \log\left(\frac{e^{it}+2}{2e^{it}+1}\right)$, where we have taken the branch of log which satisfies $\log 1 = 0$. It is easily checked that θ is smooth and $\frac{d^2\theta}{dt^2}(t) \neq 0$ for all $t \in (-\pi, \pi)\backslash\{0\}$, in particular for all $t \in [\pi/4, 3\pi/4]$. As above, define a measure μ on $\mathbb{T}^2$ by $\int_{\mathbb{T}^2} f d\mu = \int_{\pi/4}^{3\pi/4} f(t, \theta(t)) \, dt$ for all measurable f. We can now apply Proposition 6.5 with $a = \pi/4$ and $b = 3\pi/4$ to deduce that $\tilde{\mu} \in L^p(G)$ for all $p > 4$, and

Proposition 6.4 to deduce that $\alpha\tilde{\mu} = 0$. Also $\tilde{\mu} \neq 0$, so we have shown that α is a zero divisor in $L^p(G)$ for all $p > 4$. $\square$

It is interesting to actually compute $\tilde{\mu}$ explicitly, though in the above example this seems somewhat messy. We could define a measure ν on $\mathbb{T}^2$ by $\int_{\mathbb{T}^2} f \, d\nu = \int_{-\pi}^{\pi} f(t, \theta(t)) \, dt$ for all measurable f, and then as above, $\tilde{\nu} \neq 0$ and $\alpha\tilde{\nu} = 0$. Since $\frac{d^2\theta}{dt^2}(t) = 0$ when $t = 0$ or $\pm\pi$, we cannot assert from Proposition 6.5 that $\tilde{\nu} \in L^p(G)$ for $p > 4$, but we do have $\frac{d^3\theta}{dt^3}(t) \neq 0$ for $t = 0$ or $\pm\pi$, so we can assert that $\tilde{\nu} \in L^p(G)$ for all $p > 6$. We now determine $\tilde{\nu}_{mn}$, which is

$$\int_{-\pi}^{\pi} e^{-i(mt+n\theta(t))} \, dt = \int_{-\pi}^{\pi} e^{-imt} \left(e^{-i\theta(t)}\right)^n \, dt$$

$$= \int_0^{2\pi} e^{-imt} \left(\frac{2e^{it}+1}{e^{it}+2}\right)^n \, dt.$$

For $m < 0$ and $n \geq 0$ contour integration shows that $\tilde{\nu}_{mn} = 0$, and then using the substitution $t \to -t$, we see that $\tilde{\nu}_{mn} = 0$ for $m > 0$ and $n \leq 0$. Also, $\nu_{00} = 2\pi$. Now the equality $\alpha\tilde{\nu} = 0$ yields $2\tilde{\nu}_{rs} - \tilde{\nu}_{r,s+1} + \tilde{\nu}_{r+1,s} - 2\tilde{\nu}_{r+1,s+1} = 0$, so we have a recurrence relation from which to calculate the other $\tilde{\nu}_{rs}$. This determines $\tilde{\nu}$ because $\tilde{\nu} = \sum_{r,s} \tilde{\nu}_{rs} x^r y^s$.

Of course, this argument can be generalized to the case $G = \mathbb{Z}^d$ where $d > 2$. To state Puls's results in this case, we need the concept of Gaussian curvature. We shall describe this here: for more details, see [57, §8.3]. Let S be a smooth $(d-1)$-dimensional submanifold of $\mathbb{R}^d$ and let $x_0 \in S$. Then after a change of coordinates (specifically a rotation), we may assume that in a sufficiently small open neighborhood of x_0, the surface is of the form $\{(x, \theta(x)) \mid x \in U\}$, where U is a bounded open subset of $\mathbb{R}^{d-1}$ and $\theta\colon U \to \mathbb{R}$ is a smooth function. Then we say that S has nonzero Gaussian curvature at x_0 if the $(d-1)\times(d-1)$ matrix

$$\left(\frac{\partial^2\theta}{\partial x_i \partial x_j}(x_0)\right)$$

is nonsingular. Then in [52], Puls proved the following.

Theorem 6.7. *Let* $\alpha \in \mathbb{C}\mathbb{Z}^d$ *where* $2 \leq d \in \mathbb{P}$, *and suppose there exists* $x_0 \in Z(\alpha)$ *such that there is a neighborhood* S *of* x_0 *in* $Z(\alpha)$ *which is a smooth* $(d-1)$-*dimensional manifold. If* S *has nonzero Gaussian curvature at* x_0, *then* α *is a zero divisor in* $L^p(G)$ *for all* $p > \frac{2d}{d-1}$.

He uses the above theorem to give the following set of examples of zero divisors in $L^p(G)$. Let G be the free abelian group of rank d and as before let $\{x_1, \ldots, x_d\}$ be a $\mathbb{Z}$-basis for G. Let

$$\alpha = \frac{2d-1}{2} - \frac{1}{2}\sum_{i=1}^{d}(x_i + x_i^{-1}).$$

Then $\alpha \in \mathbb{C}G$ and $\hat{\alpha}(t_1, \ldots, t_d) = \frac{2d-1}{2} - \sum_{i=1}^{d} \cos t_i$. In a neighborhood of $(0, \ldots, 0, \pi/3)$, we have that $Z(\alpha)$ is of the form $\{(t, \theta(t)) \mid t \in U\}$, where

U is a bounded open neighborhood of the origin in $\mathbb{R}^{d-1}$, $t = (t_1, \ldots, t_{d-1})$, and $\theta(t) = \cos^{-1}(\frac{2d-1}{2} - \sum_{i=1}^{d-1} \cos t_i)$. A computation shows that the matrix $\left(\frac{\partial^2 \theta(t)}{\partial t_i \partial t_j}\right)$ is nonsingular at $t = 0$, hence $Z(\alpha)$ has nonzero Gaussian curvature. Therefore α is a zero divisor in $L^p(G)$ for all $p > \frac{2d}{d-1}$.

Puls has also covered many other cases in [52], in which he requires the concept of the "type" of a manifold (see [57, §8.3.2]). Let us say that M is a hyperplane in $\mathbb{T}^d$ if there exists a hyperplane N in $\mathbb{R}^d$ such that $M = N \cap [-\pi, \pi]^d$. (We have been a little sloppy here: what we really mean is that we consider $\mathbb{T}^d$ as $[-\pi, \pi]^d$ with opposite faces identified, and let M' be the inverse image of M in $[-\pi, \pi]^d$. Then we say that M is a hyperplane to mean that M' is the intersection of a hyperplane in $\mathbb{R}^d$ with $[-\pi, \pi]^d$. Perhaps this is not a very good definition because for example, it allows points to be hyperplanes.) Then the results of [52] make it seem very likely that the following conjecture is true.

Conjecture 6.8. *Let G be a free abelian group of finite rank, and let $\alpha \in \mathbb{C}G$. Then α is a nonzero divisor in $L^p(G)$ for some $p \in \mathbb{P}$ (where $p > 2$) if and only if $Z(\alpha)$ is not contained in a finite union of hyperplanes. Furthermore if α is a zero divisor in $C_0(G)$, then α is a zero divisor in $L^p(G)$ for some $p < \infty$.*

7. THE CASE $p > 2$ AND G FREE

This section also describes work of Mike Puls. It will show that when $p > 2$ and G is a nonabelian free group, then the answer to Problem 3.1 is even more in the negative than in the case of G a noncyclic free abelian group of the last section.

Let G denote the free group of rank two on the generators $\{x, y\}$, let E_n denote the words of length n on $\{x, y\}$ in G, and let $e_n = |E_n|$. Thus $E_0 = \{1\}$, $E_1 = \{x, y, x^{-1}, y^{-1}\}$, $E_2 = \{x^2, y^2, x^{-2}, y^{-2}, xy, yx, x^{-1}y^{-1}, y^{-1}x^{-1}, xy^{-1}, y^{-1}x, x^{-1}y, yx^{-1}\}$ etc. It is well known that $e_n = 4 \cdot 3^{n-1}$ for all $n \in \mathbb{P}$. We shall let χ_n denote the characteristic function of E_n, i.e.

$$\chi_n = \sum_{g \in E_n} g \in \mathbb{C}G.$$

These elements of $\mathbb{C}G$ are often called radial functions and were studied in [12], which is where some of the ideas for what follows were obtained. Let

$$\Theta = 1 - \frac{1}{3}\chi_2 + \frac{1}{3^2}\chi_4 + \cdots + \frac{1}{(-3)^n}\chi_{2n} + \cdots.$$

Then for $p > 2$,

$$\|\Theta\|_p^p = 1 + \frac{e_1}{3^p} + \frac{e_2}{3^{2p}} + \cdots + \frac{e_n}{3^{np}} + \cdots$$

$$= 1 + \frac{4}{3} \cdot 3^{-(p-1)} + \frac{4}{3} \cdot 3^{-2(p-1)} + \cdots + \frac{4}{3} \cdot 3^{-n(p-1)} + \cdots.$$

This is a geometric series with ratio between successive terms $3^{-(p-1)}$, so it is convergent when $p - 1 > 1$. It follows that $\Theta \in L^p(G)$ for all $p > 2$.

We now set $\theta = \chi_1\Theta$ and show that $\theta = 0$. If $m \in \mathbb{P}$, $g \in E_m$, and $g = g_1g_2$ with $g_1 \in E_1$, then $g_2 \in E_{m-1} \cup E_{m+1}$. Furthermore there is exactly one choice for (g_1, g_2) if $g_2 \in E_{m-1}$, and exactly three if $g_2 \in E_{m+1}$. It follows for $n \in \mathbb{N}$ that $\theta_g = 0$ for $g \in E_{2n}$, and $\theta_g = \dfrac{1}{(-3)^n} + 3 \cdot \dfrac{1}{(-3)^{n+1}} = 0$ for $g \in E_{2n+1}$.

Thus we have shown that χ_1 is a zero divisor in $L^p(G)$ for all $p > 2$.

Of course there are similar results for radial functions of free groups on more than two generators, and these are established in [53].

8. Group von Neumann Algebras

We saw in Section 6 and Theorem 6.7 that for $p > 2$, one can construct many elements in $\mathbb{C}G$ which are zero divisors in $L^p(G)$. The situation for $L^2(G)$ is different, and there is evidence that the following conjecture is true.

Conjecture 8.1. *Let G be a torsion free group. If $0 \neq \alpha \in \mathbb{C}G$ and $0 \neq \beta \in L^2(G)$, then $\alpha\beta \neq 0$.*

The reason for this is that $L^2(G)$ is a Hilbert space, whereas the spaces $L^p(G)$ are not (unless G is finite). Indeed $L^2(G)$ becomes a Hilbert space with inner product

$$\langle \sum_{g\in G} \alpha_g g, \sum_{h\in G} \beta_h h \rangle = \sum_{g\in G} a_g \bar{b}_g,$$

where $\bar{}$ denotes complex conjugation. This inner product satisfies $\langle ug, v \rangle = \langle u, vg^{-1} \rangle$ for all $g \in G$, so if U is a right $\mathbb{C}G$-submodule of $L^2(G)$, then so is $U^\perp$. In the case of right ordered groups, the argument of Theorem 4.1 can be extended to show (see [40, thèorem II])

Theorem 8.2. *Let $H \triangleleft G$ be groups such that G/H is right orderable. Suppose that nonzero elements of $\mathbb{C}H$ are nonzero divisors in $L^2(H)$. Then nonzero elements of $\mathbb{C}G$ are nonzero divisors in $L^2(G)$.*

Thus taking $H = 1$ in the above theorem, we immediately see that Problem 3.1 has an affirmative answer in the case G is right orderable.

As mentioned at the end of Section 4, a key ingredient in proving the classical zero divisor conjecture is to embed the group ring in a division ring in some nice way, and the same is true here. To accomplish this, we need the concept of the group von Neumann algebra of G.

The formula of (3.1) also yields a multiplication $L^2(G) \times L^2(G) \to L^\infty(G)$ defined by

$$\sum_{g\in G} \alpha_g g \sum_{g\in G} \beta_g g = \sum_{g\in G} \left(\sum_{x\in G} \alpha_{gx^{-1}} \beta_x \right) g.$$

Now $\mathbb{C}G$ acts faithfully and continuously by left multiplication on $L^2(G)$, so we may view $\mathbb{C}G \subseteq \mathcal{L}(L^2(G))$. Let $W(G)$ denote the group von Neumann algebra of G: thus by definition, $W(G)$ is the weak closure of $\mathbb{C}G$ in $\mathcal{L}(L^2(G))$. For $\theta \in \mathcal{L}(L^2(G))$, the following are standard facts.

(i) $\theta \in W(G)$ if and only if there exist $\theta_n \in \mathbb{C}G$ such that $\lim_{n\to\infty}\langle\theta_n u, v\rangle \to \langle\theta u, v\rangle$ for all $u, v \in L^2(G)$.

(ii) $\theta \in W(G)$ if and only if $(\theta u)g = \theta(ug)$ for all $g \in G$.

Another way of expressing (ii) above is that $\theta \in W(G)$ if and only if θ is a right $\mathbb{C}G$-map. Using (ii), we see that if $\theta \in W(G)$ and $\theta 1 = 0$, then $\theta g = 0$ for all $g \in G$ and hence $\theta\alpha = 0$ for all $\alpha \in \mathbb{C}G$. It follows that $\theta = 0$ and so the map $W(G) \to L^2(G)$ defined by $\theta \mapsto \theta 1$ is injective. Therefore the map $\theta \mapsto \theta 1$ allows us to identify $W(G)$ with a subspace of $L^2(G)$. Thus algebraically we have

$$\mathbb{C}G \subseteq W(G) \subseteq L^2(G).$$

It is not difficult to show that if $\theta \in L^2(G)$, then $\theta \in W(G)$ if and only if $\theta\alpha \in L^2(G)$ for all $\alpha \in L^2(G)$. For $\alpha = \sum_{g\in G}\alpha_g g \in L^2(G)$, define $\alpha^* = \sum_{g\in G}\bar{\alpha}_g g^{-1} \in L^2(G)$. Then for $\theta \in W(G)$, we have $\langle\theta u, v\rangle = \langle u, \theta^* v\rangle$ for all $u, v \in L^2(G)$; thus θ^* is the adjoint of the operator θ.

If $\theta = \sum_{g\in G}\theta_g g \in W(G)$, then we define the trace map $\operatorname{tr}_G : W(G) \to \mathbb{C}$ by $\operatorname{tr}_G \theta = \theta_1$. Then for $\theta, \phi \in W(G)$, we have $\operatorname{tr}_G(\theta+\phi) = \operatorname{tr}_G \theta + \operatorname{tr}_G \phi$, $\operatorname{tr}_G \theta^* = \overline{\operatorname{tr}_G\theta}$ (where the bar denotes complex conjugation), $\operatorname{tr}_G(\theta\phi) = \operatorname{tr}_G(\phi\theta)$, and $\operatorname{tr}_G \theta = \langle\theta 1, 1\rangle$. For $n \in \mathbb{P}$, this trace map extends to $\operatorname{M}_n(W(G))$ by setting $\operatorname{tr}_G \theta = \sum_{i=1}^{n}\theta_{ii}$ when $\theta \in \operatorname{M}_n(W(G))$ is a matrix with entries θ_{ij} in $W(G)$, and then $\operatorname{tr}_G \theta\phi = \operatorname{tr}_G \phi\theta$ for $\phi \in \operatorname{M}_n(W(G))$. This will be more fully described in Section 11. An important property of the trace map is given by Kaplansky's theorem (see [42] and [38, proposition 9]) which states that if $e \in \operatorname{M}_n(W(G))$ is an idempotent and $e \neq 0$ or 1, then $\operatorname{tr}_G e \in \mathbb{R}$ and $0 < \operatorname{tr}_G e < n$.

At first glance, it seems surprising that $W(G)$ is useful for proving Conjecture 8.1 because if G contains an element of infinite order, then $W(G)$ contains uncountably many idempotents, so it is very far from being a domain. However it has a classical right quotient ring $U(G)$ which we shall now describe. Let $\mathcal{U}$ denote the set of all closed densely defined linear operators [33, §2.7] considered as acting on the left of $L^2(G)$. These are maps $\theta \colon L \to L^2(G)$ where L is a dense linear subspace of $L^2(G)$ and the graph $\{(u, \theta u) \mid u \in L\}$ is closed in $L^2(G)^2$. The adjoint map $*$ extends to $\mathcal{U}$ and for $\theta \in \mathcal{U}$, it satisfies $\langle\theta u, v\rangle = \langle u, \theta^* v\rangle$ whenever θu and $\theta^* v$ are defined. We now let $U(G)$ denote the operators in $\mathcal{U}$ "affiliated" to $W(G)$ [5, p. 150]; thus for $\theta \in \mathcal{U}$, we have $\theta \in U(G)$ if and only if $\theta(ug) = (\theta u)g$ for all $g \in G$ whenever θu is defined. Then $U(G) = U(G)^*$, $U(G)$ is a $*$-regular ring [4, definition 1, p. 229] containing $W(G)$, and every element of $U(G)$ can be written in the form $\gamma\delta^{-1}$ where $\gamma \in W(G)$ and δ is a nonzero divisor in $W(G)$ (see [5], especially theorem 1 and the proof of theorem 10). On the other hand, the trace map tr_G does not extend to $U(G)$. Now a $*$-regular ring R has the property that if $\alpha \in R$, then there exists a unique projection $e \in R$ (so e is an element satisfying $e = e^2 = e^*$) such that $\alpha R = eR$, in particular every element of R is either invertible or a zero divisor. Therefore we have embedded $W(G)$ into a ring in which every element is either a zero divisor or is invertible (so $U(G)$ is a classical right quotient ring for $W(G)$), and if $0 \neq \beta \in U(G)$, then $(\beta^*\beta)^n \neq 0$ for all $n \in \mathbb{N}$. Furthermore it is obvious that if γ is an automorphism of G, then γ extends in a unique way to automorphisms of $W(G)$ and $U(G)$. Given $\alpha \in L^2(G)$, we can define an element $\hat{\alpha} \in U(G)$ by setting $\hat{\alpha}u = \alpha u$ for all $u \in \mathbb{C}G$. Then $\hat{\alpha}$ is an unbounded operator on $L^2(G)$, densely defined because $\mathbb{C}G$ is a dense subspace of $L^2(G)$ (of course $\hat{\alpha}$ does not define an element of $\mathcal{L}(L^2(G))$ in general, because the product of two elements of $L^2(G)$ does not always lie in $L^2(G)$, only in $L^\infty(G)$). It is not difficult to show that $\hat{\alpha}$ extends to a closed operator on $L^2(G)$ (see the proof of Lemma 11.3), which we shall also call $\hat{\alpha}$. Thus $\hat{\alpha}$ is an element of $\mathcal{U}$. Since $\hat{\alpha}(ug) = (\hat{\alpha}u)g$ for all $u \in \mathbb{C}G$ and $g \in G$, and $\mathbb{C}G$ is dense in $L^2(G)$, it follows (cf. [33, remark 5.6.3]) that $\hat{\alpha} \in U(G)$. Thus we have a map $L^2(G) \to U(G)$ defined by $\alpha \mapsto \hat{\alpha}$ which is obviously an injection. Algebraically, we now have

(8.1) $$\mathbb{C}G \subseteq W(G) \subseteq L^2(G) \subseteq U(G).$$

Similar properties to those of the above paragraph hold for matrix rings over $U(G)$. Let $n \in \mathbb{P}$. Then $M_n(\mathbb{C}G)$ acts continuously by left multiplication on $L^2(G)^n$, and $M_n(W(G))$ is the weak closure of $M_n(\mathbb{C}G)$ in $\mathcal{L}(L^2(G)^n)$. Also $M_n(U(G))$ is the set of closed densely defined linear operators acting on the left of $L^2(G)^n$ which are affiliated to $M_n(W(G))$. For $\theta \in M_n(U(G))$, the adjoint θ^* of θ satisfies $\langle \theta u, v \rangle = \langle u, \theta^* v \rangle$ for $u, v \in L^2(G)^n$ whenever θu and $\theta^* v$ are defined. If θ is represented by the matrix (θ_{ij}) where $\theta_{ij} \in U(G)$, then θ^* is represented by the matrix (θ_{ji}^*). Then $M_n(U(G))$ is a $*$-regular ring containing $M_n(W(G))$, and every element of $M_n(U(G))$ can be written in the form $\alpha\beta^{-1}$ where $\alpha \in M_n(W(G))$ and β is a nonzero divisor in $M_n(W(G))$. Furthermore every projection of $M_n(U(G))$ lies in $M_n(W(G))$ (use [5, theorem 1]). This means that if $\alpha \in M_n(U(G))$, then $\alpha M_n(U(G)) = e M_n(U(G))$ for a unique projection $e \in M_n(W(G))$. Thus we can define $\text{rank}_G \alpha = \text{tr}_G e$; the following lemma (see [41, Lemma 2.3]) gives some easily derived properties of rank_G; part (ii) requires Kaplansky's theorem on the trace of idempotents mentioned earlier in this section.

Lemma 8.3. *Let G be a group and let $\theta \in M_n(U(G))$. Then*

(i) $\text{rank}_G \theta\alpha = \text{rank}_G \theta = \text{rank}_G \alpha\theta$ *for all $\alpha \in GL_n(U(G))$.*
(ii) *If $0 \neq \theta \notin GL_n(U(G))$, then $0 < \text{rank}_G \theta < n$.*

Two other useful results are

Lemma 8.4. *(See [39, lemma 13].) Let G be a group, let $n \in \mathbb{P}$, and let e, f be projections in $M_n(U(G))$. If $f = ueu^{-1}$ for some unit $u \in M_n(U(G))$, then $f = vev^{-1}$ for some unit $v \in M_n(W(G))$.*

Lemma 8.5. *Let G be a group, let $n \in \mathbb{P}$, and let e, f be projections in $M_n(U(G))$. Suppose that $e M_n(U(G)) \cap f M_n(U(G)) = 0$ and $e M_n(U(G)) + f M_n(U(G)) = h M_n(U(G))$ where h is a projection in $M_n(U(G))$. Then $\text{tr}_G e + \text{tr}_G f = \text{tr}_G h$.*

Proof. This follows from the parallelogram law [4, §13]. Alternatively one could note that $e \, \mathrm{M}_n(W(G)) \cap f \, \mathrm{M}_n(W(G)) = 0$ and then apply [39, lemmas 11(i) and 12]. □

Suppose $d, n \in \mathbb{P}$, $H \leqslant G$ are groups such that $[G : H] = n$, and $\{x_1, \dots, x_n\}$ is a left transversal for H in G. Then as Hilbert spaces $L^2(G)^d = \bigoplus_{i=1}^n x_i L^2(H)^d$, hence we may view elements of $\mathcal{L}(L^2(G)^d)$ as acting on $\bigoplus_{i=1}^n x_i L^2(H)^d$ and we deduce that we have a monomorphism $\hat{} : \mathcal{L}(L^2(G)^d) \to \mathcal{L}(L^2(H)^{dn})$. It is not difficult to see that $\hat{}$ takes $\mathrm{M}_d(W(G))$ into $\mathrm{M}_{dn}(W(H))$, which yields the following result (cf. [2, (16) on p. 23])

Lemma 8.6. *Let $H \leqslant G$ be groups such that $[G : H] = n < \infty$, and let $d \in \mathbb{P}$. If $\theta \in \mathrm{M}_d(W(G))$, then $\mathrm{tr}_H \hat{\theta} = n \, \mathrm{tr}_G \theta$.*

We can now explain the usefulness of $U(G)$. Suppose we have proved Conjecture 8.1 for the torsion free group G. Then we have in particular that if $0 \neq \alpha \in \mathbb{C}G$ and $0 \neq \theta \in W(G)$, then $\alpha\theta \neq 0$. Since $U(G)$ is a classical right quotient ring for $W(G)$, it follows that α is invertible in $U(G)$. Thus in the special case that $\mathbb{C}G$ is a right order in a division ring (this will be the case when G is elementary amenable: see Theorem 4.6), we can deduce that there is a division ring D such that $\mathbb{C}G \subseteq D \subseteq U(G)$. This was exploited in [39] to obtain the following result.

Theorem 8.7. *Let G be a torsion free elementary amenable group. Then there exists a division ring D such that $\mathbb{C}G \subseteq D \subseteq U(G)$.*

Of course in the above theorem, D can be chosen so that $\mathbb{C}G$ is a right order in D, see Theorem 4.6. In view of this theorem, it seems plausible that the following conjecture is true.

Conjecture 8.8. *If G is a torsion free group, then there exists a division ring D such that $\mathbb{C}G \subseteq D \subseteq U(G)$.*

Note that the above conjecture implies Conjecture 8.1. Indeed if $0 \neq \alpha \in \mathbb{C}G$, then the above conjecture shows that α is invertible in $U(G)$, in particular $\alpha\beta \neq 0$ for all $\beta \in U(G) \backslash 0$. Then (8.1) shows that $\alpha\beta \neq 0$ for all $\beta \in L^2(G) \backslash 0$. Thus combining Theorems 8.2 and 8.7, we obtain the following.

Theorem 8.9. *Let $H \lhd G$ be groups where H is torsion free elementary amenable and G/H is right ordered. If $0 \neq \alpha \in \mathbb{C}G$ and $0 \neq \beta \in L^2(G)$, then $\alpha\beta \neq 0$.*

We conclude this section with an amusing example. Recall that the group G is of *exponential growth* (see eg. [48, p. 219]) if there is a finite subset C of G such that $\lim_{n\to\infty} |C^n|^{1/n} > 1$ (where C^n denotes the subset of G consisting of all products of at most n elements of C). We say G is exponentially bounded if it does not have exponential growth.

Example 8.10. *Let p be a prime, let $d \in \mathbb{P}$, and let H be an exponentially bounded residually finite p-group which can be generated by d elements. Write*

$H \cong F/K$ where F is the free group of rank d, and write $G = F/K'$. Then there exists a division ring D such that $\mathbb{C}G \subseteq D \subseteq U(G)$ and $\mathbb{C}G$ is a right order in D.

Of course, any finite p-group will satisfy the hypothesis for H in the above example (provided that H can be generated by d-elements), but then G will be torsion free elementary amenable and we are back in the case of Theorem 8.7. However there exist infinite periodic groups satisfying the above hypothesis for H [21, 27]; also Grigorchuk has constructed examples of such groups. Now a finitely generated elementary amenable periodic group must be finite [48, §3.11], hence H and also G cannot be elementary amenable when H is infinite. On the other hand, if H is chosen to be a periodic group, then it is easy to see that G does not contain a subgroup isomorphic to a nonabelian free group.

Proof of Example 8.10. First we show that G is right orderable. Let $\{F_i/K \mid i \in \mathcal{I}\}$ be the family of normal subgroups in H of index a power of p, and set $L = \bigcap_{i \in \mathcal{I}} F_i'$. Then F/F_i' has a finite normal series whose factors are all isomorphic to $\mathbb{Z}$ [22, §4, lemma 5], thus by the remarks just before Theorem 4.1 we see that F/F_i' is right ordered and hence so is F/L. Now $K' \leqslant L \leqslant K$, so L/K' is right ordered and we deduce that G is right ordered (again, use the remarks on right ordered groups just before Theorem 4.1). It follows from Theorem 4.1 that $\mathbb{C}G$ is a domain.

Since G is exponentially bounded, G is amenable by [48, proposition 6.8]. Now [58] tells us that if k is a field and M is an amenable group such that kM is a domain, then kM is an Ore domain. Thus $\mathbb{C}G$ is a right order in a division ring D. Since nonzero elements in $\mathbb{C}G$ are nonzero divisors in $L^2(G)$ by Theorem 8.2, it follows that the inclusion of $\mathbb{C}G$ in $L^2(G)$ extends to a ring monomorphism of D into $U(G)$, and the result follows. $\qquad\Box$

9. UNIVERSAL LOCALIZATION

The next step is to extend Theorem 8.7 to other groups. Since "most" (but not all) nonelementary amenable groups contain a nonabelian free subgroup, it is plausible to consider nonabelian free groups next. Here we come up with the problem that although $\mathbb{C}G$ is a domain, it does not satisfy the Ore condition. Indeed if G is the free group of rank two on $\{x, y\}$, then the fact that $(x-1)\mathbb{C}G \cap (y-1)\mathbb{C}G = 0$ shows that $\mathbb{C}G$ does not satisfy the Ore condition. If R is a subring of the ring S, the division closure [15, exercise 7.1.4, p. 387] of R in S, which we shall denote by $D(R, S)$, is the smallest subring of S containing R which is closed under taking inverses (i.e. $s \in D(R, S)$ and $s^{-1} \in S$ implies $s^{-1} \in D(R, S)$); perhaps a better concept is the closely related one of "rational closure" [15, p. 382], but division closure will suffice for our purposes. The division closure of $\mathbb{C}G$ in $U(G)$ will be indicated by $D(G)$. Obviously if R is an Artinian ring, then $D(R, S) = R$. In the case S is a division ring, the division closure of R is simply the smallest division subring of S containing R; thus Conjecture 8.8 could be restated as $D(G)$ is a division ring whenever G is torsion free. The following four elementary lemmas are very useful.

Lemma 9.1. *Let $R \subseteq S$ be rings, let D denote the division closure of R in S, and let $n \in \mathbb{P}$. If D is an Artinian ring, then $M_n(D)$ is the division closure of $M_n(R)$ in $M_n(S)$.*

Proof. Exercise, or see [41, lemma 4.1]. □

Lemma 9.2. *Let G be a group and let α be an automorphism of G. Then $\alpha D(G) = D(G)^* = D(G)$.*

Proof. Of course, here we have regarded α as an automorphism of $U(G)$, and $*$ as an antiautomorphism of $U(G)$; see Section 8. The result follows because $\alpha \mathbb{C}G = \mathbb{C}G^* = \mathbb{C}G$. □

Lemma 9.3. *(cf. [41, lemma 2.1].) Let $H \lhd G$ be groups, and let $D(H)G$ denote the subring of $D(G)$ generated by $D(H)$ and G. Then $D(H)G \cong D(H) * G/H$ for a suitable crossed product.*

Proof. Let T be a transversal for H in G. Since $h \mapsto tht^{-1}$ is an automorphism of H, we see that $tD(H)t^{-1} = D(H)$ for all $t \in T$ by Lemma 9.2, and so $D(H)G = \sum_{t\in T} D(H)t$. This sum is direct because the sum $\sum_{t\in T} U(H)t$ is direct, and the result is established. □

Lemma 9.4. *Let $H \lhd G$ be groups such that G/H is finite, and suppose $D(H)$ is Artinian. Then $D(G)$ is semisimple Artinian and is a crossed product $D(H) * G/H$.*

Proof. Let $D(H)G$ denote the subring generated by $D(H)$ and G. Then Lemma 9.3 shows that $D(H)G \cong D(H) * G/H$, hence $D(H)G$ is Artinian and we deduce that $D(H)G = D(G)$. Now $D(G) = D(G)^*$ by Lemma 9.2 and if $0 \neq \alpha \in D(G)$, then $(\alpha^*\alpha)^n \neq 0$ for all $n \in \mathbb{N}$. Therefore $D(G)$ has no nonzero nilpotent ideals, and the result follows. □

More generally for $n \in \mathbb{P}$, we denote the division closure of $M_n(\mathbb{C}G)$ in $M_n(U(G))$ by $D_n(G)$, and let $W_n(G) = D_n(G) \cap M_n(W(G))$. Then we have (see [41, proposition 5.1])

Proposition 9.5. *Let G be a group and let $n \in \mathbb{P}$. Then*

(i) *If e is an idempotent in $D_n(G)$, then there exists $\alpha \in GL_1(D_n(G))$ such that $eD_n(G) = \alpha eD_n(G)$ and $\alpha e\alpha^{-1}$ is a projection; in particular $eD_n(G) = fD_n(G)$ for some projection $f \in D_n(G)$.*

(ii) *If $\alpha \in D_n(G)$, then there exists a nonzero divisor $\beta \in W_n(G)$ such that $\beta\alpha \in W_n(G)$.*

The following result shows that if $D(G)$ is Artinian, then there is a bound on the length of a descending chain of right ideals in $D(G)$ in terms of the real numbers $\mathrm{tr}_G e$ for e a projection in $D(G)$.

Lemma 9.6. *Let G be a group and let $l \in \mathbb{P}$. Suppose that $D(G)$ is Artinian and that $l\,\mathrm{tr}_G e \in \mathbb{Z}$ for all projections $e \in D(G)$. If $I_0 > I_1 > \cdots > I_r$ is a strictly descending sequence of right ideals in $D(G)$, then $r \leq l$.*

Proof. Since $D(G)$ is semisimple Artinian by Lemma 9.4, the descending sequence of right ideals yields nonzero right ideals $J_1, \ldots, J_r$ of $D(G)$ such that $D(G) = J_1 \oplus \cdots \oplus J_r$. Write $1 = e_1 + \cdots + e_r$ where $e_i \in J_i$. Then $e_i^2 = e_i$ and $e_i e_j = 0$ for $i \neq j$ $(1 \leq i, j \leq r)$, hence

$$U(G) = e_1 U(G) \oplus \cdots \oplus e_r U(G).$$

In view of Proposition 9.5(i), there exist nonzero projections $f_i \in D(G)$ such that $e_i D(G) = f_i D(G)$ $(1 \leq i \leq r)$. Then $e_i U(G) = f_i U(G)$ and it now follows from Lemma 8.5 that $1 = \operatorname{tr}_G f_1 + \cdots + \operatorname{tr}_G f_r$, upon which an application of Kaplansky's theorem (see Section 8) completes the proof. $\square$

When constructing the classical right quotient ring of a ring D which satisfies the right Ore condition, one only inverts the nonzero divisors of D, but for more general rings it is necessary to consider inverting matrices. For any ring homomorphism f, we shall let f also denote the homomorphism induced by f on all matrix rings. Let Σ be any set of square matrices over a ring R. Then in [15, §7.2], Cohn constructs a ring R_Σ and a ring homomorphism $\lambda \colon R \to R_\Sigma$ such that the image of any matrix in Σ under λ is invertible. Furthermore R_Σ and λ have the following universal property: given any ring homomorphism $f \colon R \to S$ such that the image of any matrix in Σ under f is invertible, then there exists a unique ring homomorphism $\bar{f} \colon R_\Sigma \to S$ such that $\bar{f}\lambda = f$. The ring R_Σ always exists and is unique up to isomorphism, though in general λ is neither injective nor surjective. It obviously has the following useful property: if θ is an automorphism of R such that $\theta(\Sigma) = \Sigma$, then θ extends in a unique way to an automorphism of R_Σ.

Note that if R is a subring of the ring S, $D = D(R, S)$, and Σ is the set of matrices with entries in R which become invertible over D, then the inclusion $R \hookrightarrow D$ extends to a ring homomorphism $R_\Sigma \to D$. However even in the case D is a division ring, this homomorphism need not be an isomorphism.

A notable feature of the above construction of R_Σ, which is developed by Cohn in [15, §7] and Schofield in [56], is that it extends much of the classical theory of localization of Noetherian (noncommutative) rings to arbitrary rings. Indeed if S is a subset of R which contains 1, is closed under multiplication, and satisfies the Ore condition, then $RS^{-1} \cong R_S$. On the other hand, in general it is not possible to write every element of R_Σ in the form rs^{-1} with $r, s \in R$. There are "Goldie rank" versions of Conjecture 8.8. If k is a field, G is polycyclic-by-finite, and $\Delta^+(G) = 1$, then kG is a right order in a $d \times d$ matrix ring for some $d \in \mathbb{P}$. The Goldie rank conjecture states that $d = \operatorname{lcm}(G)$. This is now known to be true, and extensions of this were considered in [34]; in particular it was proved that if k is a field, G is an elementary amenable group with $\Delta^+(G) = 1$, and the orders of the finite subgroups of G are bounded, then kG is a right order in an $l \times l$ matrix ring over a division ring where $l = \operatorname{lcm}(G)$ [34, theorem 1.3]. The proof of this depends heavily on Moody's Theorem, as described in Theorem 4.4. We describe two versions of the Goldie rank conjecture.

Conjecture 9.7. *Let G be a group such that $\Delta^+(G) = 1$, and let Σ denote the matrices with entries in $\mathbb{C}G$ which become invertible over $D(G)$. Suppose the orders of the finite subgroups of G are bounded, and $l = \mathrm{lcm}(G)$. Then there is a division ring D such that $D(G) \cong \mathrm{M}_l(D) \cong \mathbb{C}G_\Sigma$.*

Conjecture 9.8. *Let G be a group such that the orders of the finite subgroups of G are bounded, and let $l = \mathrm{lcm}(G)$. If $n \in \mathbb{P}$ and $\alpha \in \mathrm{M}_n(\mathbb{C}G)$, then $l\,\mathrm{rank}_G\,\alpha \in \mathbb{N}$.*

10. C*-ALGEBRA TECHNIQUES

There is a close connection between problems related to zero divisors in $L^2(G)$ and projections in $W(G)$. Indeed Lemma 12.3 states that if $\mathrm{rank}_G\,\theta \in \mathbb{Z}$ for all $\theta \in \mathrm{M}_n(\mathbb{C}G)$ and for all $n \in \mathbb{P}$, then Conjecture 8.8 is true, and of course $\mathrm{rank}_G\,\theta$ is defined in terms of the trace of a projection in $\mathrm{M}_n(W(G))$ (Section 8). Recall that the *reduced group C*-algebra* $C^*_r(G)$ of G is the strong closure (as opposed to the weak closure for $W(G)$) of $\mathbb{C}G$ in $\mathcal{L}(L^2(G))$: thus $\mathbb{C}G \subseteq C^*_r(G) \subseteq W(G)$. There is a conjecture going back to Kaplansky and Kadison that if G is a torsion free group, then $C^*_r(G)$ has no idempotents except 0 and 1 (this is equivalent to $C^*_r(G)$ having no projections except 0 and 1). The special case G is a nonabelian free group is of particular interest, because at one time there was an open problem to as whether a simple C*-algebra was generated by its projections. Powers [51, theorem 2] proved that $C^*_r(G)$ is simple for G a nonabelian free group, so it was then sufficient to show that $C^*_r(G)$ had no nontrivial projections, but this property turned out to be more difficult to prove. However Pimsner and Voiculescu [50] established this property, thus obtaining a simple C*-algebra ($\neq \mathbb{C}$) with no nontrivial projections. Connes [16, §1] (see [20] for an exposition) gave a very elegant proof of the Pimsner-Voiculescu result, and his method was used in [41] to establish Conjecture 8.8 in the case G is a free group. For further information on this topic, see the survey article [59].

As has already been remarked, in view of Lemma 12.3 we want to prove that $\mathrm{tr}_G\,e \in \mathbb{Z}$ for certain projections e. Now in his proof that $C^*_r(G)$ has no nontrivial projections, this is exactly what Connes does. Once it is established that $\mathrm{tr}_G\,e \in \mathbb{Z}$, then the result follows from Kaplansky's theorem (§8). Of course Connes is dealing with projections in $C^*_r(G)$, while we are interested in projections which are only given to lie in $\mathrm{M}_n(W(G))$ for some $n \in \mathbb{P}$, but the Connes argument is still applicable. Connes uses a Fredholm module technique in which he constructs a "perturbation" π of $C^*_r(G)$ where G is the free group of rank two such that if $C^*_r(G)$ has a nontrivial projection, then there is a nontrivial projection $e \in C^*_r(G)$ such that the operator $e - \pi e$ on $L^2(G)$ is of trace class (though $\pi e \notin C^*_r(G)$), and it follows that the trace of $e - \pi e$ is an integer [20, lemma 4.1]. He then shows that this trace is in fact $\mathrm{tr}_G\,e$ [20, §5], thus proving that $\mathrm{tr}_G\,e \in \mathbb{Z}$ as required.

To apply Lemma 12.3 when G is the free group of rank two, we use the same perturbation π. This has the property that if $\theta \in \mathrm{M}_n(\mathbb{C}G)$ for some $n \in \mathbb{P}$, then the resulting operators $\theta, \pi(\theta)$ on $L^2(G)^n$ agree on a subspace of finite codimension. It follows that if e, e' are the projections of $L^2(G)^n$ onto $\overline{\operatorname{im} \theta}, \overline{\operatorname{im} \theta'}$ respectively, then $\operatorname{im}(e - e')$ has finite dimension and therefore has a well defined trace which is an integer. Then as in the previous paragraph, this integer turns out to be $\operatorname{tr}_G e$ and we deduce that $\operatorname{rank}_G \theta \in \mathbb{Z}$ as required. The same arguments are applied in Lemma 12.2 for the case when G is a finite direct product of free groups of rank two. For the purposes of trying to extend this to other classes of groups, it seems necessary to have that $\theta, \pi(\theta)$ agree on a subspace of finite codimension: it is not enough for $\theta - \pi(\theta)$ to have trace class, because this does not imply that $e - e'$ has trace class.

To construct the perturbation π, Connes uses the following result for free groups (see [20, section 4], [32, corollary 1.5], [18, §3]). We say that a function $\phi \colon X \to Y$ between the left G-sets X and Y is an almost G-map if for all $g \in G$, the set $\{x \in X \mid g(\phi x) \neq \phi(gx)\}$ is finite.

Theorem 10.1. *Let $\kappa \in \mathbb{N}$, let G be a free group of rank κ, let κG denote the free left G-set with κ orbits, and let $\{*\}$ denote the G-set consisting of one fixed point. Then there exists a bijective almost G-map $\phi \colon G \to \kappa G \cup \{*\}$.*

In fact the above is the only property of free groups that Connes uses, and it is also the only property of free groups used in [41] in establishing Conjecture 8.8 for G a free group. Thus it was of considerable interest to determine which other groups satisfied the conclusion of the above theorem. However Dicks and Kropholler [18] showed that free groups were the only such groups.

After proving Conjecture 8.8 for free groups, the following was established in [41] (see [41, theorem 1.5] for a generalization).

Theorem 10.2. *Let $n \in \mathbb{P}$, let $F \lhd G$ be groups such that F is free, G/F is elementary amenable, and $\Delta^+(G) = 1$, and let $D_n(G)$ denote the division closure of $\operatorname{M}_n(\mathbb{C}G)$ in $\operatorname{M}_n(U(G))$. Assume that the finite subgroups of G have bounded order, and that $l = \operatorname{lcm}(G)$. Then there exists a division ring D such that $D_n(G) \cong \operatorname{M}_{ln}(D)$.*

Of course the special case $l = n = 1$ in the above theorem yields Conjecture 8.8 for groups G which have a free subgroup F such that G/F is elementary amenable. The subsequent sections will be devoted to a proof of the following result.

Theorem 10.3. *Let $F \lhd G$ be groups, and let Σ denote the set of matrices with entries in $\mathbb{C}G$ which become invertible over $D(G)$. Suppose F is a direct product of free groups, G/F is elementary amenable, and the orders of the finite subgroups of G are bounded. Then $D(G)$ is a semisimple Artinian ring and the identity map on $\mathbb{C}G$ extends to an isomorphism $\mathbb{C}G_\Sigma \to D(G)$. Furthermore if $e \in D(G)$ is a projection, then $\operatorname{lcm}(G) \operatorname{tr}_G e \in \mathbb{Z}$ for all projections $e \in D(G)$.*

It seems very plausible that it is easy to extend the above theorem to the case when F is a subgroup of a direct product of free groups, and it certainly would be nice to establish this stronger result. However subgroups of direct products can cause more difficulty than one might intuitively expect, see for example [3]. In fact if $H \lhd G$ are groups such that G is torsion free, G/H is finite, and H is a subgroup of a direct product of free groups, then it is even unknown whether $\mathbb{C}G$ is a domain.

One can easily read off a number of related results from Theorem 10.3, for example

Corollary 10.4. *Let $F \lhd G$ be groups such that F is a direct product of free groups and G/F is elementary amenable, let $n \in \mathbb{P}$, and let $D_n(G)$ denote the division closure of $\mathrm{M}_n(\mathbb{C}G)$ in $\mathrm{M}_n(U(G))$. Suppose $\Delta^+(G) = 1$ and the orders of the finite subgroups of G are bounded, and set $l = \mathrm{lcm}(G)$. Then $D_n(G) \cong \mathrm{M}_{ln}(D)$ for some division ring D.*

For further recent information on these analytic techniques, especially in the case G is a free group, we refer the reader to the survey article [28].

11. $L^2(G)$-MODULES

We define $\mathbb{E} = \mathbb{N} \cup \{\infty\}$, where ∞ denotes the first infinite cardinal. Let G be a group, and let $L^2(G)^\infty$ denote the Hilbert direct sum of ∞ copies of $L^2(G)$, so $L^2(G)^\infty$ is a Hilbert space. Following [10, section 1], an $L^2(G)$-module $\mathcal{H}$ is a closed right $\mathbb{C}G$-submodule of $L^2(G)^n$ for some $n \in \mathbb{E}$, an $L^2(G)$-submodule of $\mathcal{H}$ is a closed right $\mathbb{C}G$-submodule of $\mathcal{H}$, an $L^2(G)$-ideal is an $L^2(G)$-submodule of $L^2(G)$, and an $L^2(G)$-homomorphism or $L^2(G)$-map $\theta \colon \mathcal{H} \to \mathcal{K}$ between $L^2(G)$-modules is a continuous right $\mathbb{C}G$-map. If X is an $L^2(G)$-ideal, then $X^\perp$ is also an $L^2(G)$-ideal, so $L^2(G) = X \oplus X^\perp$ as $L^2(G)$-modules. The following lemma shows that there can be no ambiguity in the meaning of two $L^2(G)$-modules being isomorphic.

Lemma 11.1. *Let $\mathcal{H}$ and $\mathcal{K}$ be $L^2(G)$-modules, and let $\theta \colon \mathcal{H} \to \mathcal{K}$ be an $L^2(G)$-map. If $\ker \theta = 0$ and $\overline{\mathrm{im}\,\theta} = \mathcal{K}$, then there exists an isometric $L^2(G)$-isomorphism $\phi \colon \mathcal{H} \to \mathcal{K}$.*

Proof. See [10, p. 134] and [45, §21.1]. $\square$

Lemma 11.2. *If U is an $L^2(G)$-ideal, then $U = uL^2(G)$ for some $u \in U$.*

Proof. Let e be the projection of $L^2(G)$ onto U. Then $e \in W(G)$ because U is a right $\mathbb{C}G$-module, and $eL^2(G) = U$. Thus $e1 \in U$ and we may set $u = e1$. $\square$

Lemma 11.3. *Let $n \in \mathbb{E}$, let $u \in L^2(G)^n$, and let $U = \overline{u\mathbb{C}G}$. Then U is $L^2(G)$-isomorphic to an $L^2(G)$-ideal.*

Proof. Define an unbounded operator $\theta \colon L^2(G) \to U$ by $\theta\alpha = u\alpha$ for all $\alpha \in \mathbb{C}G$. Suppose $\alpha_n \in \mathbb{C}G$, $\alpha_n \to 0$ and $\theta\alpha_n \to v$ where $v \in U \backslash 0$. Choose a standard basis element $w = (0, 0, \ldots, 0, g, 0, \ldots) \in L^2(G)^n$ where $g \in G$ such that $\langle v, w \rangle \neq 0$. Then

$$\langle v, w \rangle = \lim_{n \to \infty} \langle u\alpha_n, w \rangle = \lim_{n \to \infty} \langle u, w\alpha_n^* \rangle = 0,$$

a contradiction. Therefore θ extends to a closed operator, which we shall also call θ (see [33, p. 155]). Note that $\mathrm{im}\,\theta$ is dense in U. Using [45, §21.1, II], we may write θ uniquely in the form $\phi\psi$ where ψ is a self adjoint unbounded operator on $L^2(G)$ and $\phi\colon L^2(G) \to U$ is a partial isometry. Since θ is a right $\mathbb{C}G$-map, we see from the uniqueness of the factorization of θ that ϕ (and ψ) is also a right $\mathbb{C}G$-map. Thus ϕ induces an $L^2(G)$-isomorphism from an $L^2(G)$-ideal onto U, as required. $\qquad\square$

We shall say that an $L^2(G)$-module $\mathcal{H}$ is finitely generated if there exist $n \in \mathbb{P}$ and $u_1, \ldots, u_n \in \mathcal{H}$ such that $u_1\mathbb{C}G + \cdots + u_n\mathbb{C}G$ is dense in $\mathcal{H}$. Obviously if $\mathcal{H}$ and $\mathcal{K}$ are finitely generated, then so is $\mathcal{H} \oplus \mathcal{K}$. The next lemma gives an alternative description of this definition.

Lemma 11.4. *Let $\mathcal{H}$ be an $L^2(G)$-module. Then $\mathcal{H}$ is finitely generated if and only if $\mathcal{H}$ is isomorphic to an $L^2(G)$-submodule of $L^2(G)^n$ for some $n \in \mathbb{P}$, and in this case there exist $L^2(G)$-ideals $I_1, \ldots, I_n$ such that $\mathcal{H} \cong I_1 \oplus \cdots \oplus I_n$.*

Proof. First suppose that $\mathcal{H}$ is isomorphic to an $L^2(G)$-submodule of $L^2(G)^n$ where $n \in \mathbb{P}$. Write $L^2(G)^n = U \oplus V$ where $U \cong L^2(G)$, $V \cong L^2(G)^{n-1}$, and $U \perp V$. Let W be the orthogonal complement to $U \cap \mathcal{H}$ in $\mathcal{H}$, and let π be the projection of $L^2(G)^n$ onto V. Then the restriction of π to W is an $L^2(G)$-monomorphism, so by Lemma 11.1 W is isomorphic to an $L^2(G)$-submodule of V. Using induction, we may assume that W is finitely generated and isomorphic to a finite direct sum of $L^2(G)$-ideals. But $\mathcal{H} = U \cap \mathcal{H} \oplus W$ and $U \cap \mathcal{H}$ is finitely generated by Lemma 11.2, so $\mathcal{H}$ is finitely generated and isomorphic to a finite direct sum of $L^2(G)$-ideals.

Now suppose $\mathcal{H}$ is finitely generated, say $u_1\mathbb{C}G + \cdots + u_n\mathbb{C}G$ is dense in $\mathcal{H}$. Let $U = \overline{u_1\mathbb{C}G}$, let $V = U^\perp$, and for $i = 2, \ldots, n$, write $u_i = u_i' + v_i$ where $u_i' \in U$ and $v_i \in V$. Then $\mathcal{H} = U \oplus V$, $v_2\mathbb{C}G + \cdots + v_n\mathbb{C}G$ is dense in V, and U is isomorphic to an $L^2(G)$-ideal by Lemma 11.3. Using induction on n, we may assume that V is isomorphic to an $L^2(G)$-submodule of $L^2(G)^{n-1}$ for some $n \in \mathbb{P}$, and the result follows. $\qquad\square$

Lemma 11.5. *Let U, V and W be $L^2(G)$-modules. If $U \oplus W$ is finitely generated and $U \oplus W \cong V \oplus W$, then $U \cong V$.*

Proof. Since $U \oplus W$ is finitely generated, Lemma 11.4 shows we may assume that $U \oplus W$ is an $L^2(G)$-submodule of $L^2(G)^n$ where $n \in \mathbb{P}$. Using $U \oplus W \cong V \oplus W$, we may assume that $U \oplus W = V \oplus W_1$ where $W \cong W_1$. If X is the orthogonal complement of $U \oplus W$ in $L^2(G)^n$, then

$$U \oplus (W \oplus X) = L^2(G)^n = V \oplus (W_1 \oplus X)$$

and we need only consider the case $X = 0$.

Thus we have $U \oplus W = L^2(G)^n = V \oplus W_1$ where $W \cong W_1$. Let e and f denote the projections of $L^2(G)^n$ onto W and W_1 respectively, and let $\theta\colon W \to W_1$ be an isometric $L^2(G)$-isomorphism. Then $e, f \in \mathrm{M}_n(W(G))$ because W and W_1 are $L^2(G)$-submodules. Since

$$U \oplus L^2(G)^n = U \oplus V \oplus W_1 \cong V \oplus U \oplus W = V \oplus L^2(G)^n,$$

there is an isometric $L^2(G)$-isomorphism $\phi\colon U \oplus L^2(G)^n \to V \oplus L^2(G)^n$. If $\psi = \theta \oplus \phi$, then

$$\psi\colon L^2(G)^{2n} \to L^2(G)^{2n}$$

is a unitary operator which is also a right $\mathbb{C}G$-map, so ψ can be considered as an element of $\mathrm{M}_{2n}(W(G))$. Set

$$E = \mathrm{diag}(e, 0_n) \quad \text{and} \quad F = \mathrm{diag}(f, 0_n).$$

Then E and F are projections in $\mathrm{M}_{2n}(W(G))$ and $\psi E \psi^{-1} = F$, so E and F are equivalent [4, definition 5, §1]. By [4, proposition 8, §1],

$$\mathrm{diag}(e, 1_n) \quad \text{and} \quad \mathrm{diag}(f, 1_n)$$

are also equivalent projections. Now $\mathrm{M}_n(W(G))$ is a finite von Neumann algebra [4, definition 1, §15], [38, proposition 9], and satisfies "GC" [4, corollary 1, §14], so we may apply [4, proposition 4, §17] twice to deduce that

$$1 - \mathrm{diag}(e, 1_n) \quad \text{and} \quad 1 - \mathrm{diag}(f, 1_n)$$

are equivalent projections, and hence unitarily equivalent projections. Therefore

$$(1 - e)L^2(G)^n \cong (1 - f)L^2(G)^n$$

and the result follows. $\qquad\square$

Lemma 11.6. *Let $\mathcal{H} = L^2(G)^\infty$, let U be a finitely generated $L^2(G)$-submodule of $\mathcal{H}$, and let $V = U^\perp$. Then $V \cong \mathcal{H}$.*

Proof. Using Lemma 11.4 and induction, we may assume that U is isomorphic to an $L^2(G)$-ideal. Write $\mathcal{H} = L_1 \oplus L_2 \oplus \cdots$ where $L_i \cong L^2(G)$ for all $i \in \mathbb{P}$, let $M_n = \bigoplus_{i=1}^n L_i$, let X_n denote the orthogonal complement of $V \cap M_n$ in M_n, let T_n denote the orthogonal complement of $V \cap M_n$ in $V \cap M_{n+1}$ ($n \in \mathbb{P}$), and let π denote the projection of $\mathcal{H}$ onto U. Since $X_n \cap (V \cap M_n) = 0$ and $X_n \subseteq M_n$, we see that $X_n \cap V = 0$, hence the restriction of π to X_n is an $L^2(G)$-monomorphism and we deduce from Lemma 11.1 that X_n is isomorphic to an $L^2(G)$-submodule of U. Therefore we may write $X_n \oplus Y_n \cong L^2(G)$ for some $L^2(G)$-ideal Y_n ($n \in \mathbb{P}$). We now have

$$V \cap M_n \oplus T_n \oplus X_{n+1} = M_{n+1} = V \cap M_n \oplus X_n \oplus L_{n+1}$$
$$\cong V \cap M_n \oplus X_n \oplus X_{n+1} \oplus Y_{n+1},$$

thus by Lemma 11.5 we obtain $T_n \cong X_n \oplus Y_{n+1}$, so we may write $T_n = X'_n \oplus Y'_{n+1}$ where $X_n \cong X'_n$ and $Y_n \cong Y'_n$ ($n \in \mathbb{P}$). For $n \in \mathbb{P}$, set $F_n = V \cap M_n \oplus X'_n$. Then $F_n \subseteq F_{n+1}$, so we may define E_{n+1} to be the orthogonal complement of F_n in F_{n+1} ($n \in \mathbb{P}$); we shall set $E_1 = F_1$. Since $F_n \cong V \cap M_n \oplus X_n - M_n$, application of Lemma 11.5 yields $E_n \cong L^2(G)$ for all $n \in \mathbb{P}$. Now

$$V \cap M_n \subseteq E_1 \oplus \cdots \oplus E_n \subseteq V \cap M_{n+1}$$

for all $n \in \mathbb{P}$, hence $\bigoplus_{i=1}^\infty E_i = V$ and the result follows. $\qquad\square$

An $L^2(G)$-basis $\{e_1, e_2, \dots\}$ of the $L^2(G)$-module $\mathcal{H}$ means that there exists an isometric $L^2(G)$-isomorphism $\theta\colon \mathcal{H} \to L^2(G)^n$ for some $n \in \mathbb{E}$ such that $\theta(e_i) = (0, \dots, 0, 1, 0, \dots)$, where the 1 is in the ith position. If $\{f_1, f_2, \dots\}$ is another $L^2(G)$-basis of $\mathcal{H}$ and α is the $L^2(G)$-automorphism of $\mathcal{H}$ defined by $\alpha e_i = f_i$, then $\alpha\alpha^* = \alpha^*\alpha = 1$. Also we say that an $L^2(G)$-map θ has finite rank if $\overline{\mathrm{im}\,\theta}$ is finitely generated.

Suppose now $\mathcal{H} = L^2(G)^\infty$ and that $\theta\colon \mathcal{H} \to \mathcal{H}$ is a finite rank $L^2(G)$-map. Let $\mathcal{K} = \ker\theta$. Then the restriction of θ to $\mathcal{K}^\perp$ is an $L^2(G)$-monomorphism, so $\mathcal{K}^\perp$ is finitely generated by Lemma 11.4. Using Lemmas 11.4 and 11.6, there exists $n \in \mathbb{P}$ and an $L^2(G)$-basis $\{e_1, e_2, \dots\}$ of $\mathcal{H}$ such that $\mathrm{im}\,\theta + \mathcal{K}^\perp \subseteq \overline{U}$ where $U = e_1\mathbb{C}G + \dots + e_n\mathbb{C}G$. We may represent θ by a matrix (θ_{ij}) where $i, j \in \mathbb{P}$ and $\theta_{ij} \in W(G)$ for all i, j (so $\theta e_i = \sum_{j=1}^\infty e_j\theta_{ji}$). Then we define $\mathrm{tr}_G\,\theta = \sum_{i=1}^\infty \mathrm{tr}_G\,\theta_{ii}$, which is well defined because $\theta_{ii} = 0$ for all $i > n$. Clearly if θ_U is the restriction of θ to $\overline{U}$, then $\mathrm{tr}_G\,\theta = \mathrm{tr}_G\,\theta_U$ (where $\mathrm{tr}_G\,\theta_U$ is defined as in Section 8).

Let $\{f_1, f_2, \dots\}$ be another $L^2(G)$-basis for $\mathcal{H}$. We want to show that if (ϕ_{ij}) is the matrix of θ with respect to this basis, then $\sum_{i=1}^\infty \mathrm{tr}_G\,\phi_{ii}$ is an absolutely convergent series with sum $\mathrm{tr}_G\,\theta$. Write $f_i = \sum_{j=1}^\infty e_j\alpha_{ji}$ where $\alpha_{ij} \in W(G)$, and $\sum_{k=1}^\infty \alpha_{ik}\alpha_{kj}^*$ is an absolutely convergent series with sum δ_{ij} for all $i, j \in \mathbb{P}$. Then

$$\mathrm{tr}_G\,\phi_{ii} = \langle \theta f_i, f_i \rangle = \sum_{j,k,l=1}^n \langle e_j\theta_{jk}\alpha_{ki}, e_l\alpha_{li} \rangle$$

$$= \sum_{j,k,l=1}^n \langle e_j\theta_{jk}\alpha_{ki}\alpha_{il}^*, e_l \rangle$$

$$= \sum_{j,k=1}^n \mathrm{tr}_G(\theta_{jk}\alpha_{ki}\alpha_{ij}^*).$$

Now $\sum_{i=1}^\infty \alpha_{ki}\alpha_{ij}^*$ is absolutely convergent with sum δ_{kj}, hence $\sum_{i=1}^\infty \theta_{jk}\alpha_{ki}\alpha_{ij}^*$ is absolutely convergent with sum $\theta_{jk}\delta_{kj}$, consequently $\sum_{i=1}^\infty \mathrm{tr}_G(\theta_{jk}\alpha_{ki}\alpha_{ij}^*)$ is absolutely convergent with sum $\mathrm{tr}_G(\theta_{jk}\delta_{kj})$. Therefore $\sum_{i=1}^\infty \mathrm{tr}_G\,\phi_{ii}$ is absolutely convergent with sum $\sum_{j,k=1}^n \mathrm{tr}_G(\theta_{jk}\delta_{kj}) = \mathrm{tr}_G\,\theta$, as required.

Suppose now that $\theta, \phi\colon \mathcal{H} \to \mathcal{H}$ are finite rank $L^2(G)$-maps. Let

$$M = (\ker\theta)^\perp + (\ker\phi)^\perp + \mathrm{im}\,\theta + \mathrm{im}\,\phi.$$

Then $\overline{M}$ is finitely generated, so there exists $n \in \mathbb{P}$ and an $L^2(G)$-submodule $L \cong L^2(G)^n$ of $\mathcal{H}$ containing $\overline{M}$. Let π denote the projection of $\mathcal{H}$ onto L, and if $\alpha\colon \mathcal{H} \to \mathcal{H}$ is an $L^2(G)$-map, then α_L will denote the restriction of α to L. Then $\theta + \phi$, $\theta\phi$ and $\phi\theta$ have finite $L^2(G)$-rank, and

$$\mathrm{tr}_G(\theta + \phi) = \mathrm{tr}_G(\theta + \phi)_L = \mathrm{tr}_G\,\theta_L + \mathrm{tr}_G\,\phi_L = \mathrm{tr}_G\,\theta + \mathrm{tr}_G\,\phi,$$

$$\mathrm{tr}_G\,\theta\phi = \mathrm{tr}_G(\theta\phi)_L = \mathrm{tr}_G\,\theta_L\phi_L = \mathrm{tr}_G\,\phi_L\theta_L = \mathrm{tr}_G(\phi\theta)_L = \mathrm{tr}_G\,\phi\theta.$$

Also if α is an $L^2(G)$-automorphism of $\mathcal{H}$, then $\alpha\pi$ and $\theta\alpha^{-1}$ are finite rank $L^2(G)$-maps and $\pi\theta = \theta = \theta\pi$, hence by the above

$$\operatorname{tr}_G \alpha\theta\alpha^{-1} = \operatorname{tr}_G(\alpha\pi)(\theta\alpha^{-1}) = \operatorname{tr}_G(\theta\alpha^{-1})(\alpha\pi) = \operatorname{tr}_G \theta.$$

Suppose M is a finitely generated $L^2(G)$-submodule of $L^2(G)^m$ where $m \in \mathbb{E}$. Then $\dim_G M$ is defined to be $\operatorname{tr}_G e$ where e is the projection of $L^2(G)^m$ onto M ($\dim_G$ is precisely d_G of [10, p. 134]). In view of Kaplansky's theorem (see Section 8), $\dim_G M \geq 0$ and $\dim_G M = 0$ if and only if $M = 0$. Let N be an $L^2(G)$-submodule of $L^2(G)^n$ where $n \in \mathbb{E}$ and $N \cong M$. Then

$$M^\perp \oplus L^2(G)^n \cong N^\perp \oplus L^2(G)^m,$$

hence there is a unitary $L^2(G)$-map α of $L^2(G)^m \oplus L^2(G)^n$ which takes M to N. Therefore if f is the projection of $L^2(G)^n$ onto N, then $0 \oplus f = \alpha(e \oplus 0)\alpha^{-1}$ and it follows that $\operatorname{tr}_G f = \operatorname{tr}_G e$. Thus $\dim_G N = \dim_G M$, in other words $\dim_G M$ depends only on the isomorphism type of M. If $n \in \mathbb{P}$ and $\phi \in \mathrm{M}_n(W(G))$, we may view ϕ as an $L^2(G)$-map $L^2(G)^n \to L^2(G)^n$, and then $\dim_G \operatorname{im} \phi = \operatorname{rank}_G \phi$. We need the following technical result.

Lemma 11.7. *Let $\mathcal{H} = L^2(G)^\infty$, let $\theta \colon \mathcal{H} \to \mathcal{H}$ be an $L^2(G)$-homomorphism, and let $\{e_1, e_2, \dots\}$ be an $L^2(G)$-basis for $\mathcal{H}$. For $r, s \in \mathbb{E}$, $r \leq s$, let $\mathcal{H}_{r,s} = \overline{e_r \mathbb{C}G + \dots + e_s \mathbb{C}G}$ ($s \neq \infty$), let $\mathcal{H}_{r,\infty} = \overline{e_r \mathbb{C}G + e_{r+1}\mathbb{C}G + \dots}$, and let $U_{r,s} = \overline{\theta\mathcal{H}_{r,s}}$. Suppose for all $i \in \mathbb{P}$ we can write θe_i as a finite sum of elements $\sum_{j=1}^r e_j \alpha_j$ where $\alpha_j \in \mathbb{C}G$ for all j (where r depends on i). If $\operatorname{rank}_G \phi \in \mathbb{Z}$ for all $\phi \in \mathrm{M}_r(\mathbb{C}G)$ and for all $r \in \mathbb{P}$, then $\dim_G U_{1,m} \cap U_{n,\infty} \in \mathbb{Z}$ for all $m, n \in \mathbb{P}$.*

Proof. Suppose $a, b, c, d \in \mathbb{P}$ with $a \leq b$ and $b, c \leq d$. Using the hypothesis that θe_i can be written as a finite sum of elements of the form $e_j \alpha_j$ where $\alpha_j \in \mathbb{C}G$, there exists $r \in \mathbb{P}$, $r \geq d$ such that $U_{1,d} \subseteq \mathcal{H}_{1,r}$. Define an $L^2(G)$-map $\phi \colon \mathcal{H}_{1,r} \to \mathcal{H}_{1,r}$ by $\phi e_i = \theta e_i$ if $a \leq i \leq b$ or $c \leq i \leq d$, and $\phi e_i = 0$ otherwise. Then with respect to the $L^2(G)$-basis $\{e_1, \dots, e_r\}$ of $\mathcal{H}_{1,r}$ the matrix of ϕ is in $\mathrm{M}_r(\mathbb{C}G)$, so $\operatorname{rank}_G \phi \in \mathbb{Z}$. But $\operatorname{im} \phi = \theta(\mathcal{H}_{a,b} + \mathcal{H}_{c,d})$ and it follows that $\dim_G \overline{U_{a,b} + U_{c,d}} \in \mathbb{Z}$.
Let $s \in \mathbb{P}$ with $s \geq m, n$. Using Lemma 11.1 we can obtain standard isomorphism theorems, in particular

$$\overline{(U_{1,m} + U_{n,s})} \oplus (U_{1,m} \cap U_{n,s}) \cong U_{1,m} \oplus U_{n,s}.$$

Therefore $\dim_G \overline{U_{1,m} + U_{n,s}} + \dim_G U_{1,m} \cap U_{n,s} = \dim_G U_{1,m} + \dim_G U_{n,s}$ and we deduce from the previous paragraph that $\dim_G U_{1,m} \cap U_{n,s} \in \mathbb{Z}$. Thus as s increases, $\dim_G U_{1,m} \cap U_{n,s}$ forms an increasing sequence of integers bounded above by $\dim_G U_{1,m}$, hence there exists $t \in \mathbb{P}$ such that $\dim_G U_{1,m} \cap U_{n,s} = \dim_G U_{1,m} \cap U_{n,t}$ for all $s \geq t$. Therefore $U_{1,m} \cap U_{n,s} = U_{1,m} \cap U_{n,t}$ for all $s \geq t$, and it follows that $U_{1,m} \cap U_{n,\infty} = U_{1,m} \cap U_{n,t}$. We conclude that $\dim_G U_{1,m} \cap U_{n,\infty} = \dim_G U_{1,m} \cap U_{n,t} \in \mathbb{Z}$ as required. $\square$

12. THE SPECIAL CASE OF A DIRECT PRODUCT OF FREE GROUPS

Here we generalize the theory of [41, section 3]. If $\mathcal{H}$ is a Hilbert space and G is a group of operators acting on the right of $\mathcal{H}$, then we define

$$\mathcal{L}_G(\mathcal{H}) = \{\theta \in \mathcal{L}(\mathcal{H}) \mid \theta(ug) = (\theta u)g \text{ for all } u \in \mathcal{H} \text{ and } g \in G\}.$$

Note that von Neumann's double commutant theorem [1, theorem 1.2.1] (or see (ii) after Theorem 8.2) tells us that

$$\mathcal{L}_G(L^2(G)) = \{\theta \in \mathcal{L}(L^2(G)) \mid \theta(ug) = (\theta u)g \text{ for all } g \in G\} = W(G).$$

Suppose now that H and A are groups, $G = H \times A$, $n \in \mathbb{P}$, $\theta \in \mathcal{L}_A(L^2(G))$, $\phi \in \mathrm{M}_n(\mathcal{L}_A(L^2(G)))$, and ϕ is represented by the matrix (ϕ_{ij}) where $\phi_{ij} \in \mathcal{L}_A(L^2(G))$ for all i, j. We make θ act on $L^2(G) \oplus L^2(G) \oplus L^2(A)$ by $\theta(u, v, x) = (\theta u, \theta v, 0)$, and ϕ act on $L^2(G)^n \oplus L^2(G)^n \oplus L^2(A)^n$ by $\phi(u, v, x) = (\phi u, \phi v, 0)$ $(u, v \in L^2(G)$ or $L^2(G)^n$, $x \in L^2(A)$ or $L^2(A)^n)$. Note that the actions of θ and ϕ on $L^2(G) \oplus L^2(G) \oplus L^2(A)$ and $L^2(G)^n \oplus L^2(G)^n \oplus L^2(A)^n$ are right $\mathbb{C}A$-maps.

Now let H be the free group on two generators, and let A act on the right of A by right multiplication as usual; i.e. $ab = ab$ for all $a \in A$ and $b \in A$. We also make H act trivially on A: thus $ha = a$ for all $a \in A$ and $h \in H$ (though $h1_H = h$). Theorem 10.1 shows that there is a bijection $\pi \colon H \to H \cup H \cup \{1_A\}$ (where 1_A is the identity of A) such that

(12.1) $\pi 1_H = 1_A$,

(12.2) $\{k \in H \mid h(\pi k) \neq \pi(hk)\}$ is finite for all $h \in H$.

We extend π to a right A-map

(12.3) $\pi \colon G \to G \cup G \cup A$

by setting $\pi(ha) = (\pi h)a$ for all $h \in H$ and $a \in A$. This in turn defines a unitary operator $\alpha \colon L^2(G) \to L^2(G) \oplus L^2(G) \oplus L^2(A)$, and hence also a unitary operator (equal to the direct sum of n copies of α)

(12.4) $\beta \colon L^2(G)^n \to L^2(G)^n \oplus L^2(G)^n \oplus L^2(A)^n$.

We note that α and β are right $\mathbb{C}A$-maps. Suppose $\phi \in \mathrm{M}_n(W(G))$ and $\phi - \beta^{-1}\phi\beta$ has finite $L^2(A)$-rank. Then we have

Lemma 12.1. $\mathrm{tr}_G \phi = \mathrm{tr}_A(\phi - \beta^{-1}\phi\beta)$.

Proof. (cf. [20, section 5].) Let (ϕ_{ij}) denote the matrix of ϕ. Since $\phi_{ij} - \alpha^{-1}\phi_{ij}\alpha$ has finite $L^2(A)$-rank for all i, j, $\mathrm{tr}_G \phi = \sum_{i=1}^n \mathrm{tr}_G \phi_{ii}$ and $\mathrm{tr}_A(\phi - \beta^{-1}\phi\beta) = \sum_{i=1}^n \mathrm{tr}_A(\phi_{ii} - \alpha^{-1}\phi_{ii}\alpha)$, it will be sufficient to show that $\mathrm{tr}_G \theta = \mathrm{tr}_A(\theta - \alpha^{-1}\theta\alpha)$ for all $\theta \in W(G)$ such that $\theta - \alpha^{-1}\theta\alpha$ has finite $L^2(A)$-rank. If $\theta = \sum_{g \in G} \theta_g g$ where $\theta_g \in \mathbb{C}$ for all $g \in G$, then $\mathrm{tr}_G \theta = \theta_1$ and $\langle \theta g, g \rangle = \theta_1$ for all $g \in G$. Using (12.1), we see that $\langle \theta \pi h, \pi h \rangle = \theta_1$ for all $h \in H \backslash 1$ and $\langle \theta \pi 1, \pi 1 \rangle = 0$, hence

$$\langle (\theta - \alpha^{-1}\theta\alpha)h, h \rangle = 0 \quad \text{if } h \in H \backslash 1,$$
$$\langle (\theta - \alpha^{-1}\theta\alpha)1, 1 \rangle = \theta_1.$$

Since H is an $L^2(A)$-basis for $L^2(G)$, we can calculate $\mathrm{tr}_A \theta$ with respect to this basis and the result follows. □

Let G be a group, let $n \in \mathbb{P}$, let $\theta \in \mathrm{M}_n(\mathbb{C}G)$, and let $X \subseteq G$. If $\theta = \sum_{g \in G} \theta_g g$ where $\theta_g \in \mathrm{M}_n(\mathbb{C})$, then $\mathrm{supp}\,\theta$ is defined to be $\{g \in G \mid \theta_g \neq 0\}$, a finite subset of G. Also $L^2(X)$ will indicate the closed subspace of $L^2(G)$ with Hilbert basis X.

Lemma 12.2. *Let H be the free group of rank two, let A be a group, let $G = H \times A$, let $n \in \mathbb{P}$, and let $\theta \in \mathrm{M}_n(\mathbb{C}G)$. If $\mathrm{rank}_A \phi \in \mathbb{Z}$ for all $\phi \in \mathrm{M}_r(\mathbb{C}A)$ and for all $r \in \mathbb{P}$, then $\mathrm{rank}_G \theta \in \mathbb{Z}$.*

Proof. Let $\pi : G \to G \cup G \cup A$ be the bijection given by (12.3), and let $\beta : L^2(G)^n \to L^2(G)^n \oplus L^2(G)^n \oplus L^2(A)^n$ be the unitary operator given by (12.4). Let

$$K = \{k \in H \mid g(\pi k) = \pi(gk) \quad \text{for all } g \in \mathrm{supp}\,\theta\},$$

let $J = H \backslash K$, let $L_1 = \theta L^2(G)^n$, let $L_2 = \beta^{-1}\theta\beta L^2(G)^n$, and let λ denote the projection of $L^2(G)^n$ onto $\overline{L_1}$. Then $|J| < \infty$ by (12.2), and $\beta^{-1}\lambda\beta$ is the projection of $L^2(G)^n$ onto $\overline{L_2}$. Since $\mathrm{rank}_G \theta = \mathrm{tr}_G \lambda$, we want to prove that $\mathrm{tr}_G \lambda \in \mathbb{Z}$.

Let $M = \theta L^2(KA)^n$, and let μ denote the projection of $L^2(G)^n$ onto $\overline{M}$. Note that $M = \beta^{-1}\theta\beta L^2(KA)^n$ because $\beta^{-1}\theta\beta u = \theta u$ for all $u \in L^2(KA)^n$. Let N_1 and N_2 denote the orthogonal complements of $\overline{M}$ in $\overline{L_1}$ and $\overline{L_2}$ respectively, and let η_1 and η_2 denote the projections of $L^2(G)^n$ onto N_1 and N_2 respectively. Let $P_1 = \theta L^2(JA)^n$, let $P_2 = \beta^{-1}\theta\beta L^2(JA)^n$, and for $i = 1,2$, let Q_i denote the orthogonal complement of $\overline{P_i} \cap \overline{M}$ in $\overline{P_i}$. Note that $\overline{M} \cap Q_i = 0$ and $\overline{M} + Q_i$ is dense in $\overline{L_i}$ $(i = 1,2)$. Thus if π_i is the projection of $\overline{L_i}$ onto N_i, then the restriction of π_i to Q_i is an $L^2(A)$-monomorphism with dense image, so $N_i \cong Q_i$ by Lemma 11.1 $(i = 1,2)$. Therefore

(12.5) $$N_i \oplus (\overline{P_i} \cap \overline{M}) \cong \overline{P_i}.$$

Using Lemma 11.4, we see that N_i is finitely generated, hence $\eta_1 - \eta_2$ has finite $L^2(A)$-rank. Also $\lambda = \mu + \eta_1$ and $\beta^{-1}\lambda\beta = \mu + \eta_2$, hence $\lambda - \beta^{-1}\lambda\beta = \eta_1 - \eta_2$. Therefore $\mathrm{tr}_G \lambda = \mathrm{tr}_A(\lambda - \beta^{-1}\lambda\beta)$ by Lemma 12.1, and since $\mathrm{tr}_A(\eta_1 - \eta_2) = \mathrm{tr}_A \eta_1 - \mathrm{tr}_A \eta_2$, it will suffice to prove that $\mathrm{tr}_A \eta_1$ and $\mathrm{tr}_A \eta_2 \in \mathbb{Z}$. Now $\mathrm{tr}_A \eta_i = \dim_A N_i$ so in view of (12.5), we require that $\dim_A \overline{P_i} \cap \overline{M}$ and $\dim_A \overline{P_i} \in \mathbb{Z}$. We apply Lemma 11.7: note that with respect to the standard $L^2(A)$-basis H^n of $L^2(G)^n$, the matrices of θ and $\beta^{-1}\theta\beta$ have the required form for this lemma. But $P_1 = \theta L^2(JA)^n$, $P_2 = \beta^{-1}\theta\beta L^2(JA)^n$, and $M = \theta L^2(KA)^n = \beta^{-1}\theta\beta L^2(KA)^n$, and the result follows. □

The proof of the following lemma is identical to the proof of [41, lemma 3.7].

Lemma 12.3. *Let G be a group. If $\mathrm{rank}_G \theta \in \mathbb{Z}$ for all $\theta \in \mathrm{M}_n(\mathbb{C}G)$ and for all $n \in \mathbb{P}$, then $D(G)$ is a division ring.*

Proof. We shall use the theory of [15, section 7.1]. Let R denote the rational closure [15, p. 382] of $\mathbb{C}G$ in $U(G)$, and let $\alpha \in D(G) \backslash 0$. By [15, exercise 7.1.4]

$D(G) \subseteq R$, so we can apply Cramer's rule [15, proposition 7.1.3] to deduce that α is stably associated over R to a matrix in $M_m(\mathbb{C}G)$ for some $m \in \mathbb{P}$. Therefore there exists $n \geq m$ such that $\text{diag}(\alpha, 1_{n-1})$ is associated over R to a matrix $\theta \in M_n(\mathbb{C}G)$, which means that there exist $X, Y \in GL_n(U(G))$ such that $X \text{diag}(\alpha, 1_{n-1})Y = \theta$.

Suppose α is not invertible in $U(G)$. Using Lemma 8.3, we see that $0 < \text{rank}_G \alpha < 1$ and thus $n - 1 < \text{rank}_G \theta < n$. This contradicts Lemma 12.2, hence α is invertible in $U(G)$. Since $D(G)$ is closed under taking inverses, $D(G)$ must be a division ring. □

Lemma 12.4. *Let $n \in \mathbb{P}$, and let $G = H_1 \times \cdots \times H_n$ where H_i is isomorphic to the free group of rank two for all i. Then $D(G)$ is a division ring.*

Proof. By induction on n and Lemma 12.2, $\text{rank}_G \theta \in \mathbb{Z}$ for all $\theta \in M_n(\mathbb{C}G)$ and for all $n \in \mathbb{P}$. Now use Lemma 12.3. □

Lemma 12.5. *Let $H \lhd G$ be groups such that G/H is free, let Φ denote the matrices over $\mathbb{C}H$ which become invertible over $D(H)$, and let Σ denote the matrices over $\mathbb{C}G$ which become invertible over $D(G)$. Suppose $D(G)$ is a division ring. If the identity map on $\mathbb{C}H$ extends to an isomorphism $\phi \colon \mathbb{C}H_\Phi \to D(H)$, then the identity map on $\mathbb{C}G$ extends to an isomorphism $\sigma \colon \mathbb{C}G_\Sigma \to D(G)$.*

Proof. By Lemma 9.3, we may view $\sum_{g \in G} D(H)g$ as $D(H) * [G/H]$. Suppose $H \subseteq N \lhd K \subseteq G$ and $K/N \cong \mathbb{Z}$ with generator Nt where $t \in K$. Then if $d_i \in D(N)$ and $\sum_i d_i t^i = 0$, it follows that $d_i = 0$ for all i, which means in the terminology of [31, §2] that $D(G)$ is a free division ring of fractions for $D(H) * [G/H]$. Therefore $D(G)$ is the universal field of fractions for $D(H) * [G/H]$ by the theorem of [31] and the proof of [37, proposition 6]. Since $D(H) * [G/H]$ is a free ideal ring [14, theorem 3.2], the results of [15, §7.5] show that $D(G) = D(H) * [G/H]_\Psi$ for a suitable set of matrices Ψ with entries in $D(H) * [G/H]$. The proof is completed by applying [56, proof of theorem 4.6] and [15, exercise 7.2.8]. □

Lemma 12.6. *Let $G \leqslant F$ be groups such that F is a direct product of finitely generated free groups, and let Σ denote the set of matrices over $\mathbb{C}G$ which become invertible over $D(G)$. Then $D(G)$ is a division ring, and the identity map on $\mathbb{C}G$ extends to an isomorphism $\mathbb{C}G_\Sigma \to D(G)$.*

Proof. We may write $F = F_1 \times \cdots \times F_n$ where $n \in \mathbb{P}$ and the F_i are finitely generated free groups, and since any finitely generated free group is isomorphic to a subgroup of the free group of rank 2, we may assume that each F_i is free of rank 2. Then $D(F)$ is a division ring by Lemma 12.4, hence $D(G)$ is a division ring. Write $H_i = F_1 \times \cdots \times F_i$ for $0 \leq i \leq n$ (so $H_0 = 1$). Then $(G \cap H_i)/(G \cap H_{i-1})$ is isomorphic to a free group for all i, so we can now use Lemma 12.5 and induction on n to complete the proof. □

13. PROOF OF THEOREM 10.3

To simplify the notation in the following lemma, we assume that $1, 2, \ldots \in \mathcal{I}$.

Lemma 13.1. *Let $\{H_i \mid i \in \mathcal{I}\}$ be a family of nonabelian free groups, let $G = H_1 \times H_2 \times \cdots$, and let θ be an automorphism of G. Then $\theta H_1 = H_i$ for some $i \in \mathcal{I}$.*

Proof. Suppose $g = (g_1, g_2, \dots) \in G$ where $g_i \in H_i$ for all i. Then

$$C_G(g) = C_{H_1}(g_1) \times C_{H_2}(g_2) \times \cdots$$

and $C_{H_i}(g_i) \cong \mathbb{Z}$ if $g_i \neq 1$, and $C_{H_i}(g_i) = H_i$ if $g_i = 1$. It follows that $Z(C_G(g)) \cong \mathbb{Z}^r$, where $Z(C_G(g))$ denotes the center of $C_G(g)$ and $r = |\{i \mid g_i \neq 1\}|$.

Let $x, y \in H_1 \backslash 1$. Then by the above we have $\theta x \in H_i$ and $\theta y \in H_j$ for some $i, j \in \mathcal{I}$. If $i \neq j$, then $\langle x, y \rangle \cong \mathbb{Z} \times \mathbb{Z}$ which is not possible. Therefore $i = j$ and the result follows. $\qquad\square$

Lemma 13.2. *Let $H \lhd G$ be groups such that H is a direct product of non-abelian free groups and G/H is finite. Let X be a finite subset of G. Then there exists a finitely generated subgroup G_0 of G such that $X \subseteq G_0$ and $G_0 \cap H$ is a direct product of nonabelian free groups.*

Proof. By enlarging X if necessary, we may assume that $HX = G$. Let H be the direct product of the nonabelian free groups H_i. Using Lemma 13.1 we see that G permutes the H_i by conjugation, so we may write $H = \times_i K_i$ where $K_i = K_{i1} \times \cdots \times K_{im_i}$ with the K_{ij} nonabelian free groups (so each K_{ij} is an H_k for some k), and for each i the set $\{K_{i1}, \dots, K_{im_i}\}$ is permuted transitively by conjugation by G. For each i, let N_i denote the normalizer of K_{i1} in G, and then choose right transversals $S_i \subseteq X$ for H in N_i, and $T_i \subseteq X$ for N_i in G; thus $|T_i| = m_i$ and we may write $T_i = \{t_{i1}, \dots, t_{im_i}\}$ where $t_{ij}^{-1} K_{i1} t_{ij} = K_{ij}$. Set $H_0 = H \cap \langle X \rangle$ and note that since it is a subgroup of finite index in a finitely generated group, it is also finitely generated, so we may write $H_0 \subseteq K_1 \times \cdots \times K_n$ for some $n \in \mathbb{P}$, and then there are finite subsets $Y_{ij} \subseteq K_{ij}$ ($1 \leq i \leq n$ and $1 \leq j \leq m_i$) such that $H_0 \subseteq \langle \bigcup_{ij} Y_{ij} \rangle$. Then we may choose finitely generated nonabelian free subgroups $\tilde{K}_{i1}$ of K_{i1} such that

$$Y_{ij} \subseteq t_{ij}^{-1} \tilde{K}_{i1} t_{ij} \quad \text{for } j = 1, \dots, m_i.$$

Set $L_{i1} = \langle s^{-1} \tilde{K}_{i1} s \mid s \in S_i \rangle$ and

$$L_i = t_{i1}^{-1} L_{i1} t_{i1} \times t_{i2}^{-1} L_{i1} t_{i2} \times \cdots \times t_{im_i}^{-1} L_{i1} t_{im_i}.$$

Then L_{i1} and hence also L_i is a finitely generated subgroup. Also if $i, j \in \mathbb{P}$ with $i \leq n$, $j \leq m_i$ and $x \in X$, then we may write $t_{ij} x = h s t_{ik}$ for some $h \in H_0$, $s \in S_i$ and $k \in \mathbb{P}$, and then $x^{-1} t_{ij}^{-1} L_{i1} t_{ij} x = t_{ik}^{-1} L_{i1} t_{ik}$ and we deduce that X normalizes L_i. Therefore if $L = L_1 \times \cdots \times L_n$, then L is a finitely generated subgroup and X normalizes L. Moreover L_{i1} is a free group because it is a subgroup of the free group K_{i1}, and it is nonabelian because it contains the nonabelian subgroup $\tilde{K}_{i1}$, hence L is a direct product of nonabelian free groups. Thus we may set $G_0 = L \langle X \rangle$ for the required subgroup. $\qquad\square$

For the purposes of the next two lemmas, given a group G and $n \in \mathbb{P}$, we shall define $\mathcal{S}_n G$ to be the intersection of normal subgroups of index at most n in G. Note that $\mathcal{S}_n G$ is a characteristic subgroup of G and that $\mathcal{S}_n G \supseteq \mathcal{S}_{n+1} G$ for all $n \in \mathbb{P}$. Furthermore if G is finitely generated, then $G/\mathcal{S}_n G$ is finite.

Lemma 13.3. *Let $F \lhd G$ be groups such that F is finitely generated free and G/F is finite. Suppose for all $n \in \mathbb{P}$, there exists $H_n \leqslant G$ such that $H_n F = G$ and $H_n \cap F = \mathcal{S}_n F$. Then there exists $H \leqslant G$ such that $HF = G$ and $H \cap F = 1$.*

Proof. Since $\mathcal{S}_n F$ is a normal subgroup of finite index in G, there are only finitely many subgroups of G which contain $\mathcal{S}_n F$, hence an application of the König graph theorem shows we may assume that $H_n \supseteq H_{n+1}$ for all $n \in \mathbb{P}$. It follows that if $\hat{G}$ denotes the profinite completion of G, then $\hat{G}$ has a subgroup K isomorphic to G/F.

We shall now use the notation and results of [60]. Since G has a free subgroup of finite index, we see from [17, theorem IV.3.2] that G is isomorphic to the fundamental group of a graph of groups $\pi_1(\mathcal{G}, \Gamma)$ with respect to some tree T, where Γ is a finite graph of groups, and the vertex groups $G(v)$ are finite for all vertices v of Γ. Then we can form the fundamental group $\Pi_1(\mathcal{G}, \Gamma, T)$ in the category of profinite groups, and by construction, $\Pi_1(\mathcal{G}, \Gamma, T) \cong \hat{G}$ [60, p. 418]. Of course the vertex groups $G(v)$ are the same as the vertex groups $\hat{G}(v)$. By [60, theorem 3.10] and the fact that K is a finite subgroup, we see that $K \subseteq g\hat{G}(v)g^{-1}$ for some vertex v of Γ and some $g \in \hat{G}$. Thus G has a subgroup isomorphic to G/F and the result follows. $\qquad\square$

Lemma 13.4. *Let $l \in \mathbb{P}$, and let $H \lhd G$ be groups such that G is finitely generated, G/H is finite, and H is a direct product of nonabelian free groups. Assume that whenever $K \lhd G$ such that $K \subseteq H$ and G/K is abelian-by-finite, then G/K has a subgroup of order l. Then G has a subgroup of order l.*

Proof. Write $H = H_1 \times \cdots \times H_t$ where the H_i are nonabelian free groups, and set $H_{(n)} = \mathcal{S}_n H_1 \times \cdots \times \mathcal{S}_n H_t$ for $n \in \mathbb{P}$. Note that if $K \lhd G$ and G/K is abelian-by-finite, then $H'_{(n)} \subseteq K$ for some $n \in \mathbb{P}$ and that $H/H'_{(n)}$ is torsion free.

First we reduce to the case $|G/H| = l$. We know by hypothesis that $G/H'_{(n)}$ has a subgroup $L_n/H'_{(n)}$ of order l for all $n \in \mathbb{P}$. Since $H/H'_{(n)}$ is torsion free, we see that $L_n \cap H = H'_{(n)}$ and therefore $|L_n H/H| = l$. Now G/H has only finitely many subgroups of order l, hence there exists a subgroup G_0/H of order l in G/H with $L_n \subseteq G_0$ for infinitely many n. Thus replacing G with G_0, we may assume that $|G/H| = l$.

We now use induction on t, the case $t = 1$ being a consequence of Lemma 13.3. Suppose we can write $H = F_1 \times F_2$ where $F_1, F_2 \lhd G$ and each F_i is a direct product of a proper subset of $\{H_1, \ldots, H_t\}$. Then by induction on t there exists $G_1 \leqslant G$ such that $F_1 \subseteq G_1$ and $|G_1/F_1| = l$. The natural injection $G_1 \hookrightarrow G$ induces an isomorphism $G_1 \hookrightarrow G/F_2$, so again using induction we see that G_1, and hence also G, has a subgroup of order l. Therefore we may assume that no such decomposition $H = F_1 \times F_2$ as above exists. It now follows from Lemma 13.1 that G permutes the H_i transitively by conjugation. Let D_n be the normalizer of H_1 in L_n, and let $Z = H_2 \times \cdots \times H_t$. Then $D_n H$ is the normalizer of both H_1 and Z in G for all $n \in \mathbb{P}$, so we may set $D = D_n H$ for all n. Since $D_n H = D$ and $D_n \cap H = H_{(n)}$ for all $n \in \mathbb{P}$, we have from the case $t = 1$ that D/Z has a subgroup of order $|D/H|$. Thus D/Z is isomorphic to a semidirect product of H_1 and D/H, so we may apply [30, theorem 3] to obtain a subgroup of G isomorphic to G/H, which is what is required. □

Lemma 13.5. *Let $G = \bigcup_{i \in \mathcal{I}} G_i$ be groups such that given $i, j \in \mathcal{I}$, there exists $l \in \mathcal{I}$ such that $G_i, G_j \subseteq G_l$, let Σ denote the matrices with entries in $\mathbb{C}G$ which become invertible over $D(G)$, and let Σ_i denote the matrices with entries in $\mathbb{C}G_i$ which become invertible over $D(G_i)$. Assume that the orders of the finite subgroups of G are bounded, and that $D(G_i)$ is an Artinian ring for all $i \in \mathcal{I}$. Suppose $\mathrm{lcm}(G_i) \, \mathrm{tr}_{G_i} e \in \mathbb{Z}$ whenever e is a projection in $D(G_i)$, for all $i \in \mathcal{I}$. Then*

(i) *$D(G) = \bigcup_{i \in \mathcal{I}} D(G_i)$ and $\mathrm{lcm}(G) \, \mathrm{tr}_G e \in \mathbb{Z}$ for all projections $e \in D(G)$.*

(ii) *$D(G)$ is a semisimple Artinian ring.*

(iii) *Suppose the identity map on $\mathbb{C}G$ extends to an isomorphism $\lambda_i \colon \mathbb{C}G_{i\Sigma_i} \to D(G_i)$ for all $i \in \mathcal{I}$. Then the identity map on $\mathbb{C}G$ extends to an isomorphism $\lambda \colon \mathbb{C}G_\Sigma \to D(G)$.*

Proof. (i) This is obvious.

(ii) If $I_0 > I_1 > \cdots > I_r$ is a strictly descending sequence of right ideals in $D(G)$, then

$$I_0 \cap D(G_i) > I_1 \cap D(G_i) > I_2 \cap D(G_i) > \cdots > I_r \cap D(G_i)$$

is a strictly descending sequence of right ideals in $D(G_i)$ for some $i \in \mathcal{I}$, hence $r \leq \mathrm{lcm}(G)$ by (i) and Lemma 9.6. This shows that $D(G)$ is Artinian, and the result now follows from Lemma 9.4.

(iii) Since every matrix in Σ_i becomes invertible over $\mathbb{C}G_\Sigma$, we see that there are maps $\mu_i \colon \mathbb{C}G_{i\Sigma_i} \to \mathbb{C}G_\Sigma$ which extend the inclusion map $\mathbb{C}G_i \to \mathbb{C}G$. Now λ_i is an isomorphism for all $i \in \mathcal{I}$, hence there are maps $\nu_i \colon D(G_i) \to \mathbb{C}G_\Sigma$ defined by $\nu_i = \mu_i \lambda_i^{-1}$, which extend the inclusion map $\mathbb{C}G_i \to \mathbb{C}G$. If $G_i \subseteq G_j$ and $\psi_{ij} \colon D(G_i) \to D(G_j)$ is the inclusion map, then $\nu_j \psi_{ij} = \nu_i$ and it follows that the ν_i fit together to give a map $\nu \colon \bigcup_{i \in \mathcal{I}} D(G_i) \to \mathbb{C}G_\Sigma$ such that $\nu \psi_i = \nu_i$, where $\psi_i \colon D(G_i) \to \bigcup_{i \in \mathcal{I}} D(G_i)$ is the natural inclusion. But $\bigcup_{i \in \mathcal{I}} D(G_i) = D(G)$ by (i), and we deduce that $\nu \colon D(G) \to \mathbb{C}G_\Sigma$ is a map which extends the identity on $\mathbb{C}G$. By the universal property of $\mathbb{C}G_\Sigma$, there is a map $\lambda \colon \mathbb{C}G_\Sigma \to D(G)$ which also extends the identity on $\mathbb{C}G$. Then $\nu\lambda$ is the identity on $\mathbb{C}G_\Sigma$ and $\lambda\nu$ is the identity on $D(G)$, and we deduce that λ is an isomorphism, as required. □

We need the following three technical lemmas.

Lemma 13.6. *(cf. [41, lemma 4.4].) Let Q be a semisimple Artinian ring, let $G = \langle x \rangle$ be an infinite cyclic group, let $Q * G$ be a crossed product, and let S be the set of nonzero divisors of $Q * G$. Let R be a ring containing $Q * G$, and let D be the division closure of $Q * G$ in R. Suppose every element of $Q * G$ of the form $1 + q_1 x + \cdots + q_t x^t$ with $q_i \in Q$ and $t \in \mathbb{P}$ is invertible in R. Then $Q * G$ is a semiprime Noetherian ring and D is an Artinian ring. Furthermore every element of S is invertible in D, and the identity map on $Q * G$ extends to an isomorphism $Q * G_S \to D$.*

Lemma 13.7. *Let $H \lhd G$ be groups, let $D(H)G$ denote the subring generated by $D(H)$ and G in $D(G)$, and let Σ denote the matrices with entries in $\mathbb{C}H$ which become invertible over $D(H)$. If the identity map on $\mathbb{C}H$ extends to an isomorphism $\mathbb{C}H_\Sigma \to D(H)$, then the identity map on $\mathbb{C}G$ extends to an isomorphism $\mathbb{C}G_\Sigma \to D(H)G$.*

Proof. This follows from Lemma 9.3 and [41, 4.5] □

Lemma 13.8. *(See [41, lemma 4.7].) Let D be a $*$-ring, let $\mathcal{R}$ be a set of subrings of D, let $n \in \mathbb{P}$, and let $e \in M_n(D)$ be an idempotent. Assume that whenever $R \in \mathcal{R}$ and P is a finitely generated projective R-module, there exist projections $f_i \in R$ such that $P \cong \bigoplus f_i R$. If the natural induction map*

$$\bigoplus_{R \in \mathcal{R}} K_0(R) \to K_0(D)$$

is onto, then there exist $r, s \in \mathbb{P}$, $R_1, \ldots, R_s \in \mathcal{R}$, and projections $f_i \in R_i$ $(1 \le i \le s)$ such that

$$\mathrm{diag}(e, 1_r, 0_s) = u \, \mathrm{diag}(f_1, \ldots, f_s, 0_{n+r}) u^{-1},$$

where $u \in \mathrm{GL}_{n+r+s}(D)$.

The essence of the next two lemmas is to show that if Theorem 10.3 holds for the group G_0 and G/G_0 is finitely generated abelian-by-finite, then it also holds for G. This is to prepare for an induction argument to follow.

Lemma 13.9. *Let $H \lhd G$ be groups such that G/H is free abelian of finite rank, let $D(H)G$ denote the subring of $D(G)$ generated by $D(H)$ and G, and let S denote the nonzero divisors in $D(H)G$. Suppose $D(H)$ is an Artinian ring. Then $D(H)G$ is a semiprime Noetherian ring and $D(G)$ is an Artinian ring. Furthermore every element of S is invertible in $D(G)$, and the identity map on $D(H)G$ extends to an isomorphism from $D(H)G_S$ to $D(G)$.*

Proof. By induction on the rank of G/H, we immediately reduce to the case G/H is infinite cyclic, say $G = \langle Hx \rangle$ where $x \in G$. Since $D(H)$ is semisimple by Lemma 9.4 and $D(H)G \cong D(H) * G/H$ by Lemma 9.3, we are in a position to apply Lemma 13.6. If $\alpha = 1 + q_1 x + \cdots + q_t x^t \in D(H)G$ where $t \in \mathbb{P}$ and $q_i \in D(H)$, then by Proposition 9.5(ii) there is a nonzero divisor β in $W(H)$ such that $\beta q_i \in W(H)$ for all i. Using [40, theorem 4], we see that $\beta \alpha \gamma \ne 0$ for all $\gamma \in W(G) \backslash 0$, and we deduce that α is invertible in $U(G)$. The result now follows from Lemma 13.6. □

Lemma 13.10. *Let* $N \lhd H \lhd G$ *be groups such that* $N \lhd G$, H/N *is free abelian of finite rank, and* G/H *is finite. Let* $D(N)G$ *denote the subring of* $D(G)$ *generated by* $D(N)$ *and* G, *and let* S *denote the nonzero divisors of* $D(N)G$. *Suppose* $D(N)$ *is an Artinian ring. Then*

(i) *$D(N)G$ is a semiprime Noetherian ring and $D(G)$ is a semisimple Artinian ring. Furthermore every element of S is invertible in $D(G)$, and the identity map on $D(N)G$ extends to an isomorphism from $D(N)G_S$ to $D(G)$.*

(ii) *Let Φ denote the matrices of $\mathbb{C}N$ which become invertible over $D(N)$, and let Σ denote the matrices of $\mathbb{C}G$ which become invertible over $D(G)$. If the identity map on $\mathbb{C}N$ extends to an isomorphism $\mathbb{C}N_\Phi \to D(N)$, then the identity map on $\mathbb{C}G$ extends to an isomorphism $\mathbb{C}G_\Sigma \to D(G)$.*

(iii) *Suppose $m, n \in \mathbb{P}$ and the orders of the finite subgroups of G are bounded. If $m \operatorname{lcm}(F) \operatorname{tr}_F e \in \mathbb{Z}$ whenever $F/N \in \mathcal{F}(G/N)$ and e is a projection in $D(F)$, then $m \operatorname{lcm}(G) \operatorname{tr}_G e \in \mathbb{Z}$ for all projections e in $\mathrm{M}_n(D(G))$.*

Proof. (i) This follows from Lemmas 9.4 and 13.9.

(ii) Lemma 13.7 shows that the identity map on $\mathbb{C}G$ extends to an isomorphism $D(N)G \to \mathbb{C}G_\Phi$. We now see from (i) and the proof of [56, theorem 4.6] that $D(G)$ is $\mathbb{C}G_\Psi$ for a suitable set of matrices Ψ with entries in $\mathbb{C}G$. An application of [15, exercise 7.2.8] completes the proof.

(iii) Using (i), we see that $D(N)G$ is Noetherian and that $D(G) \cong D(N)GS^{-1}$, so it follows from [34, lemma 2.2] that the natural inclusion $D(N)G \to D(G)$ induces an epimorphism $G_0(D(N)G) \to G_0(D(G))$. Now $D(F) \cong D(N)*F/N$ whenever $F/N \in \mathcal{F}(G/N)$ by Lemma 9.4, and $D(N)G \cong D(N) * G/N$ by Lemma 9.3, so we can apply Moody's induction theorem (Lemma 4.4) to deduce that the natural map

$$\bigoplus_{F/N \in \mathcal{F}(G/N)} G_0(D(F)) \longrightarrow G_0(D(G))$$

is also onto. Since $D(G)$ and $D(F)$ are semisimple Artinian by (i), we have natural isomorphisms $K_0(D(G)) \cong G_0(D(G))$ and $K_0(D(F)) \cong G_0(D(F))$ for all F such that $F/N \in \mathcal{F}(G/N)$, and we conclude that the natural induction map

$$\bigoplus_{F/N \in \mathcal{F}(G/N)} K_0(D(F)) \longrightarrow K_0(D(G))$$

is onto. When $F/N \in \mathcal{F}(G/N)$, we see from Lemma 9.4, that $D(F)$ is semisimple Artinian, hence every indecomposable $D(F)$-module is of the form $eD(F)$ for some idempotent $e \in D(F)$ and in view of Proposition 9.5(i), we may assume that e is a projection. We are now in a position to apply Lemma 13.8, so we obtain $r, s \in \mathbb{P}$, $F_1/N, \ldots, F_s/N \in \mathcal{F}(G/N)$, and projections $f_i \in D(F_i)$ such that

$$\operatorname{diag}(e, 1_r, 0_s) = u \operatorname{diag}(f_1, \ldots, f_s, 0_{n+r})u^{-1}$$

where $u \in \mathrm{GL}_{n+r+s}(D(G))$. Applying Lemma 8.4, we may assume that $u \in \mathrm{GL}_{n+r+s}(W(G))$, hence

$$\mathrm{tr}_G\, e + r = \mathrm{tr}_G\, f_1 + \cdots + \mathrm{tr}_G\, f_s$$

and the result follows.   □

The following result could easily be proved directly, but is also an immediate consequence of the above Lemma 13.10(iii) (use the case $G = N$ and note that the orders of the finite subgroups of G all divide l).

Corollary 13.11. *Let G be a group such that $D(G)$ is Artinian, and let $l, n \in \mathbb{P}$. If $l\,\mathrm{tr}_G\, e \in \mathbb{Z}$ for all projections $e \in D(G)$, then $l\,\mathrm{tr}_G\, e \in \mathbb{Z}$ for all projections $e \in \mathrm{M}_n(D(G))$.*

Lemma 13.12. *Let $H \lhd G$ be groups such that $|G/H| < \infty$ and H is a direct product of nonabelian free groups, let $l = \mathrm{lcm}(G)$, and let Σ denote the set of matrices with entries in $\mathbb{C}G$ which become invertible over $D(G)$. Then*

(i) *$D(G)$ is a semisimple Artinian ring.*
(ii) *The identity map on $\mathbb{C}G$ extends to an isomorphism $\mathbb{C}G_\Sigma \to D(G)$.*
(iii) *If $e \in D(G)$ is a projection, then $l\,\mathrm{tr}_G\, e \in \mathbb{Z}$.*

Proof. Let $\{X_i \mid i \in \mathcal{I}\}$ denote the family of finite subsets of G. For each $i \in \mathcal{I}$, there is by Lemma 13.2 a finitely generated subgroup G_i containing X_i such that $G_i \cap H$ is a direct product of nonabelian free groups. Let Σ_i denote the matrices over $\mathbb{C}G_i$ which become invertible over $D(G_i)$. If (i), (ii) and (iii) are all true for all $i \in \mathcal{I}$ when G is replaced by G_i and Σ by Σ_i, then the result follows from Lemma 13.5 so we may assume that G is finitely generated.

Lemma 13.4 now shows that there exists $K \lhd G$ such that $K \subseteq H$, G/K is abelian-by-finite, and $\mathrm{lcm}(G/K) = l$. Using Lemma 12.6, we see that $D(K)$ is a division ring and that the identity map on $\mathbb{C}K$ extends to an isomorphism $\mathbb{C}K_\Phi \to D(K)$, where Φ denotes the matrices with entries in $\mathbb{C}K$ which become invertible over $D(K)$. Therefore the only projections of $D(K)$ are 0 and 1, so $\mathrm{tr}_K\, e \in \mathbb{Z}$ for all projections $e \in D(K)$.

Let $F/K \in \mathcal{F}(G/K)$, let $[F : K] = f$, let $\{x_1, \ldots, x_f\}$ be a transversal for K in F, let $e \in D(F)$ be a projection, and let $\hat{\ }: W(F) \to \mathrm{M}_f(W(K))$ denote the monomorphism of Lemma 8.6. In view of the previous paragraph, Corollary 13.11 tells us that $\mathrm{tr}_K\, h \in \mathbb{Z}$ for all projections $h \in \mathrm{M}_f(D(K))$. Since $e \in W(F)$, we may write $e = \sum \epsilon_i x_i$ where $\epsilon_i \in W(K)$ for all i. Using Lemma 9.3, we deduce that $\epsilon_i \in D(K)$ for all i, and it is now not difficult to see that $\hat{e} \in \mathrm{M}_f(D(K))$. Therefore $\mathrm{tr}_K\, \hat{e} \in \mathbb{Z}$ by Corollary 13.11, and we conclude from Lemma 8.6 that $f\,\mathrm{tr}_F\, e \in \mathbb{Z}$. But $f \mid l$ and the result follows from Lemma 13.10.   □

Proof of Theorem 10.3. Replacing F with F', we may assume that F is a direct product of *nonabelian* free groups. We now use a transfinite induction argument, and since this is standard when dealing with elementary amenable groups, we will only sketch the details. If $\mathcal{Y}$ is a class of groups, then $H \in \mathrm{L}\,\mathcal{Y}$

means that every finite subset of the group H is contained in a $\mathcal{Y}$-subgroup, and $\mathcal{B}$ denotes the class of finitely generated abelian-by-finite groups. For each ordinal α, define $\mathcal{X}_\alpha$ inductively as follows:

$$\mathcal{X}_0 = \text{all finite groups},$$
$$\mathcal{X}_\alpha = (\mathrm{L}\,\mathcal{X}_{\alpha-1})\mathcal{B} \quad \text{if } \alpha \text{ is a successor ordinal},$$
$$\mathcal{X}_\alpha = \bigcup_{\beta < \alpha} \mathcal{X}_\beta \quad \text{if } \alpha \text{ is a limit ordinal}.$$

Then $\bigcup_{\alpha \geq 0} \mathcal{X}_\alpha$ is the class of elementary amenable groups [34, lemma 3.1(i)]. Let α be the least ordinal such that $G/F \in \mathcal{X}_\alpha$. If $\alpha = 0$, the result follows from Lemma 13.12. The use of transfinite induction now means that we have two cases to consider.

Case (i)   The result is true with H in place of G whenever H/F is a finitely generated subgroup of G/F. Here we use Lemma 13.5.

Case (ii)   There exists $H \lhd G$ such that $F \subseteq H$ and G/H is finitely generated abelian-by-finite, and the result is true with E in place of G whenever E/H is a finite subgroup of G/H. Here we use Lemma 13.10. $\square$

Proof of Corollary 10.4. By Theorem 10.3, we know that $D(G)$ is semisimple Artinian so if $D(G)$ is not simple Artinian, then there is a central idempotent $e \in D(G)$ such that $0 \neq e \neq 1$. Using Proposition 9.5(i), we deduce that $e \in W(G)$. Since $geg^{-1} = e$ for all $g \in G$, we see that $\{gxg^{-1} \mid g \in G\}$ is finite whenever $x \in G$ and $e_x \neq 0$, hence $e \in D(\Delta(G))$ where $\Delta(G)$ denotes the finite conjugate center of G [47, §5]. But $\Delta^+(G) = 1$, hence $\Delta(G)$ is torsion free abelian by [47, lemma 5.1(ii)] and it now follows from Theorem 10.3 that $\mathrm{tr}_G\, e \in \mathbb{Z}$. Therefore $e = 0$ or 1 by Kaplansky's theorem (§8), a contradiction, thus $D(G)$ is simple Artinian and we may write $D(G) = \mathrm{M}_m(D)$ for some $m \in \mathbb{P}$ and some division ring D.

It remains to prove that $m = l$. Using Lemma 9.6 and Theorem 10.3, we see that $m \leq l$. Now let $F \in \mathcal{F}(G)$ and set $f = \frac{1}{|F|}\sum_{g \in F} g$, a projection in $\mathbb{C}F$. Write $1 = e_1 + \cdots + e_r + \cdots + e_m$ where the e_i are primitive idempotents of $D(G)$, $1 \leq r \leq m$, and $f = e_1 + \cdots + e_r$. By Lemma 9.5(i), there are projections $f_i \in D(G)$ such that $f_i D(G) = e_i D(G)$ $(1 \leq i \leq m)$, and then application of Lemma 8.5 shows that $\mathrm{tr}_G\, f_1 + \cdots + \mathrm{tr}_G\, f_m = 1$. Also for each i, there exists a unit $u_i \in D(G)$ such that $u_i f_i u_i^{-1} = f_1$, and by Lemma 8.4 we may assume that $u_i \in W(G)$ for all i. Therefore $\mathrm{tr}_G\, f_i = \mathrm{tr}_G\, f_1$ for all i and we deduce that $\mathrm{tr}_G\, f_i = 1/m$ for all i. Another application of Lemma 8.5 shows that $\mathrm{tr}_G\, f = \mathrm{tr}_G\, f_1 + \cdots + \mathrm{tr}_G\, f_r$ and we conclude that $1/|F| = r/m$. Therefore $|F|$ divides m for all $F \in \mathcal{F}(G)$, hence $l | m$ and we have proven the result in the case $n = 1$. The case for general n follows from Lemma 9.1 and Corollary 13.11. $\square$

REFERENCES

[1] W. Arveson, *An invitation to C*-algebra*, Graduate Texts in Mathematics, vol. 39, Springer-Verlag, Berlin-New York, 1976.

246 *P. A. Linnell*

[2] H. Bass, *Traces and Euler characteristics*, Homological Group Theory (C.T.C. Wall, ed.), London Math. Soc. Lecture Note Series, vol. 36, Cambridge University Press, 1979, pp. 1–26.

[3] G. Baumslag and J. E. Roseblade, *Subgroups of direct products of free groups*, J. London Math. Soc. **30** (1984), 44–52.

[4] S. K. Berberian, *Baer *-rings*, Grundlehren, vol. 195, Springer-Verlag, Berlin-New York, 1972.

[5] _____, *The maximal ring of quotients of a finite von Neumann algebra*, Rocky Mountain J. Math. **12** (1982), 149–164.

[6] K. A. Brown, *On zero divisors in group rings*, Bull. London Math. Soc. **8** (1976), 251–256.

[7] J. Cheeger and M. Gromov, *L_2-cohomology and group cohomology*, Topology **25** (1986), 189–215.

[8] G. H. Cliff, *Zero divisors and idempotents in group rings*, Canadian J. Math. **32** (1980), 596–602.

[9] G. H. Cliff and A. Weiss, *Moody's induction theorem*, Illinois J. Math. **32** (1988), 489–500.

[10] J. M. Cohen, *Von Neumann dimension and the homology of covering spaces*, Quart. J. Math. Oxford **30** (1979), 133–142.

[11] _____, *L^2-cohomology in topology, algebra and analysis*, Rend. Circ. Mat. Palmero (2) suppl. no. 18 (1988), 31–36.

[12] J. M. Cohen and L. de Michele, *The radial Fourier-Stieltjes algebra of free groups*, Operator algebras and K-theory (R. G. Douglas and C. Schochet, eds.), Contemporary Mathematics, vol. 10, Amer. Math. Soc., 1982, pp. 33–40.

[13] P. M. Cohn, *On the free product of associative rings II*, Math. Z. **73** (1960), 433–456.

[14] _____, *On the free product of associative rings III*, J. Algebra **8** (1968), 376–383.

[15] _____, *Free rings and their relations (second edition)*, London Math. Soc. monographs, vol. 19, Academic Press, London-New York, 1985.

[16] A. Connes, *Non-commutative differential geometry*, Publ. Math. IHES **62** (1985), 257–360.

[17] W. Dicks, *Groups, trees and projective modules*, Lecture Notes in Mathematics, vol. 790, Springer-Verlag, Berlin-New York, 1980.

[18] W. Dicks and P. H. Kropholler, *Free groups and almost equivariant maps*, Bull. London Math. Soc. **27** (1995), 319–326.

[19] G. A. Edgar and J. M. Rosenblatt, *Difference equations over locally compact abelian groups*, Trans. Amer. Math. Soc. **253** (1979), 273–289.

[20] E. G. Effros, *Why the circle is connected: an introduction to quantized topology*, Math. Intelligencer **11** (1989), 27–34.

[21] J. Fabrykowski and N. Gupta, *On groups with exponential growth*, J. Indian Math. Soc. **49** (1985), 249–256.

[22] D. R. Farkas, *Miscellany on Bieberbach group algebras*, Pacific J. Math. **59** (1975), 427–435.

[23] D. R. Farkas and P. A. Linnell, *Zero divisors in group rings: something old, something new*, Representation theory, group rings, and coding theory (M. Isaacs, A. I. Lichtman, D. S. Passman, and S. K. Sehgal, eds.), Contemporary Mathematics, vol. 93, Amer. Math. Soc., 1989, pp. 155–166.

[24] D. R. Farkas and R. L. Snider, *K_0 and Noetherian group rings*, J. Algebra **42** (1976), 192–198.

[25] E. Formanek, *The zero divisor question for supersolvable groups*, Bull. Austral. Math. Soc. **9** (1973), 69–71.

[26] M. Gromov, *Asymptotic invariants of infinite groups*, London Math. Soc. Lecture Notes Series, vol. 182, Cambridge University Press, Cambridge-New York, 1993.

[27] N. Gupta, *On groups in which every element has finite order*, Amer. Math. Monthly **96** (1989), 297–308.

[28] P. de la Harpe, *Operator algebras, free groups and other groups*, Recent advances in Operator Algebras, Orléans, 1992, Astérisque, vol. 232, Société Mathématique de France, 1995, pp. 121–153.

[29] G. Higman, *The units of group rings*, Proc. London Math. Soc. (2) **46** (1940), 231–248.

[30] D. F. Holt, *Embeddings of group extensions into Wreath products*, Quart. J. Math. Oxford **29** (1978), 463–468.

[31] I. Hughes, *Division rings of fractions for group rings*, Comm. Pure Appl. Math. **23** (1970), 181–188.

[32] P. Julg and A. Valette, *K-theoretic amenability for $SL_2(\mathbb{Q}_p)$ and the action on the associated tree*, J. Functional Analysis **58** (1984), 194–215.

[33] R. V. Kadison and J. R. Ringrose, *Fundamentals of the theory of operator algebras, volume 1, elementary theory*, Pure and Applied Mathematics Series, vol. 100, Academic Press, London-New York, 1983.

[34] P. H. Kropholler, P. A. Linnell, and J. A. Moody, *Applications of a new K-theoretic theorem to soluble group rings*, Proc. Amer. Math. Soc. **104** (1988), 675–684.

[35] M. Lazard, *Groupes analytiques p-adiques*, I.H.E.S. Pub. Math. **26** (1965), 389–603.

[36] J. Lewin, *A note on zero divisors in group-rings*, Proc. Amer. Math. Soc. **31** (1972), 357–359.

[37] _____, *Fields of fractions for group algebras of free groups*, Trans. Amer. Math. Soc. **192** (1974), 339–346.

[38] P. A. Linnell, *On accessibility of groups*, J. Pure Appl. Algebra **30** (1983), 39–46.

[39] _____, *Zero divisors and group von Neumann algebras*, Pacific J. Math. **149** (1991), 349–363.

[40] _____, *Zero divisors and $L^2(G)$*, C. R. Acad. Sci. Paris Sér. I Math. **315** (1992), 49–53.

[41] _____, *Division rings and group von Neumann algebras*, Forum Math. **5** (1993), 561–576.

[42] M. S. Montgomery, *Left and right inverses in group algebras*, Bull. Amer. Math. Soc. **75** (1969), 539–540.

[43] J. A. Moody, *Brauer induction for G_0 of certain infinite groups*, J. Algebra **122** (1989), 1–14.

[44] R. B. Mura and A. H. Rhemtulla, *Orderable groups*, Lecture Notes in Pure and Appl. Math., vol. 27, Marcel Dekker, New York, 1977.

[45] M. A. Naimark, *Normed algebras*, Wolters-Noordhoff Publishing, Groningen, 1972.

[46] A. Neumann, *Completed group algebras without zero divisors*, Arch. Math. **51** (1988), 496–499.

[47] D. S. Passman, *Infinite crossed products*, Pure and Applied Mathematics, vol. 135, Academic Press, London-New York, 1989.

[48] A. L. T. Paterson, *Amenability*, Mathematical Surveys and Monographs, vol. 29, American Mathematical Society, Providence R.I., 1988.

[49] J.-P. Pier, *Amenable locally compact groups*, Wiley-Interscience, New York, 1984.

[50] M. Pimsner and D. Voiculescu, *K-groups of reduced crossed products by free groups*, J. Operator Theory **8** (1982), 131–156.

[51] R. T. Powers, *Simplicity of the C^*-algebra associated with the free group on two generators*, Duke Math. J. **42** (1975), 151–156.

[52] M. J. Puls, *Zero divisors and $L^p(G)$*, Proc. Amer. Math. Soc., to appear.

[53] _____, *Analytic versions of the zero divisor conjecture*, Ph.D. thesis, VPISU, Blacksburg, VA, 1995.

[54] J. M. Rosenblatt, *Linear independence of translations*, J. Austral. Math. Soc. Ser. A **59** (1995), 131–133.

[55] W. Rudin, *Fourier analysis on groups*, Interscience Tracts in Pure and Applied Mathematics, vol. 12, John Wiley and Sons, London-New York, 1962.

[56] A. H. Schofield, *Representations of rings over skew fields*, London Math. Soc. Lecture Note Series, vol. 92, Cambridge University Press, Cambridge-New York, 1985.

[57] E. M. Stein, *Harmonic analysis: real-variable methods, orthogonality, and oscillatory integrals*, Princeton Mathematical Series, vol. 43, Princeton University Press, Princeton, N.J., 1993.

[58] D. Tamari, *A refined classification of semi-groups leading to generalized polynomial rings with a generalized degree concept*, Proceedings of the International Congress of Mathematicians, Amsterdam, 1954 (Johan C. H. Gerretsen and Johannes de Groot, eds.), vol. 3, Groningen, 1957, pp. 439–440.

[59] A. Valette, *The conjecture of idempotents: a survey of the C*-algebraic approach*, Bull. Soc. Math. Belg. Ser. A **41** (1989), 485–521.

[60] P. A. Zalesskii and O. V. Mel'nikov, *Subgroups of profinite groups acting on trees*, Math. USSR Sb. **63** (1989), 405–424.

On the Geometric Invariants of Soluble Groups of Finite Prüfer Rank

Holger Meinert

Fachbereich Mathematik der Johann Wolfgang Goethe-Universität, Robert-Mayer-Str. 6–10, D-60054 Frankfurt a.M., Germany

1. Introduction

Given a finitely generated group G, R. Bieri and B. Renz introduced in 1988 two descending chains

$$
\begin{array}{ccccccccc}
Hom(G,\mathbb{R}) & \supseteq & \Sigma^1(G,\mathbb{Z}) & \supseteq & \Sigma^2(G,\mathbb{Z}) & \supseteq & \Sigma^3(G,\mathbb{Z}) & \supseteq & \cdots \\
\| & & \| & & |\cup & & |\cup & & \\
Hom(G,\mathbb{R}) & \supseteq & \Sigma^1(G) & \supseteq & \Sigma^2(G) & \supseteq & \Sigma^3(G) & \supseteq & \cdots
\end{array}
$$

of conical subsets of the real vector space $Hom(G,\mathbb{R})$, the *homological and homotopical geometric invariants* $\Sigma^*(G,\mathbb{Z})$ and $\Sigma^*(G)$ [BiRe, Re2]. They contain rather detailed information on the structure of the group G, for example:

(i) $\Sigma^m(G,\mathbb{Z})$ and $\Sigma^m(G)$ characterize the normal subgroups of G with abelian quotient that are of type FP_m and of type F_m, respectively.

(ii) If G is metabelian, $\Sigma^1(G) = \Sigma^1(G,\mathbb{Z})$ carries the information whether G is finitely presented or not. Moreover, it is conceivable that in this case $\Sigma^1(G)$ also determines the higher finiteness properties "type FP_m" and the higher geometric invariants $\Sigma^m(G) \subseteq \Sigma^m(G,\mathbb{Z})$ for $m \geq 2$.

In general, the geometric invariants are difficult to compute and not much is known about the higher ones. It is the aim of this article to present the progress that has been made in the study of the invariants for soluble-by-finite groups of finite Prüfer rank.

The paper is organized as follows. In Section 2 we introduce the geometric invariants, and Section 3 contains a method for "computing" $\Sigma^1(G)$ for soluble-by-finite groups of finite Prüfer rank. In Section 4 we will be concerned with the relationship between $\Sigma^1(G)$ and finiteness properties of metabelian (or more general soluble) groups. Finally, Section 5 is devoted to the higher invariants.

1.1. The basic settings

Throughout this article G always denotes a finitely generated group, R a commutative ring with non-trivial unity, M a (left) module over the group ring RG, and m a non-negative integer or ∞.

If we endow $Hom(G, \mathbb{R})$ with the compact-open topology, where G carries the discrete and $\mathbb{R}$ the usual topology, then $Hom(G, \mathbb{R})$ becomes a real topological vector space, its dimension d being the torsion-free rank of the abelianization G/G'. A concrete model is given by choosing an epimorphism $\vartheta : G \twoheadrightarrow \mathbb{Z}^d \subseteq \mathbb{R}^d$ and identifying the vector $x \in \mathbb{R}^d$ with the homomorphism $G \to \mathbb{R}$, $g \mapsto \langle x, \vartheta(g) \rangle$, where $\langle \cdot, \cdot \rangle$ is the standard scalar product in $\mathbb{R}^d$.

A *rationally defined open half space of* $Hom(G, \mathbb{R})$ is a subset of $Hom(G, \mathbb{R})$ of the form $\{\chi \mid \chi(g) > 0\}$, defined by some $g \in G$.

2. The geometric invariants

In this section we define the geometric invariants and state some of their main features. We start by recalling some general definitions.

2.1 Finiteness properties A module over a ring Λ (with $1 \neq 0$) is said to be *of type FP_m over* Λ if it admits a projective resolution over Λ with finitely generated modules in all dimensions $< m + 1$. A monoid Γ is said to be *of type FP_m over* R if the trivial $R\Gamma$-module R is of type FP_m over the monoid ring $R\Gamma$. If $R = \mathbb{Z}$ we merely say that Γ is *of type FP_m*. Following C.T.C. Wall we say that a CW-complex is *of type F_m* if it has finitely many cells in each dimension $< m + 1$. By definition, a group Γ is *of type F_m* if there exists a $K(\Gamma, 1)$-complex of type F_m.

Any group is of type F_0 and FP_0 over any R. Moreover, the implications (finitely generated $\Leftrightarrow$ F_1 $\Leftrightarrow$ FP_1 over R), and (finitely presented $\Leftrightarrow$ F_2 $\Rightarrow$ FP_2 over R), and (F_m $\Rightarrow$ FP_m $\Rightarrow$ FP_m over R) are true for all groups and all R. For finitely presented groups F_m and FP_m are equivalent. It is an open question whether FP_2 implies F_2, but non-finitely-presented groups which are FP_2 over any field exist (see Section 4). For more details the reader is referred to [Bi1, Br1].

2.2 The homological invariants For a given homomorphism $\chi : G \to \mathbb{R}$ we consider the submonoid $G_\chi = \{g \in G \mid \chi(g) \geq 0\}$ and ask whether the RG-module M is of type FP_m over the monoid ring RG_χ. The answers are codified in the homological invariants

$$\Sigma^m_R(G, M) = \{\chi \in Hom(G, \mathbb{R}) \mid M \text{ is of type } FP_m \text{ over } RG_\chi\}.$$

We set $\Sigma^m(G, M) = \Sigma^m_{\mathbb{Z}}(G, M)$ and denote by $\Sigma^m_R(G, M)^c$ and $\Sigma^m(G, M)^c$ the complements of the indicated sets in $Hom(G, \mathbb{R})$, respectively. By $\Sigma^{\bullet}_R(G, R)$ we shall always mean the invariants of the trivial RG-module R.

In contrast to the original definition in [BiRe] we do not restrict to the integers as coefficient ring, this will be crucial later on. Moreover, we have endowed $\Sigma_R^m(G, M)$ with a "singular point": if $\Sigma_R^m(G, M) \neq \emptyset$ then M is of type FP_m over RG (see [BiRe]) which in turn is equivalent to $0 \in \Sigma_R^m(G, M)$.

2.3 The homotopical invariants There seems to be no convincing definition of "type F_m" for monoids, in general, but if one restricts to monoids of the form G_χ one can proceed as follows. Let $\chi : G \to \mathbb{R}$ be a homomorphism, and let $\tilde{K}$ be the universal covering complex of a $K(G, 1)$-complex of type F_m. By a *regular χ-equivariant height function on $\tilde{K}$* we shall mean a continuous map $h : \tilde{K} \to \mathbb{R}$ such that

(i) $h(gx) = \chi(g) + h(x)$ for all $g \in G$ and all $x \in \tilde{K}$,

(ii) $h(\tilde{K}^0) \subseteq \chi(G)$, where $\tilde{K}^0$ is the set of vertices of $\tilde{K}$,

(iii) For each closed cell e of $\tilde{K}$, $h|_e$ attains its minimum on the boundary of e.

It is not difficult to see that such height functions always exist. Moreover, the cellular chain complex of the maximal subcomplex $\tilde{K}_h$ contained in $h^{-1}([0, \infty))$ is a free $\mathbb{Z}G_\chi$-chain complex with finitely generated modules in all dimensions $< m + 1$ [Me1, Me5].

We say that G_χ is *of type F_m* if there is a $K(G, 1)$-complex K of type F_m and a regular χ-equivariant height function $h : \tilde{K} \to \mathbb{R}$ on the universal covering complex such that $\tilde{K}_h$ is $(m - 1)$-connected (resp. contractible if $m = \infty$). Finally we set

$$\Sigma^m(G) = \{\chi \in Hom(G, \mathbb{R}) \mid G_\chi \text{ is of type } F_m\},$$

and denote by $\Sigma^m(G)^c$ its complement in $Hom(G, \mathbb{R})$. Clearly, $0 \in \Sigma^m(G)$ if and only if G is of type F_m if and only if $\Sigma^m(G) \neq \emptyset$.

From the discussions above one can see that $\Sigma^0(G) = \Sigma_R^0(G, R) = Hom(G, \mathbb{R})$ and that $\Sigma^m(G) \subseteq \Sigma^m(G, \mathbb{Z}) \subseteq \Sigma_R^m(G, R)$ for all $m \geq 1$ and all R.

Theorem 2.4. [Re2] *Let R be a commutative ring with non-trivial unity. Then*

(i) $\Sigma^1(G) = \Sigma_R^1(G, R)$,

(ii) $\Sigma^m(G) = \Sigma^2(G) \cap \Sigma^m(G, \mathbb{Z})$ for all $m \geq 2$.

2.5 Some remarks on $\Sigma^1(-)$

(i) Monoids of type FP_1 are not necessarily finitely generated (also see [Co]). Let $G = \langle a, t \mid t^{-1}at = a^2 \rangle$ and $\chi : G \to \mathbb{Z}$ given by $\chi(a) = 0$ and $\chi(t) = 1$. Then G_χ is, in fact, of type FP_∞ but not finitely generated.

(ii) Let $\Gamma(G, X)$ be the Cayley graph of G with respect to a finite generating set X. Given $\chi \in Hom(G, \mathbb{R})$, let $\Gamma(G, X)_\chi$ be the full subgraph of $\Gamma(G, X)$ with vertex set G_χ. Then $\chi \in \Sigma^1(G)$ if and only if $\Gamma(G, X)_\chi$ is *connected* [BiRe, BiSt4, Re1].

(iii) The invariant $\Sigma^1(G)$ coincides up to sign with the Bieri-Neumann-Strebel invariant $\Sigma_{G'}$ studied in [BiNeSt] (see [BiRe, Re2]).

(iv) K.S. Brown has given a powerful and elegant characterization of $\Sigma^1(G)$ (resp. $\Sigma_{G'}$) in terms of $\mathbb{R}$-tree actions [Br3] (also see [BiSt4, Ge1, Ge2, Le]). Applications are to be found in [BiSt4, Ge1, Ge2, Ho, Me4].

In the introduction we already hinted at the following interesting properties of our invariants. The proofs in the homological setting have been given only for $R = \mathbb{Z}$ but they work perfectly for arbitrary non-trivial commutative rings R with unity.

Theorem 2.6. [BiNeSt, BiRe, Re2] *Let G be a finitely generated group, $N \trianglelefteq G$ a normal subgroup containing the derived subgroup G', and M an RG-module.*

(i) M is of type FP_m over RN if and only if $\{\chi \in Hom(G, \mathbb{R}) \mid \chi(N)=0\} \subseteq \Sigma_R^m(G, M)$.

(ii) N is of type F_m if and only if $\{\chi \in Hom(G, \mathbb{R}) \mid \chi(N) = 0\} \subseteq \Sigma^m(G)$.

Theorem 2.7. [BiNeSt, BiRe, Re2] $\Sigma_R^m(G, M) - \{0\}$ and $\Sigma^m(G) - \{0\}$ are open subsets of $Hom(G, \mathbb{R}) - \{0\}$.

There is no shortage of examples for $\Sigma_R^0(G, M)$ if G is abelian [BiSt1, BiSt2] and for $\Sigma^1(G)$ [BiNeSt, BiSt4, Br3]. The higher invariants $\Sigma^m(G) \subseteq \Sigma^m(G, \mathbb{Z})$ have been computed for 1-relator groups [BiRe, Re2], fundamental groups of compact 3-manifolds [BiSt4], or direct products of these [Ge1, Ge2]. In [Me2] we determined the higher invariants for direct products of virtually free groups and, using Theorem 2.6, also the finiteness properties "FP_m" and "F_m" of all normal subgroups with abelian quotient.

In the following we often want to pass to sub- or supergroups of finite index. We will see later on that this can be harmful although we have:

Theorem 2.8. [BiNeSt, BiSt4, Sch] *Let $H \leq G$ be a subgroup of finite index, and let $\chi : G \to \mathbb{R}$ be a homomorphism. Then $\chi \in \Sigma_R^m(G, M)$ (resp. $\chi \in \Sigma^m(G)$) if and only if $\chi|_H \in \Sigma_R^m(H, M)$ (resp. $\chi|_H \in \Sigma^m(H)$).*

3. On $\Sigma^1(G)$ for soluble groups of finite Prüfer rank

For a metabelian group G, given by an extension

(E) $\qquad 1 \longrightarrow A \longrightarrow G \longrightarrow Q \longrightarrow 1$, A and Q abelian groups,

Bieri and Strebel introduced in their 1980 paper [BiSt1] an invariant Σ_A and proved that it contains the full information whether G is finitely presented or not (cf. Theorem 4.1 (ii)). By definition, $\Sigma_A = \Sigma^0(Q, A)$, and the following result shows that $\Sigma^1(-)$ can be thought of as a generalization of Σ_A to the class of all finitely generated groups.

Theorem 3.1. [BiSt4] *Let N be a normal subgroup of a finitely generated group G, $\pi : G \twoheadrightarrow G/N = Q$ the canonical projection, inducing the map $\pi^* : Hom(Q, \mathbb{R}) \hookrightarrow Hom(G, \mathbb{R})$, and A the abelianization N/N' viewed as a $\mathbb{Z}Q$-module via conjugation. If N is nilpotent and Q' finitely presentable then $\Sigma^1(G)^c = \pi^* (\Sigma^0(Q, A)^c)$.*

Remark. The inclusion $\Sigma_R^0(Q, R \otimes_{\mathbb{Z}} A)^c \subseteq \Sigma^0(Q, A)^c$ can fail to be an equality. If $Q = \langle q \mid - \rangle$ is infinite cyclic and q acts by multiplication by 2 on $A = \mathbb{Z}[\frac{1}{2}]$ then $\Sigma_K^0(Q, K \otimes_{\mathbb{Z}} A)^c = \emptyset$ for any field K, but $\Sigma^0(Q, A)^c \neq \emptyset$.

Although Theorem 3.1 tells us that, in the above setting, $\Sigma^1(G)$ and $\Sigma^0(Q, A)$ are essentially the same invariants, we find it convenient to state the results in terms of $\Sigma^1(G)$, if possible, avoiding the reference to a certain extension. However, almost all the results below have been obtained for the invariant $\Sigma^0(Q, A)$ (one exception is the first part of Theorem 4.1 which has a beautiful proof using the Cayley graph approach to $\Sigma^1(-)$ as explained in 2.5(ii); see [BiSt4]). Indeed, there are very strong methods, using valuations on fields, for "computing" $\Sigma^0(Q, A)$ (see [BiSt2, BiGr2]). We shall not be concerned with the general case here but restrict ourselves to groups of finite Prüfer rank.

3.2 Finite Prüfer rank A group has *finite Prüfer rank* if there is a uniform bound on the minimal numbers of generators of all finitely generated subgroups. Given an extension (E) with a finitely generated group G in the middle, this group has finite Prüfer rank if and only if the torsion subgroup T of A is finite and A/T has finite torsion-free rank (see, e.g., [Bol]). By a result of Mal'cev, finitely generated soluble-by-finite groups of finite Prüfer rank are nilpotent-by-abelian-by-finite (see, e.g., [Ro], Proof of Theorem 10.38).

Theorem 3.3. [BiSt2] *Let G be a finitely generated soluble-by-finite group of finite Prüfer rank. Then there exists a finite set $S \subseteq Hom(G, \mathbb{Z}) - \{0\}$ of non-trivial homomorphisms such that $\Sigma^1(G)^c = \{r \cdot \chi \mid 0 < r \in \mathbb{R}, \chi \in S\}$.*

In the remainder of this section we shall outline a procedure for "computing" $\Sigma^1(-)^c$ for soluble-by-finite groups of finite Prüfer rank.

3.4 Computation of $\Sigma^1(G)^c$ in the finite Prüfer rank case Given a finitely generated soluble-by-finite group G of finite Prüfer rank, one proceeds in two steps.

1) Reduction step. We first choose a subgroup $H \leq G$ of finite index containing a nilpotent normal subgroup $N \trianglelefteq H$ with $Q = H/N$ free abelian. By Theorem 2.8 it suffices to "compute" $\Sigma^1(H)^c$. Next, Theorem 3.1 asserts that $\Sigma^1(H)^c$ is completely determined by the invariant $\Sigma^0(Q, A)^c$ of the finitely generated $\mathbb{Z}Q$-module $A = N/N'$. Since the metabelian group H/N' is also of finite Prüfer rank, the torsion subgroup T of A is finite, hence $\Sigma^0(Q, A)^c =$

$\Sigma^0(Q, A/T)^c$, and A/T has finite torsion-free rank.

We have now reduced the computation of $\Sigma^1(G)^c$ to the computation of $\Sigma^0(Q, M)^c$, where Q is a finitely generated free abelian group and M is a finitely generated $\mathbb{Z}Q$-module which is torsion-free as abelian group and has finite torsion-free rank $n = \dim_{\mathbb{Q}}(M \otimes_{\mathbb{Z}} \mathbb{Q})$. Then our second step can be found in [Åbg], I.4.

2) The core. We first extend the action of Q on M to an action $\psi_1 : Q \to \mathrm{GL}(M \otimes_{\mathbb{Z}} \mathbb{Q})$. Let k be the finite field extension of $\mathbb{Q}$ obtained by adjoining all eigenvalues of a finite generating set of $\psi_1(Q)$. Let V be the k-vector space $M \otimes_{\mathbb{Z}} k$ of dimension n, and extend ψ_1 to a homomorphism $\psi_2 : Q \to \mathrm{GL}(V)$. Then one trigonalizes the image of Q in $\mathrm{GL}(V)$ (see [Bor], Chap. I, (4.6)). This gives us a homomorphism $\psi : Q \to U_n(k)$ into the group of all upper triangular matrices in $\mathrm{GL}_n(k)$. Let $p_i : U_n(k) \to k^*$ be the projection onto the (i, i)-coordinate, $i = 1, \ldots, n$. In this way we obtain homomorphisms $\phi_i = p_i \circ \psi : Q \to k^*$ for $1 \le i \le n$.

Theorem 3.5. [Åbg] *With the notation above, $\Sigma^0(Q, M)^c$ is the set of all non-zero homomorphisms $r \cdot (Q \xrightarrow{\phi_i} k^* \xrightarrow{v} \mathbb{Z}) \in Hom(Q, \mathbb{R})$, where v ranges over all normalized discrete valuations $k^* \twoheadrightarrow \mathbb{Z}$, r over all positive real numbers, and i over $\{1, \ldots, n\}$.*

Recall that a *normalized discrete valuation* is a group epimorphism $v : k^* \twoheadrightarrow \mathbb{Z}$ with the additional property $v(x + y) \ge \min\{v(x), v(y)\}$ for all $x, y \in k^*$. Since k is an algebraic number field, all valuations on k are "known" (see, e.g., [Bou]). So the theorem above allows one to determine $\Sigma^0(Q, M)^c$ and hence $\Sigma^1(G)^c$.

3.6 Examples

(i) Let Q be a finitely generated abelian group that acts via a homomorphism $\kappa : Q \to \mathbb{Q}^*$ on the field of rational numbers, and let A be the cyclic Q-module $\mathbb{Z}Q \cdot 1 \subseteq \mathbb{Q}$. If $A \rightarrowtail G \xrightarrow{\pi} Q$ is an arbitrary extension then G is metabelian of finite Prüfer rank, and $\Sigma^1(G)^c$ is made up of all non-trivial homomorphisms $r \cdot (v_p \circ \kappa \circ \pi) \in Hom(G, \mathbb{R})$, where $r > 0$, p is a prime number, and $v_p : \mathbb{Q}^* \twoheadrightarrow \mathbb{Z}$ is the p-adic valuation.

(ii) Fix a natural number s. Then the group $H = H_s$ consisting of all upper triangular matrices $(a_{i,j}) \in \mathrm{GL}_{s+1}(\mathbb{Z}[\frac{1}{2}])$ with $a_{i,i} > 0$ for all i, is of finite Prüfer rank since it is constructible (see Section 4 below) by [Ki]. If N denotes the nilpotent normal subgroup of all matrices with ones on the diagonal then $Q = H/N$ is free abelian on the $s + 1$ cosets $q_1 = \mathrm{diag}(2, 1, \ldots, 1) \cdot N$, $\ldots$, $q_{s+1} = \mathrm{diag}(1, \ldots, 1, 2) \cdot N$. Using the procedure above it is easy to compute $\Sigma^1(H)^c$: it consists of all homomorphisms $r \cdot (\chi_i \circ \pi) \in Hom(H, \mathbb{R})$, where $r > 0$, π denotes the projection $H \twoheadrightarrow Q$, $1 \le i \le s$, and the homomorphism $\chi_i : Q \to \mathbb{Z}$ is given by

$$\chi_i(q_j) = \begin{cases} 0 & \text{if } i \notin \{j-1, j\} \\ -1 & \text{if } i = j - 1 \\ 1 & \text{if } i = j. \end{cases}$$

(iii) Abels' group $\Gamma = \Gamma_s = \{(a_{i,j}) \in H \mid a_{1,1} = a_{s+1,s+1} = 1\}$ is a normal subgroup of H with abelian quotient. It is of finite Prüfer rank, of type F_{s-1}, but not of type FP_s [AbBr, Br2]. By the recipe above, one finds $\Sigma^1(\Gamma)^c = \{\chi|_\Gamma \mid \chi \in \Sigma^1(H)^c\}$.

4. The FP_m-Conjecture

As mentioned above, the starting point of the "Σ-theory" was the characterization of the finitely presented metabelian groups in terms of the Bieri-Strebel invariant Σ_A. Using the invariant $\Sigma^1(G)$ along with the notation $-\Sigma^1(G) = \{-\chi \mid \chi \in \Sigma^1(G)\}$ we state it as the second part of the following theorem.

Theorem 4.1. *Let R be a non-trivial commutative ring with unity.*
(i) [BiNeSt, BiSt4] Let G be a group without free subgroups of rank 2. If G is of type FP_2 over R then $\Sigma^1(G) \cup -\Sigma^1(G) = Hom(G, \mathbb{R})$.
(ii) [BiSt1] A metabelian group G is finitely presented if and only if it is of type FP_2 over R if and only if $\Sigma^1(G) \cup -\Sigma^1(G) = Hom(G, \mathbb{R})$.

Let G be the quotient of Abels' group Γ_3, the group of all upper triangular matrices $(a_{i,j}) \in GL_4(\mathbb{Z}[\frac{1}{2}])$ with $a_{i,i} > 0$ and $a_{1,1} = a_{4,4} = 1$, modulo its centre $Z(\Gamma_3)$. This nilpotent-by-abelian group of finite Prüfer rank satisfies $\Sigma^1(G) \cup -\Sigma^1(G) = Hom(G, \mathbb{R})$, and it is of type FP_2 over any field but not of type FP_2 [BiSt1].

4.2 Tameness We say that $\Sigma^1(G)^c$ (or $\Sigma^0_R(G, M)^c$) is *m-tame*, where $m \in \mathbb{N}$, if every subset of at most m homomorphisms is contained in a rationally defined open half space of $Hom(G, \mathbb{R})$. By some well-known separation theorem this is equivalent to saying that $0 \notin \text{conv}_{\leq m} \Sigma^1(G)^c$, where the latter set is the union of the convex hulls of all subsets of $\Sigma^1(G)^c$ of at most m elements. Note that the condition on $\Sigma^1(G)^c$ in Theorem 4.1 is equivalent to the 2-tameness condition.

Theorem 4.3. [BiGr1] *Let G be a metabelian group given by an extension (E). If G is of type FP_m over a field K ($m \in \mathbb{N}$) then $\Sigma^0_K(Q, K \otimes_{\mathbb{Z}} A)^c$ is m-tame.*

The two preceding theorems have led to the following

FP$_m$-Conjecture. *Let G be a metabelian group, and $m \in \mathbf{N}$. Then G is of type FP_m if and only if $\Sigma^1(G)^c$ is m-tame.*

Notice that the case $m = 1$ is trivial and that Theorem 4.1 gives an affirmative answer for $m = 2$. Moreover, Theorem 4.3 can be thought of as "one implication over a field". Recently G.A. Noskov showed that if G, given by a *split extension (E) with torsion-free A*, is of type FP$_m$ then $\Sigma^1(G)^c$ is m-tame [No]. And K.-U. Bux established the FP$_m$-conjecture for semi-direct products $O_S \rtimes O_S^*$, where O_S is an S-arithmetic subring of a global function field [Bu]. In his proof he used Theorem 4.3 to deduce that $\Sigma^1(G)^c$ is m-tame if, in the extension (E), A is an elementary abelian p-group (p prime) and G is of type FP$_m$ over $\mathbb{F}_p$.

Thanks to the deep work of H. Abels and H. Åberg much more is known about the finiteness properties of soluble-by-finite groups of finite Prüfer rank.

Theorem 4.4. [Åbg] *Let R be a commutative ring with non-trivial unity, and $m \in \mathbf{N}$.*

 (i) *Suppose G is a soluble-by-finite group of finite Prüfer rank. If G is of type FP_m over R then $\Sigma^1(G)^c$ is m-tame.*

 (ii) *A metabelian group G of finite Prüfer rank is of type FP_m over R if and only if $\Sigma^1(G)^c$ is m-tame.*

We have seen that the 2-tameness condition is necessary but, in general, not sufficient for a nilpotent-by-abelian group to be of type FP$_2$. There is a second (easy) necessary condition. Suppose $N \rightarrowtail H \twoheadrightarrow Q$ is an extension with an abelian group Q. If H is of type FP$_2$ then the Schur multiplier $H_2(N, \mathbb{Z})$ of N is a finitely generated $\mathbb{Z}Q$-module [BiGr1]. Now, Abels proved that the two necessary conditions are also sufficient for S-arithmetic nilpotent-by-abelian groups (which are of finite Prüfer rank!).

Theorem 4.5. [Ab] *Let G be an S-arithmetic soluble group, and let $H \leq G$ be a subgroup of finite index containing a nilpotent normal subgroup $N \trianglelefteq H$ with $Q = H/N$ abelian. Then G is finitely presented if and only if G is of type FP_2 if and only if both $\Sigma^1(H)^c$ is 2-tame and $H_2(N, \mathbb{Z})$ is finitely generated as $\mathbb{Z}Q$-module.*

Finally, we want to give a group theoretical characterization of the soluble-by-finite groups of type FP$_\infty$. Recall that a soluble-by-finite group G is *constructible (in the sense of [BaBi])* if it admits a finite chain $1 = H_0 \leq H_1 \leq \cdots \leq H_k = G$ of subgroups such that either H_i has finite index in H_{i+1} or $H_{i+1} = \langle H_i, t \mid tH_i t^{-1} \leq H_i \rangle$ is an ascending HNN-extension with base group H_i. These groups are of finite Prüfer rank and of type F$_\infty$. Obviously all polycyclic-by-finite groups are constructible but the class of constructible soluble-by-finite groups is much bigger. Recently P.H. Kropholler proved:

Theorem 4.6. [Kr] *Every soluble-by-finite group of type* FP_∞ *is constructible.*

This result closes the remaining gap in the "FP_∞-conjecture for nilpotent-by-abelian groups":

Theorem 4.7. *Let G be a nilpotent-by-abelian group, and d the torsion-free rank of its abelianization G/G'. Then the following assertions are equivalent:*

(i) *G is of type FP_{d+1}.*

(ii) *G is of type FP_∞.*

(iii) *G is constructible.*

(iv) *$\Sigma^1(G)^c$ is contained in a rationally defined open half space of $Hom(G, \mathbb{R})$.*

(v) *$\Sigma^1(G)^c$ is $(d+1)$-tame.*

Proof (i) implies (ii) is an observation in [Bi2], (ii) implies (iii) follows from the preceeding theorem, the equivalence (iii) if and only if (iv) is discussed in [BiSt3], and (v) implies (iv) is a general fact about convex subsets (see, e.g., [Va]).

Now, let G be a nilpotent-by-abelian-by-finite group. If G is constructible, it follows from Theorem 4.7 and Theorem 2.8 that $\Sigma^1(G)^c$ is contained in a rationally defined open half space of $Hom(G, \mathbb{R})$. The converse is false: take a non-constructible nilpotent-by-abelian group H and form the semi-direct product $G = (H \times H) \rtimes \mathbb{Z}_2$, where the cyclic group of order two permutes the factors. Then G is not constructible but $\Sigma^1(G)^c = \emptyset$.

5. The higher invariants for soluble groups of finite Prüfer rank

In Section 4 we pointed out that for a metabelian group G, given by an extension (E), the invariants $\Sigma^1(G)$ and $\Sigma^0(Q, A)$ are expected to carry the whole information on the finiteness properties of G. Then there are two obvious questions concerning the higher invariants:

(i) What can be said about $\Sigma^m(G)$ and $\Sigma^m(G, \mathbb{Z})$ for $m \geq 2$?

(ii) What can be said about $\Sigma^m(Q, A)$ for $m \geq 1$?

It turns out that only the first question is of interest. It is well-known that the group ring $\mathbb{Z}Q$ of an abelian group Q is noetherian, so the second question is answered by the following

Proposition 5.1. [Me3] *Let G be a group, and M an RG-module. If the group ring RG is left noctherian then $\Sigma^n_R(G, M) = \Sigma^\infty_R(G, M)$.*

Our first question seems to be much more difficult. It is conceivable that the complements $\Sigma^m(G, \mathbb{Z})^c \subseteq \Sigma^m(G)^c$ of the higher invariants are also completely determined by $\Sigma^1(G)^c$, though not in the very strong sense of the preceeding proposition. Recall that $\mathrm{conv}_{\leq m} \Sigma^1(G)^c = \bigcup_\sigma \mathrm{conv}(\sigma)$, where the union is taken over all $\sigma \subseteq \Sigma^1(G)^c$ with $|\sigma| \leq m$.

Σ^m-Conjecture. *Let G be a metabelian group, and $m \in \mathbb{N}$. If G is of type FP_m then $\mathrm{conv}_{\leq m} \Sigma^1(G)^c = \Sigma^m(G, \mathbb{Z})^c = \Sigma^m(G)^c$.*

Although this conjecture is connected somehow with the FP_m-conjecture (see [Me5]), none of them seems to imply the other. R. Gehrke proved that $\mathrm{conv}_{\leq 2} \Sigma^1(G)^c \subseteq \Sigma^2(G, \mathbb{Z})^c$ holds for all finitely presented metabelian groups [Ge1, Ge2].

Using Åberg's above result and, in particular, the CW-complex constructed by him in his proof of the "if"-part of Theorem 4.4(ii), the author established the Σ^m-conjecture for metabelian groups of finite Prüfer rank ([Me5, Me6], which are revised versions of [Me1]). Although we only considered the integers as coefficient ring, the arguments in [Me5] work for any R.

Theorem 5.2. [Me5, Me6] *Let G be a metabelian group of finite Prüfer rank, and R a non-trivial commutative ring with unity. If G is of type FP_m then*

$$\mathop{\mathrm{conv}}_{\leq m} \Sigma^1(G)^c = \Sigma^m_R(G, R)^c = \Sigma^m(G, \mathbb{Z})^c = \Sigma^m(G)^c \ .$$

Using this result together with work of Åberg and Strebel we are going to prove:

Theorem 5.3. *Let G be a soluble-by-finite group of finite Prüfer rank, and let K be a field. If G is of type FP_m over K, then*

$$\mathop{\mathrm{conv}}_{\leq m} \Sigma^1(G)^c \subseteq \Sigma^m_K(G, K)^c \subseteq \Sigma^m(G, \mathbb{Z})^c \subseteq \Sigma^m(G)^c \ .$$

Proof We only have to prove the first inclusion, and by Theorem 2.8 we can pass to the situation where G is nilpotent-by-abelian (and of finite Prüfer rank).

Let us assume for a moment that $p : G \twoheadrightarrow \tilde{G}$ is an epimorphism with abelian kernel $A \leq G''$. It follows easily from Theorem 3.1 that $\Sigma^1(G)^c = p^*(\Sigma^1(\tilde{G})^c)$, where $p^* : Hom(\tilde{G}, \mathbb{R}) \rightarrowtail Hom(G, \mathbb{R})$ is the induced linear map. Now, Åberg has shown that the homology groups $H_i(A, K)$ are finite dimensional K-vector spaces for all $i \in \mathbb{N}_0$ ([Åbg], IV.2, Proof of Proposition 2.1). By Theorem B in [St] one concludes that $\tilde{G}$ is of type FP_m over K. Moreover, from Strebel's proof one can also infer that $p^*(\Sigma^m_K(\tilde{G}, K)^c) \subseteq \Sigma^m_K(G, K)^c$. Summarizing we find: if $\mathrm{conv}_{\leq m} \Sigma^1(\tilde{G})^c \subseteq \Sigma^m_K(\tilde{G}, K)^c$ holds then we have the inclusion $\mathrm{conv}_{\leq m} \Sigma^1(G)^c \subseteq \Sigma^m_K(G, K)^c$.

Finally, we consider the epimorphism $G \twoheadrightarrow G/G'' = \tilde{G}$. Since G'' is soluble we may as well assume that it is abelian. Then the discussion above shows that $\tilde{G}$ is of type FP_m over K which implies $\mathrm{conv}_{\leq m} \Sigma^1(\tilde{G})^c = \Sigma^m_K(\tilde{G}, K)^c$ by Theorem 5.2. Hence we obtain the desired inclusion by the argument above.

Question. *Let G be a nilpotent-by-abelian group of finite Prüfer rank. If G is of type F_m is it true that $\Sigma^m(G)^c$ is contained in the convex hull $\text{conv}\,\Sigma^1(G)^c$ of $\Sigma^1(G)^c$?*

We know that the answer is "yes" if G is constructible.

Theorem 5.5.   [Me5, Me6] *Let G be a constructible nilpotent-by-abelian group. For any commutative ring R with non-trivial unity and any $m \in \mathbf{N}$ we have*

$$\underset{\leq m}{\text{conv}}\,\Sigma^1(G)^c \subseteq \Sigma^m_R(G,R)^c \subseteq \Sigma^m(G,\mathbb{Z})^c \subseteq \Sigma^m(G)^c \subseteq \text{conv}\,\Sigma^1(G)^c\ ,$$

and $\Sigma^m_R(G,R)^c = \Sigma^m(G,\mathbb{Z})^c = \Sigma^m(G)^c = \text{conv}\,\Sigma^1(G)^c$ for all $m \geq d$, where d is the torsion-free rank of G/G'.

Conjecture 5.6.   *If G is a constructible nilpotent-by-abelian group and $m \in \mathbf{N}$ then $\text{conv}_{\leq m}\,\Sigma^1(G)^c = \Sigma^m(G)^c$.*

It would be of interest to prove this conjecture. In conjunction with Theorem 2.6 it would yield results on finiteness properties of certain nilpotent-by-abelian groups of finite Prüfer rank, for example, a new proof that the group Γ_s of Example 3.6(iii) is of type F_{s-1}.

Theorem 5.5 does not hold for soluble constructible groups: let $H = \langle\, a, t \mid t^{-1}at = a^2 \,\rangle$ and $G = (H \times H) \rtimes \mathbb{Z}_2$, where the cyclic group of order 2 acts by permuting the factors. Then G is constructible and $\Sigma^1(G)^c = \emptyset$, but $\Sigma^2_R(G,R)^c = \Sigma^\infty_R(G,R)^c = \Sigma^\infty(G)^c = \Sigma^2(G)^c \neq \emptyset$ for all R.

Although we cannot completely determine the invariants of constructible soluble-by-finite groups, we can characterize the groups where the complements of the invariants are empty:

Theorem 5.7.   [Me3] *A soluble-by-finite group G is polycyclic-by-finite if and only if $\Sigma^\infty(G)^c = \emptyset$ if and only if $\Sigma^\infty(G,\mathbb{Z})^c = \emptyset$.*

Note added in proof

In 1995 M. Bestvina and N. Brady [*Morse theory and finiteness properties of groups*, to appear in Invent. Math.] proved that there exist groups of type FP_∞ which are not finitely presented. Their examples also imply that the invariants $\Sigma^m(G,R)$ and $\Sigma^m(G)$ can differ for all $m \geq 2$ (and any non-zero ring R). However, it is still unknown whether such examples can occur within the class of soluble groups.

References

[Ab]   H. Abels, *Finite presentability of S-arithmetic groups - Compact presentability of solvable groups*, Lecture Notes in Mathematics, **1261** Springer, Berlin (1987).

[AbBr]   H. Abels and K.S. Brown, *Finiteness properties of solvable S-arithmetic groups: an example*, J. Pure Appl. Algebra **44** (1987), pp. 77–83.

[Åbg]   H. Åberg, *Bieri-Strebel valuations (of finite rank)*, Proc. London Math. Soc. (3) **52** (1986), pp. 269–304.

[BaBi]   G. Baumslag and R. Bieri, *Constructable solvable groups*, Math. Z. **151** (1976), pp. 249–257.

[Bi1]   R. Bieri, *Homological dimension of discrete groups*, Queen Mary College Mathematics Notes, London, 1976, 2nd ed. (1981).

[Bi2]   R. Bieri, *A connection between the integral homology and the centre of a rational linear group*, Math. Z. **170** (1980), pp. 263–266.

[BiGr1]   R. Bieri and J.R.J. Groves, *Metabelian groups of type* FP$_\infty$ *are virtually of type* FP, Proc. London Math. Soc. (3) **45** (1982), pp. 365–384.

[BiGr2]   R. Bieri and J.R.J. Groves, *The geometry of the set of characters induced by valuations*, J. Reine Angew. Math. **347** (1984), pp. 168–195.

[BiNeSt] R. Bieri, W.D. Neumann and R. Strebel, *A geometric invariant for discrete groups*, Invent. Math. **90** (1987), pp. 451–477.

[BiRe]   R. Bieri and B. Renz, *Valuations on free resolutions and higher geometric invariants of groups*, Comment. Math. Helv. **63** (1988), pp. 464–497.

[BiSt1]   R. Bieri and R. Strebel, *Valuations and finitely presented metabelian groups*, Proc. London Math. Soc. (3) **41** (1980), pp. 439–464.

[BiSt2]   R. Bieri and R. Strebel, *A geometric invariant for modules over an abelian group*, J. Reine Angew. Math. **322** (1981), pp. 170–189.

[BiSt3]   R. Bieri and R. Strebel, *A geometric invariant for nilpotent-by-abelian-by-finite groups*, J. Pure Appl. Algebra **25** (1982), pp. 1–20.

[BiSt4]   R. Bieri and R. Strebel, *Geometric Invariants for Discrete Groups*, Preprint of a monograph, in preparation, Universität Frankfurt a.M. (1994).

[Bol]   J. Boler, *Subgroups of finitely presented metabelian groups of finite rank*, J. Austral. Math. Soc., Ser. A **22** (1979), pp. 501–508.

[Bor]   A. Borel, *Linear Algebraic Groups*, W.A. Benjamin Inc., New York (1969).

[Bou]   N. Bourbaki, *Commutative algebra*, Hermann, Paris (1972), and Addison-Wesley, Reading, MA (1972).

[Br1]   K.S. Brown, *Cohomology of groups*, Grad. Texts in Math., **87** Springer (1982).

[Br2]   K.S. Brown, *Finiteness properties of groups*, J. Pure Appl. Algebra **44** (1987), pp. 45–75.

[Br3]   K.S. Brown, *Trees, valuations, and the Bieri-Neumann-Strebel invariant*, Invent. Math. **90** (1987), pp. 479–504.

[Bu]   K-U. Bux, *Eine Serie metabelscher S-arithmetischer Gruppen*, Diplomarbeit, Universität Frankfurt a.M. (1993).

[Co]   D.E. Cohen, *A monoid which is right FP_∞ but not left FP_1*, Bull. London Math. Soc. **24** (1992), pp. 340–342.

[Ge1]   R. Gehrke, *Die höheren geometrischen Invarianten für Gruppen mit Kommutatorrelationen*, Dissertation, Universität Frankfurt a.M. (1992).

[Ge2]   R. Gehrke, *The higher geometric invariants for groups with sufficient commutativity*, Preprint (to appear in Comm. in Algebra), Universität Frankfurt a.M., (1993).

[Ho]   T. Holm, *Über einen Zusammenhang zwischen der Invariante Σ^1 und Darstellungen von Gruppen*, Diplomarbeit, Universität Frankfurt a.M. (1991).

[Ki]   D. Kilsch, *On minimax groups which are embeddable in constructible groups*, J. London Math. Soc. (2) **18** (1978), pp. 472–474.

[Kr]   P.H. Kropholler, *On groups of type FP_∞*, J. Pure Appl. Algebra **90** (1993), pp. 55–67.

[Le]   G. Levitt, *$\mathbb{R}$-trees and the Bieri-Neumann-Strebel invariant*, Preprint Centre de Recerca Matemàtica, Institut d'Estudis Catalans, Barcelona (to appear in Publicacions Matemàtiques) (1993).

[Me1]   H. Meinert, *Die höheren geometrischen Invarianten Σ^m von Gruppen via Operationen auf CW-Komplexen und der Beweis der Σ^m-Vermutung für metabelsche Gruppen endlichen Prüfer-Ranges*, Dissertation, Universität Frankfurt a.M. (1993).

[Me2]   H. Meinert, *The higher geometric invariants of direct products of virtually free groups*, Comment. Math. Helv. **69** (1994), pp. 39–48.

[Me3]   H. Meinert, *The higher geometric invariants of modules over noetherian group rings*, in: Combinatorial and Geometric Group Theory, Edinburgh 1993 (A. J. Duncan, N. D. Gilbert, J. Howie, eds.), London Math. Soc. Lecture Note Series, **204** Cambridge University Press (1995), pp. 247–254.

[Me4]   H. Meinert, *The Bieri-Neumann-Strebel invariant for graph products of groups*, J. Pure Appl. Algebra **103** (1995), pp. 205–210.

[Me5]   H. Meinert, *The homological invariants for metabelian groups of finite Prüfer rank: a proof of the Σ^m-conjecture*, Proc. London Math. Soc. (3) **72** (1996), pp. 385–424.

[Me6]   H. Meinert, *Actions on 2-complexes and the homotopical invariant*
         Σ^2 *of a group*, Preprint (to appear in J. Pure Appl. Algebra), Uni-
         versität Frankfurt a.M. (1994).

[No]   G. A. Noskov, *Bieri-Strebel invariant and homological finiteness*
         properties of metabelian groups, SFB-Preprint 93–028 Universität
         Bielefeld (1993).

[St]   R. Strebel, *On quotients of groups having finite homological type,*
         Arch. Math. **41** (1983), pp. 419–426.

[Re1]   B. Renz, *Geometric invariants and HNN-extensions,* in: Group The-
         ory, Proceedings of the Singapore Group Theory Conference, June
         1987 (K.N. Cheng and Y.K. Leong, eds.), W. de Gruyter, Berlin–
         New York (1989), pp. 465–484.

[Re2]   B. Renz, *Geometrische Invarianten und Endlichkeitseigenschaften*
         von Gruppen, Dissertation, Universität Frankfurt a.M. (1988).

[Ro]   D.J.S. Robinson, *Finiteness conditions for soluble groups,* (2 Vols.)
         Springer-Verlag, Berlin–Heidelberg–New York (1972).

[Sch]   S. Schmitt, *Über den Zusammenhang der geometrischen Invarianten*
         von Gruppe und Untergruppe mit Hilfe von variablen Modulkoef-
         fizienten, Diplomarbeit, Universität Frankfurt a.M. (1991).

[Va]   F.A. Valentine, *Convex sets,* McGraw-Hill, New York (1964).

Some Constructions Relating to Hyperbolic Groups

K.V. Mikhajlovskii and A.Yu. Ol'shanskii

Abstract. The purpose of this paper is to construct certain quotients, HNN-extensions, amalgamated products and inductive limits of hyperbolic groups, and to apply the results to construct finitely generated verbally complete and divisible groups.

0. Introduction

The first examples of infinite non-abelian groups, all of whose proper subgroups are cyclic [9], were constructed as inductive limits of hyperbolic groups, although the notion of hyperbolicity was not exploited explicitly. An explicit application of hyperbolic properties for such constructions was proposed in [1] by M.Gromov. This approach was realized in [2].

In the present paper we focus on a method for constructing divisible and verbally complete groups by means of hyperbolic group theory. Recall that a group G is said to be divisible if for any element g of G and any nonzero integer n the equation $x^n = g$ has a solution in G.

The groups $\mathbb{Q}$ and $\mathbb{C}_{p^\infty}$ are natural examples of divisible groups. For a long period of time, it was unknown whether or not there exist non-trivial finitely generated divisible groups. The first examples were constructed by V.S.Guba [10]. These groups are torsion free. Later on, periodic examples have been given by S.V.Ivanov [7]. Until now it was unknown whether or not there exist non-trivial finitely generated verbally complete groups. Recall that a group G is verbally complete if for any non-trivial word $v(x_1, \ldots, x_n)$ of the free group $F(x_1, x_2, \ldots)$ with countable set of generators, and for any element $g \in G$, the equation $v(x_1, \ldots, x_n) = g$ has a solution in G. We have obtained the following results.

Partly supported by Russian Fund of Fundamental Research Grant 010-15-41. The second author was also supported by ISF Grant MID 000.

Theorem 1. *For every non-cyclic torsion free hyperbolic group G there exists a non-trivial torsion free verbally complete quotient H of G.*

Theorem 2. *Every non-elementary hyperbolic group G has a non-trivial verbally complete torsion quotient group $\tilde{G}$.*

Corollary 1. *There exist non-trivial finitely generated torsion free verbally complete groups.*

Corollary 2. *There exist non-trivial finitely generated verbally complete torsion groups.*

Corollary 3. *For every non-cyclic torsion free hyperbolic group G there exists a non-abelian torsion free divisible quotient H of G.*

Corollary 4. *Every non-elementary hyperbolic group G has a non-trivial divisible torsion quotient group $\tilde{G}$.*

Corollary 5 (Guba). *There exists a non-trivial finitely generated torsion free divisible group.*

Corollary 6 (Ivanov). *There exists a non-trivial finitely generated divisible torsion group.*

Our proof of Theorems 1,2 uses results of [2] and criteria for hyperbolicity of HNN-extensions and free products with amalgamated subgroups, which are given below. For their formulation, recall that in M.Gromov's terminology an elementary group is any cyclic-by-finite group. It is known [1,3] that every non-elementary subgroup of a hyperbolic group G contains a 2-generated free subgroup, and every element g of infinite order in G belongs to a unique maximal elementary subgroup $E(g) \subset G$.

Theorem 3. *Let G be a hyperbolic group with isomorphic infinite elementary subgroups A and B, and let ψ be an isomorphism from A to B. The HNN-extension $\overline{G} = \langle G, t \mid t^{-1}at = \psi(a), a \in A \rangle$ of G with associated subgroups A and B is hyperbolic if and only if the following two conditions hold:*
1) either A or B is a maximal elementary subgroup of G;
2) for all $g \in G$ the subgroup $gAg^{-1} \cap B$ is finite.

Corollary 7. *Let G and H be hyperbolic groups, A and B be infinite elementary subgroups of G, H respectively. Then the free product of the groups G and H with amalgamated subgroups A and B is hyperbolic if and only if either A is a maximal elementary subgroup of G or B is a maximal elementary subgroup of H.*

In the case when the elementary subgroups A and B in Theorem 3 and Corollary 7 are abelian, these criteria were obtained simultaneously and independently by O.Kharlampovich and A.Myasnikov [5].The authors are grateful to professors O.Kharlampovich and A.Myasnikov for communicating their results. Notice also, that Corollary 7 was earlier proved by different methods in the work of M.Bestvina and M.Feighn [4], and in the case of maximal cyclic subgroups A and B, the hyperbolic property of free constructions was established by M.Gromov [1] (without details). The above results were partially included in the Abstracts of the Durham symposium on geometrical and cohomological methods in group theory (July 1994).

1. General notions and definitions

There are several equivalent definitions of a hyperbolic group [1,3]. We will use the following one. Let

$$G = \langle g_1, \ldots, g_k \,|\, r_1, \ldots, r_l \rangle \tag{1}$$

be a finitely presented group. Every word W, which is equal to 1 in G, can be expressed in the free group $F(g_1, \ldots, g_k)$ as

$$W = \prod_{i=1}^{n} u_i r_{s_i}^{\pm 1} u_i^{-1} \tag{2}$$

where the number $n = n(W)$ can be assumed the minimal possible one. The group G is said to be *hyperbolic* (word hyperbolic or negatively curved), if there exists a linear function bounding the number of factors $n = n(W)$ in (2) depending on the length $\|W\|$ of the word W. In other words, there is a constant $\beta = \beta(G)$ such that $n(W) \leq \beta \|W\|$ for every word W representing the identity in G. The definition does not depend on the choice of the presentation (1) of G.

To prove the theorems, we will use the geometric language of diagrams over groups [6].

Recall that a *map* is a finite planar connected and simply-connected 2-complex. A *diagram* Δ over an alphabet $\mathcal{A}$ is a map whose edges e are labeled by letters $\phi(e) \in \mathcal{A}^{\pm 1}$ such that $\phi(e)^{-1} = \phi(e^{-1})$. A diagram over $\mathcal{A}$ is called a *diagram over the group G* given by the presentation (1), where $\mathcal{A} = \{g_1, g_2, \ldots, g_k\}$, if the label of the boundary path of every face of Δ is a cyclic permutation of some relator. In view of van Kampen's lemma [6,7], to prove that a group G, given by the presentation (1), is hyperbolic, it suffices to find a constant $C = C(G) > 0$ such that $n(\Delta) < C\|\partial\Delta\|$ for every minimal circular diagram Δ over G, where $n(\Delta)$ is the number of faces in Δ and $\|\partial\Delta\|$ coincides with the perimeter of $\partial\Delta$.

We first prove the sufficiency of the conditions of Theorem 3. For definiteness we will assume the subgroup A to be maximal elementary in G. Let the group

G and its subgroups A and B from Theorem 3 be given by

$$G = \langle g_1, \ldots, g_\nu \mid r_1, \ldots, r_\tau \rangle \tag{3}$$

$$A = \langle a \rangle \cup x_1 \langle a \rangle \cup \ldots \cup x_n \langle a \rangle \tag{4}$$

$$B = \langle b \rangle \cup y_1 \langle b \rangle \cup \ldots \cup y_n \langle b \rangle \tag{5}$$

where the elements a, b, x_i, y_i $i = 1, \ldots, n$ are chosen in accordance with the given isomorphism ψ, i.e., $\psi : a \mapsto b, \psi : x_i \mapsto y_i$ $i = 1, \ldots, n$ and $\langle a \rangle$, $\langle b \rangle$ are infinite cyclic subgroups of finite index in A, B respectively. We may assume $\langle a \rangle$, $\langle b \rangle$ to be normal subgroups of A and B. Then the group $\overline{G}$ has the presentation

$$\overline{G} = \langle g_1, \ldots, g_\nu, t \mid r_1, \ldots, r_\tau, t^{-1}atb^{-1}, t^{-1}x_1 t y_1^{-1}, \ldots, t^{-1}x_n t y_n^{-1} \rangle \tag{6}.$$

Define further our alphabet as a set $\mathcal{A}$, where

$$\mathcal{A} = \{g_1, \ldots, g_\nu, a, b, x_1, \ldots, x_n, y_1, \ldots, y_n, t\}^{\pm 1}.$$

For elements of the groups G and $\overline{G}$, we introduce the following length functions $\|\ \|$ and $|\ |$. If W is a word in the alphabet $\mathcal{A}$, then $\|W\|$ is the length of W. Another length $|W|$ is the minimal length among all words representing the same element of the group $G * \langle t \rangle_\infty$ as the word W (we consider words in the alphabet $\mathcal{A}$ to be elements of $G * \langle t \rangle_\infty$), that is $|W| = \min\{\|V\| \mid V = W, V \in G * \langle t \rangle_\infty\}$. By the above definition, we have that $\|a\| = \|b\| = 1$ and $\|x_i\| = \|y_i\| = 1$ for $i = 1, \ldots, n$. The notation $X \equiv Y$ will be used for *graphic* equality of the words X and Y.

Consider a path $p = e_1 \ldots e_m$ in a diagram Δ over $\overline{G}$ or over G. We will use the notation p_-, p_+ for the initial and the terminal vertices of a path p, respectively. The length $\|p\|$ of p is, by definition, the number of edges in its presentation and the length $|p|$ of p is the length $|\phi(p)|$ of its label as an element of the group $G * \langle t \rangle_\infty$, which was defined above.

Further we assume the notions of the *Cayley graph* $C(G)$ of the group G and *hyperbolic space* to be known [1,3]. Let us endow each edge e of $C(G)$ with the metric of the unit segment $[0,1]$ and define a geodesic metric on $C(G)$ by extending the metrics of all edges. (Recall [1,3], that a group G is hyperbolic if and only if its Cayley graph $C(G)$ is a hyperbolic space.)

Consider a path p in $C(G)$ with the natural parametrization by length. The path p is called (λ, c)-*quasigeodesic* for some $\lambda > 0$ and $c \geq 0$, if for any points $p(s)$ and $p(u)$

$$|p(s) - p(u)| \geq \lambda|s - u| - c.$$

A word W in the alphabet $\mathcal{A}$ is called (λ, c)-quasigeodesic, if W is the label of a (λ, c)-quasigeodesic path in the Cayley graph of the group $G * \langle t \rangle_\infty$. Fixing a vertex o of a circular diagram Δ over the group G, one can define a natural mapping γ of Δ into $C(G)$. Let $\gamma(o) = O_1$ be the unity vertex of $C(G)$. For an arbitrary vertex O the image $\gamma(O)$ is, by definition, the element of G represented by the word $\phi(p)$, where $p_- = o$ and $p_+ = O$ for a path p in Δ. In view of van Kampen's lemma, $\phi(p) = \phi(q)$ in G if $p_- = q_-$ and $p_+ = q_+$ since Δ is a simply-connected diagram. Then for an edge e in Δ the image $\bar{e} = \gamma(e)$ is by definition the edge with the starting point $\gamma(e_-)$ and with the same label as e. Clearly, the mapping γ can be extended onto the set of paths in Δ. Obviously, the mapping γ preserves the lengths $|p|$ and $\|p\|$ of any path p in Δ.

Call a path p in a diagram (λ, c)-*quasigeodesic* if its γ-image in $C(G)$ is (λ, c)-quasigeodesic. The following lemmas will be useful. Consider a hyperbolic group G, its Cayley graph $C(G)$ and denote the distance between two points $x, y \in C(G)$ by $\rho(x, y)$.

Lemma 1 ([1;3, p.90]). *There exists a constant $H = H(G, \lambda, c)$ such that for any (λ, c)-quasigeodesic path p in $C(G)$ and any geodesic path q with the conditions $p_- = q_-$ and $p_+ = q_+$, the inequalities $\rho(u, p) < H$ and $\rho(v, q) < H$ hold for any points $u \in q$ and $v \in p$.*

Call two paths p and q in $C(G)$ K-*bound* for some $K > 0$ if

$$\max(\rho(p_-, q_-), \rho(p_+, q_+)) \leq K.$$

A geodesic n-gon $[x_1, x_2, \ldots, x_n]$ in $C(G)$ is a closed broken line $x_1 - x_2 - \ldots - x_n - x_1$ where each path $x_i - x_j$ is a geodesic segment $[x_i, x_j]$.

Lemma 2 ([8, Lemma 25]). *There are positive constants $c_1 = c_1(G)$ and $c_2 = c_2(G)$ such that for any geodesic n-gon P in the $C(G)$ the following property holds. If the set of all segments of P is divided into three subsets N_1, N_2, N_3 with the length sums σ_1, σ_2, σ_3, respectively and $\sigma_1 > cn$, $\sigma_3 < 10^{-3}cn$ for some $c \geq c_2$, then there exist distinct segments $p_1 \in N_1$ and $p_2 \in N_1 \cup N_2$ having c_1-bound subsegments of length greater than $10^{-3}c$.*

2. Planar diagrams over HNN-extensions

For circular diagrams over the group G or over $\overline{G}$, given by presentations (3) and (6), we will partition the edges into two systems:
1) g-edges with labels in the group G;
2) t-edges with labels $t^{\pm 1}$.
When considering diagrams over $\overline{G}$, it will sometimes be convenient to admit faces with boundary labels from the set of all relators (not only defining) of

the group G. A face of a diagram Δ is called a g-face if its boundary label belongs to the set of all relators of the group G. A face with boundary label from the set $\{t^{-1}atb^{-1}, t^{-1}x_ity_i^{-1}\ i = 1, \ldots, n\}$ is called a t-face.

Now we introduce *elementary transformations* for circular diagrams over G or $\overline{G}$.

1. Assume a vertex o has degree 1 in a diagram Δ, i.e., $o = e_-$ for a single edge e. Then one may delete e except for the vertex e_+.

2. Let $o_1 = f_+ = (f_1)_-, o_2 = (f_1)_+ = (f_2)_-$ for some edges f, f_1, f_2, where o_1 and o_2 are vertices of degree 2 in Δ and $\phi(f_1) = \phi(f_2)^{-1}$. Then one can delete o_1, o_2 from the set of vertices by declaring $f f_1 f_2$ to be a single edge e with the label $\phi(e) = \phi(f)$.

3. If two different g-faces Π_1 and Π_2 have a common edge e in their boundaries then one may delete it (except for e_- and e_+) making Π_1 and Π_2 into a single g-face Π.

Let $1', 2', 3'$ be the inverses of the elementary transformations 1, 2, 3 respectively (where $3'$ is permitted only when the new faces Π_1 and Π_2 arising from Π correspond to relators of G).

Definition.   Let Δ be a diagram over the group $\overline{G}$ given by presentation (6). A sequence of t-faces $\pi_1, \ldots, \pi_m$ in Δ is called a *t-strip* if for any $i = 1, \ldots, m - 1$ the t-faces π_i and π_{i+1} have a common t-edge (Fig.1).

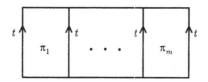

Figure 1.

A t-strip Π is called *cyclic* if it consists of $m \geq 1$ t-faces $\pi_1, \ldots, \pi_m$ such that π_1 and π_m have a common t-edge (Fig.2).

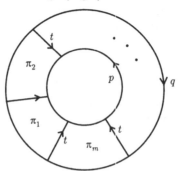

Figure 2.

Consider a circular diagram Δ over the group $\overline{G}$ which has τ t-faces. Then the τ is called the *type* $\tau(\Delta)$ of Δ. All diagrams over $\overline{G}$ will be ordered by their types.

Definition. A circular diagram Δ over the group $\overline{G}$ with boundary label W is said to be *minimal* if for any circular diagram Δ' over $\overline{G}$ with the same boundary label the following inequality $\tau(\Delta) \leq \tau(\Delta')$ holds.

Lemma 3. *A minimal circular diagram Δ over $\overline{G}$ has no cyclic t-strips.*

Proof Assume that a minimal diagram Δ over $\overline{G}$ has a cyclic t-strip Π and denote the inner and outer contours of Π by p and q (Fig.2). Let us consider the circular subdiagram Δ' of Δ with $\partial\Delta' = q$. The boundary label of Δ' is, by definition, an element from A or B (in Fig.2 $\phi(q) \in A$) and since Δ' is a circular diagram over $\overline{G}$, we obtain $\phi(q) = 1$ in $\overline{G}$. Recall that by the definition of HNN-extensions (see [6]), the groups A, B are embedded into $\overline{G}$ by the natural maps $a \mapsto a$, $b \mapsto b$ for $a \in A$, $b \in B$. Therefore we will have that $\phi(q) = 1$ either in the group A or in B. So one can cut out Δ' from Δ and paste in its place a diagram with the same boundary label consisting of g-faces only, thus reducing the number of t-faces in the resulting diagram. But this contradicts the minimality of Δ. Hence our assumption is false and the lemma is proved.

A system of t-strips in a diagram Δ over $\overline{G}$ is called a *distinguished system* if:
1) different t-strips have no common t-edges (that is they are disjoint);
2) any t-face of Δ belongs to some t-strip.
A circular diagram Δ over $\overline{G}$ is said to be *simple* if $\partial\Delta$ is a cyclically reduced path and $\Delta \neq \Delta_1 \cup \Delta_2$, where Δ_1, Δ_2 are circular subdiagrams of Δ with non-empty sets of edges such that $\Delta_1 \cap \Delta_2$ consists of a vertex.
Since cyclic t-strips are not contained in minimal diagrams, the boundary t-edges of any distinguished t-strip in such diagrams belong to $\partial\Delta$. Therefore a minimal simple diagram Δ over $\overline{G}$ can be presented as the union of subdiagrams

$$\Delta = \cup_i \Delta_i \cup_j \Pi_j$$

where the Δ_i are maximal circular subdiagrams over G and the Π_j are distinguished t-strips in Δ (Fig.3).

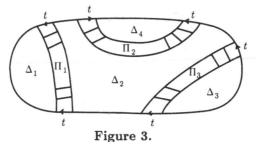

Figure 3.

When considering diagrams over $\overline{G}$, we will always assume them to be: 1) circular; 2) simple; 3) with a fixed distinguished system of t-strips.

3. Choice of constants

Our proofs and some definitions will be based on a fixed system of constants. It is convenient to introduce all of them and to indicate in which order they will be chosen. All of the constants, that will be used below, will be chosen sequentially, one after another. Each of the following constants is chosen after its predecessor

$$\beta, \lambda, c, H, c_1, \varepsilon, \theta_1, M, \rho, \theta, \alpha_1, \alpha. \tag{7}$$

Each of them is positive and depends on the group G only (not on a fixed diagram over G or $\overline{G}$).

The first constant is β which determines the linear isoperimetric inequality for the hyperbolic group G. Then we choose $\lambda > 0$ and $c \geq 0$ using Lemma 2.4 [2] such that any word from the set

$$\{x_i a^m x_j, \ y_i b^m y_j \mid i, j = 1, \ldots, n, \ m \in \mathbf{Z}\}$$

is (λ, c)-quasigeodesic.

The next constant is $H = H(G, \lambda, c) > 0$ which we select using Lemma 1. The constant $c_1 = c_1(G)$ is chosen in accordance with Lemma 2. The remaining parameters will be introduced below.

4. Contiguity subdiagrams

Let Π_1 and Π_2 be distinct t-strips in a diagram Δ over $\overline{G}$. Consider a simple closed path $w = p_1 q_1 p_2 q_2$ in Δ such that q_1 and q_2 are subpaths of the boundary cycles of Π_1 and Π_2 with the properties: 1) $\|p_1\|, \|p_2\| \leq \varepsilon$ (see (7)); 2) $\phi(p_1), \phi(p_2)$ consist of g-edges only; 3) a subdiagram Γ of Δ, bounded by w, has no t-faces; 4) $\min(\|q_1\|, \|q_2\|) \geq 1$.

Then we call Γ a *contiguity subdiagram* between Π_1 and Π_2. We will also consider a contiguity subdiagram Γ of some t-strip Π_1 to a section q of $\partial\Delta$, if q_2 is a subpath of q.

The notation $\partial(\Pi_1, \Gamma, \Pi_2)$ (or $\partial(\Pi_1, \Gamma, q)$)$= p_1 q_1 p_2 q_2$ will define the above partition of the contour w of Γ. The subpaths q_1 and q_2 are called the *contiguity arcs* while p_1, p_2 are called the *side arcs* of Γ.

We choose the constant ε with the property $\varepsilon > 2H + c_1$, where H and c_1 are introduced above.

A system $\mathcal{M}$ of contiguity subdiagrams of t-strips to t-strips or to a contour $\partial\Delta$ in a diagram Δ over $\overline{G}$ is called a *distinguished system* if:

1) distinct subdiagrams in $\mathcal{M}$ have no common faces and no common edges in their contiguity arcs (i.e., these subdiagrams are disjoint);

2) the sum of lengths $\|p\|$ of all contiguity arcs p of all subdiagrams of $\mathcal{M}$ is not less than the similar sum for any other system $\mathcal{M}'$ with the property 1);

3) the number of subdiagrams in $\mathcal{M}$ is minimal among all systems with the properties 1) and 2).

If Γ is a distinguished contiguity subdiagram of a t-strip Π to the contour $\partial\Delta$ and $\partial(\Pi,\Gamma,\partial\Delta) = p_1 q_1 p_2 q_2$, then the arc q_1 of Π (and all of its edges) is said to be *outer*. If $\partial(\Pi,\Gamma,\Pi_1) = p_1 q_1 p_2 q_2$ for a t-strip Π_1 then q_1 is an *inner* arc. The other edges of $\partial\Pi$, that are neither outer no inner, are called *unbound*. Every maximal subpath of the boundary of Π consisting of unbound edges will be called an *unbound arc* of t-strip Π. An unbound arc of $\partial\Delta$ can be defined similary.

In order to prove hyperbolicity of the group $\overline{G}$, i.e., for obtaining a linear isoperimetric inequality for the minimal circular diagrams over $\overline{G}$, we can restrict ourselves to considering only such diagrams over $\overline{G}$, which consist of g-faces and the following special t-strips. A t-strip Π in a minimal diagram Δ over $\overline{G}$ is said to be *special* if $\phi(\partial\Pi) \equiv t^{-1} x_i a^k x_j t y_j^{-1} b^{-k} y_i^{-1}$ (Fig.4), where $\psi(x_i) = y_i, \psi(x_j) = y_j$ are defined above, $k \in \mathbf{Z}$ and $i, j = 0, 1, \ldots, n$. In the above definition the case $k = 0$ is just the case, when t-faces with boundary label $t^{-1} a t b^{-1}$ do not occur in Π, and similarly, if $i = 0$ then a t-face with boundary label $t^{-1} x_i t y_i^{-1}$ does not occur in Π.

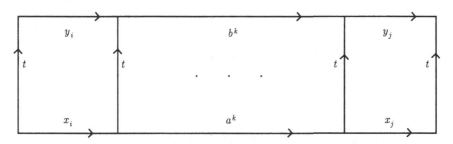

Figure 4.

In particular, let us recall that we assume the subgroups $\langle a \rangle$ and $\langle b \rangle$ to be normal in the groups A and B respectively (see Section 1), that is for any $i = 1, \ldots, n$ we have $x_i a x_i^{-1} = a^{\pm 1}$, $y_i b y_i^{-1} = b^{\pm 1}$ in G. This means that if we have a t-strip Π in a diagram Δ over $\overline{G}$, which is not special, then using the elementary transformations of Δ one can obtain a t-strip Π' from a t-strip Π with $\phi(\partial\Pi') \equiv t^{-1} x_{i_1} \ldots x_{i_k} a^l t b^{-l} y_{i_k}^{-1} \ldots y_{i_1}^{-1}$, where $x_{i_j} \in \{x_1, \ldots, x_n\}$, $y_{i_j} \in \{y_1, \ldots, y_n\}$. Then by the decompositions (4) and (5) of the groups A, B we have $x_{i_1} \ldots x_{i_k} = x_{i_0} a^m$, $y_{i_1} \ldots y_{i_k} = y_{i_0} b^m$, where $\psi(x_{i_0}) = y_{i_0}$ and $x_{i_0} \in \{x_1, \ldots, x_n\}$, $y_{i_0} \in \{y_1, \ldots, y_n\}$.

Thus, by elementary transformations of Δ one can obtain a t-strip Π'' from Π' which will be special. Hence, starting from a minimal diagram Δ, we obtain a diagram Δ' with the same boundary label consisting of g-faces and t-strips of special type. In the general case, the number of t-faces in Δ' may increase. Therefore, if we prove a linear isoperimetric inequality for such minimal special (i.e., with special t-strips) diagrams, then the hyperbolicity of $\overline{G}$ will be established. Hence, from now on we will consider minimal special diagrams only (i.e. diagrams having minimal type among all special diagrams with the same boundary label).

To prove Theorem 3, we shall exploit the notion of estimating graphs Φ and Φ' (see also [7,8]). Namely, choose a point $o = o(\Pi)$ inside each t-strip Π of a diagram Δ over $\overline{G}$ and define the set of all $o(\Pi)$ to be the set of vertices of the graph Φ. Let Γ be a distinguished contiguity subdiagram of a t-strip Π_1 to a t-strip Π_2, then the vertices $o_1 = o_1(\Pi_1)$ and $o_2 = o_2(\Pi_2)$ are connected in Φ by a non-oriented edge of Φ through the subdiagram Γ.

In our considerations, a contour of a minimal diagram Δ over $\overline{G}$ will be regarded as a single section whose maximal subpaths consisting of g-edges are (λ, c)-quasigeodesic. To define the graph Φ', choose a vertex O outside of Δ and regard the set of vertices of Φ' consisting of the vertex O and the vertices of the graph Φ. For each contiguity subdiagram Γ of a t-strip Π to a contour $\partial\Delta$ the vertex $o(\Pi)$ is connected with O in Φ' by an edge of Φ' passing through Γ.

Lemma 4. *Let Δ be a minimal diagram over $\overline{G}$, then the estimating graphs Φ and Φ' satisfy the following condition: there is a vertex of Φ inside every 2-gon of Φ'.*

Proof The statement follows immediately from the definition of distinguished (maximal) contiguity subdiagrams and the fact that the boundary t-edges of any distinguished t-strip in the minimal diagram Δ lie on $\partial\Delta$.

5. Compatibility of paths and reduction of type

If g is an element of infinite order of a hyperbolic group G, then by $E(g)$ we will denote the unique maximal elementary subgroup of G containing g (the elementarizer of g) [2].

Considering the circular diagrams over $\overline{G}$, we will partition the t-strips into two sets. A t-strip Π will be called *long* if $\|\partial\Pi\| > \rho$, where ρ depends on the group G only and will be chosen below. Otherwise Π will be called *short*.

Consider two paths u_1, u_2 in the Cayley graph $C(G)$ of the group G and suppose $\phi(u_1) = W_1^{m_1}$, $\phi(u_2) = W_2^{m_2}$, where W_1 and W_2 have infinite order in G. The vertices $o_1, o_2, \ldots$ of the path u_1 such that labels of subpaths $o_i - o_{i+1}$ are equal to W_1 in G, will be called the *phase vertices* of u_1. Similarly, choose phase vertices $\bar{o}_1, \bar{o}_2, \ldots$ on u_2. We call the paths u_1 and u_2 *compatible* (see [2]), if there is a path v in $C(G)$ joining some phase vertices o_i and $\bar{o}_j$ of paths u_1, u_2 such that $VW_2V^{-1} \in E(W_1)$ in G for the label $V = \phi(v)$ and there are nonzero m, l such that $(VW_2V^{-1})^m = W_1^l$ in G.

For a t-strip Π in a diagram Δ over $\overline{G}$, we define the *A-section* and the *B-section* of the contour $\partial\Pi$ as the maximal subpaths of the boundary with labels in A and in B respectively.

Let us consider a circular diagram Δ over $\overline{G}$ and a contiguity subdiagram Γ between A-sections of long t-strips Π_1 and Π_2 in Δ. We put $\partial(\Pi_1, \Gamma, \Pi_2) = p_1 q_1 p_2 q_2$, where $\|p_1\|, \|p_2\| \leq \varepsilon$,

$$\phi(\partial\Pi_1) \equiv t^{-1} x_i^1 a^k x_j^1 t(y_j^1)^{-1} b^{-k}(y_i^1)^{-1},$$
$$\phi(\partial\Pi_2) \equiv t^{-1} x_i^2 a^l x_j^2 t(y_j^2)^{-1} b^{-l}(y_i^2)^{-1},$$

where $x_i^1, x_i^2, x_j^1, x_j^2 \in \{x_1, \ldots, x_n\}$, $y_i^1, y_i^2, y_j^1, y_j^2 \in \{y_1, \ldots, y_n\}$.

For simplicity, we assume the initial and terminal vertices of paths p_1, p_2 to be phased with period a (Fig.5). We recall that the subgroup A of G is supposed to be maximal elementary, that is $A = E(a)$ in G.

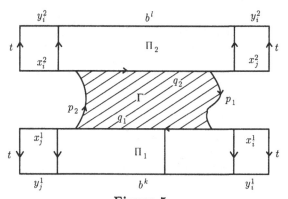

Figure 5.

Let $\phi(q_1) \equiv a^{k_1}$, $\phi(q_2) \equiv a^{l_1}$, $|k_1| \leq |k|$, $|l_1| \leq |l|$. By the definition of a contiguity subdiagram, Γ is a circular diagram over G. Consider the γ-image of Γ in $C(G)$ defined in Section 1. Then by Lemma 2.5 [2] there exists a constant $M = M(a, b, \varepsilon) > 0$ such that either $|k_1|, |l_1| < M$ or the paths q_1 and q_2 are compatible.

Let us show how to reduce the number of t-faces in Δ in the second case, i.e. when the diagram Δ is not minimal.

The compatibility of the paths q_1 and q_2 yields that there exists a path p in Γ joining some vertices o_1 and o_2 on q_1, q_2, respectively, such that $\phi(p)a\phi(p)^{-1} \in E(a)$. Therefore, by Lemma 1.16 [2] we obtain $\phi(p) \in E(a)$ and as $A = E(a)$, finally $\phi(p) \in A$. Then from Γ one can obtain $\phi(p_1), \phi(p_2) \in A$ as $\phi(p) \in A$.

As we regard the case when inequalities $|k_1|$, $|l_1| < M$ are not satisfied, we assume $|l_1| \geq M$. Then using the (λ, c)-quasigeodesity of the paths q_1 and q_2 and the equality $\phi(q_1) = \phi(p_1)^{-1}\phi(q_2)^{-1}\phi(p_2)^{-1}$ in G, we obtain from Γ $|k_1|\|a\| = \|q_1\| \geq |q_1| = |p_1^{-1}q_2^{-1}p_2^{-1}| \geq |q_2| - 2\varepsilon \geq \lambda\|q_2\| - c - 2\varepsilon = \lambda|l_1|\|a\| - c - 2\varepsilon$. Since $a \in \mathcal{A}$, one obtains

$$|k_1| \geq \lambda M - c - 2\varepsilon = \overline{M}. \tag{8}$$

Let us define the substrips π_1 and π_2 of Π_1 and Π_2, respectively, with contours $\partial\pi_1 = e_1q_1e_2q_1'$, $\partial\pi_2 = e_1'q_2e_2'q_2'$, where e_1, e_2, e_1', e_2' are t-edges, $\phi(q_1') = b^{-k_1}$, $\phi(q_2') = b^{-l_1}$ (Fig.6).

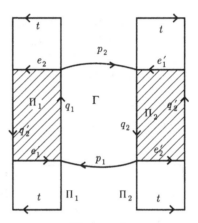

Figure 6.

Then we cut out a subdiagram $\overline{\Gamma}$ with contour

$$\partial\overline{\Gamma} = p_1e_1^{-1}(q_1')^{-1}e_2^{-1}p_2(e_1')^{-1}(q_2')^{-1}(e_2')^{-1}$$

from Δ (Fig.7). $\overline{\Gamma}$ consists of the union of the t-strips π_1, π_2 and the subdiagram Γ.

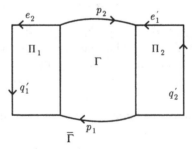

Figure 7.

As mentioned above, $\phi(p_1), \phi(p_2) \in A$, so by the decomposition (4) of A, $\phi(p_1) = \bar{x}_1a^{k_0}$, $\phi(p_2) = \bar{x}_2a^{l_0}$, where $\bar{x}_1, \bar{x}_2 \in \{x_1, \ldots, x_n\}$. Since the words

$\bar{x}_1 a^{k_0}, \bar{x}_2 a^{l_0}$ are (λ, c)-quasigeodesic, we obtain

$$\|p_1\| \geq |p_1| = |\bar{x}_1 a^{k_0}| \geq \lambda \|\bar{x}_1 a^{k_0}\| - c \geq |k_0| \lambda \|a\| - c.$$

Then, using the inequality $\|p_1\| \leq \varepsilon$ and the fact that $\|a\| = 1$ (recall that $a \in \mathcal{A}$), we obtain

$$|k_0| \leq (\varepsilon + c)\lambda^{-1}. \tag{9}$$

Similarly,

$$|l_0| \leq (\varepsilon + c)\lambda^{-1}. \tag{10}$$

Now let us consider our circular diagram $\bar{\Gamma}$ separately (Fig.7) and carry out a series of transformations with it. Eventually, we will obtain a new diagram Γ' with the same boundary label, but with a smaller number of t-faces than in $\bar{\Gamma}$. Finally, we will paste Γ' in Δ instead of $\bar{\Gamma}$, reducing the number of t-faces in Δ.

As $\phi(p_1) = \bar{x}_1 a^{k_0}$, $\phi(p_2) = \bar{x}_2 a^{l_0}$ in G, we can glue to paths p_1, p_2 of $\bar{\Gamma}$ diagrams Δ_1, Δ_2 over G with contours $\partial \Delta_1 = p_1 v_1^{-1}, \partial \Delta_2 = p_2 v_2^{-1}$, where $\phi(v_1) = \bar{x}_1 a^{k_0}, \phi(v_2) = \bar{x}_2 a^{l_0}$. We construct a diagram $\bar{\Gamma}'$ over $\bar{G}$ with the contour

$$\partial \bar{\Gamma}' = v_1 e_1^{-1}(q_1')^{-1} e_2^{-1} v_2 (e_1')^{-1}(q_2')^{-1}(e_2')^{-1}.$$

Then one can glue to paths $(e_2')^{-1} v_1 e_1^{-1}$, $e_1^{-1} v_2 (e_1')^{-1}$ t-strips π_1', π_2' with boundary labels $\phi(\partial \pi_1') \equiv t^{-1} \bar{x}_1 a^{k_0} t b^{-k_0} \psi(\bar{x}_1)^{-1}$, $\phi(\partial \pi_2') \equiv t^{-1} \bar{x}_2 a^{l_0} t b^{-l_0} \psi(\bar{x}_2)^{-1}$, where $\psi : A \longrightarrow B$ is our isomorphism.

We obtain a circular diagram $\bar{\Gamma}''$ over $\bar{G}$ with a boundary label from the group B, where $\partial \bar{\Gamma}'' = \bar{v}_1 (q_1')^{-1} \bar{v}_2 (q_2')^{-1}$ and $\phi(\bar{v}_1) = \psi(\bar{x}_1) b^{k_0}, \phi(\bar{v}_2) = \psi(\bar{x}_2) b^{l_0}$ (Fig.8).

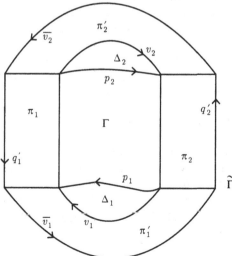

Figure 8.

Since $\phi(\partial\overline{\Gamma}'') \in B$ and the group B is embedded in $\overline{G}$ by the natural map $b \mapsto b$, $b \in B$ [6], one can replace $\overline{\Gamma}''$ by a circular diagram with the same boundary label consisting of g-faces only. Thus we can consider $\overline{\Gamma}''$ to consist of g-faces only.

In order to construct the diagram Γ' replacing $\overline{\Gamma}$, we have to carry out inverse transformations. Namely, glue to paths $\bar{v}_1$, $\bar{v}_2$ the mirror copies of the t-strips π'_1 and π'_2. Then glue to v_1 and v_2 copies of the diagrams Δ_1 and Δ_2. Eventually, one can obtain a diagram Γ' with the same boundary label as $\overline{\Gamma}$. Then we paste Γ' in Δ instead of $\overline{\Gamma}$ and get a new diagram Δ' with the same boundary label as Δ. We prove below that $\tau(\Delta') < \tau(\Delta)$, i.e. the diagram Δ is not minimal.

Notice, that after our transformations the t-strips Π_1, Π_2 are replaced by t-strips Π'_1 and Π'_2, where Π'_1 consists of the union of $\bar{\pi}_1, \pi'_1, \tilde{\pi}_1$ and Π'_2 consists of the union of $\tilde{\pi}_1, \pi'_2, \tilde{\pi}_2$, where $\tilde{\pi}_1, \bar{\pi}_1$ are substrips of Π_1 such that $\Pi_1 = \bar{\pi}_1 \cup \pi_1 \cup \tilde{\pi}_1$ and, similarly $\Pi_2 = \bar{\pi}_2 \cup \pi_2 \cup \tilde{\pi}_2$.

In the general case, Π'_1 and Π'_2 are not the special t-strips, as there are g-edges with labels $\bar{x}_1$ and $\bar{x}_2$ in substrips π'_1, π'_2. But since the subgroups $\langle a \rangle, \langle b \rangle$ are normal in the groups A and B, then after a series of elementary transformations over Δ' we can assume that the initial pair of edges in the contours $\partial\Pi'_1$ and $\partial\Pi'_2$ have the labels $x^1_i(\bar{x}_1)^{-1}$, $(x^1_j)^{-1}\bar{x}_2$, respectively. Finally, as above, these products belong to A. Therefore we obtain $x^1_i(\bar{x}_1)^{-1} = x'a^{n_1}$ and $(x^1_j)^{-1}\bar{x}_2 = x''a^{n_2}$, where $x', x'' \in \{x_1, \ldots, x_n\}$. Thus, after elementary transformations over Δ' one can regard the t-strips Π'_1 and Π'_2 to be special. Recall that the elements $x_1, \ldots, x_n$ belong to the alphabet $\mathcal{A}$. Therefore we have $\|x^1_i(\bar{x}_1)^{-1}\|, \|(x^1_j)^{-1}\bar{x}_2\| \leq 2$ and using the (λ, c)-quasigeodesity of the words $x'a^{n_1}$, $x''a^{n_2}$ we obtain

$$|n_1|, |n_2| \leq (2+c)\lambda^{-1} \qquad (11)$$

So, the total number of t-faces in t-strips Π'_1 and Π'_2 is equal to $\bar{n} = |k_0| + |l_0| + |n_1| + |n_2| + |k - k_1| + |l - l_1| + 4$. Therefore, using the inequalities (9),(10),(11) one obtains

$$\bar{n} \leq 2(\varepsilon + c)\lambda^{-1} + 2(2 + c)\lambda^{-1} + |k - k_1| + |l - l_1| + 4$$

Moreover, the total number of t-faces in the t-strips Π_1 and Π_2 is equal to $\bar{m} = |k| + |l| + 4$, but as was mentioned above $|l_1| \geq M$ and $|k_1| \geq \overline{M}$ by (8). In other words, $|l_1|$ and $|k_1|$ are sufficiently large and so the inequality $\bar{n} < \bar{m}$ holds. This means that the number of t-faces in the t-strips Π'_1 and Π'_2 is less than that in Π_1 and Π_2. Thus, we have shown that the diagram Δ was not minimal in the case of compatibility of paths q_1 and q_2.

6. Outer arcs of long t-strips

Let a group $\overline{G}$ be given by the presentation (6). We prove in this section that the sum of the lengths of all outer arcs of long t-strips in a minimal special diagram Δ over $\overline{G}$ is almost equal to the sum of the perimeters of all t-strips in Δ.

Consider a minimal diagram Δ over $\overline{G}$ and construct the estimating graphs Φ and Φ' for it (as in Section 4). We will use the notations $V(\Phi)$ and $E(\Phi')$ for the number of vertices of Φ and for the number of edges of Φ', respectively. Taking into account Lemma 4, it is easy to restrict the number $E(\Phi')$ in terms of $V(\Phi)$. For example, Lemmas 10.3,10.4 [7] give the following (rather rough) estimate:

Lemma 5. *For any minimal diagram Δ over $\overline{G}$ the inequality $E(\Phi') \le 10V(\Phi)$ holds.*

Extracting all t-strips and all distinguished contiguity subdiagrams from Δ, one can obtain a set of diagrams $\Delta_1, \ldots, \Delta_l$ over G. As mentioned in Section 2, the boundary t-edges of any distinguished t-strip in a minimal special diagram Δ lie on $\partial\Delta$. Therefore the diagrams $\Delta_1, \ldots, \Delta_l$ are circular. Let every boundary $\partial\Delta_i$ consists of n_i arcs, where each arc is either (1) an unbound arc of a t-strip, or (2) an unbound arc of $\partial\Delta$, or (3) a side arc of some distinguished contiguity subdiagram in Δ.

Lemma 6. $\sum_{i=1}^{l} n_i < 50m$ *for any minimal diagram Δ over $\overline{G}$, where m is the number of t-strips in Δ ($m \ge 1$).*

Proof If the contour of a t-strip Π (or of $\partial\Delta$) consists of i contiguity arcs and j unbound arcs, then $j \le i+1$ since unbound arcs have to be separated by contiguity arcs. So the total number of unbound arcs in Δ is not greater than $r + m + 1$, where r is the number of distinguished contiguity arcs in Δ (recall that we regard $\partial\Delta$ as a single section). On the other hand, r is not greater than twice the number of edges of the estimating graph Φ' for Δ. Hence, by Lemma 5 we have $r \le 20m$. Similarly, the number of arcs of type (3) is not greater than $20m$. Therefore the total number of all arcs satisfies the inequality $\sum_{i=1}^{l} n_i \le (20m + m + 1) + 20m < 50m$.

Now we consider a minimal special diagram Δ over $\overline{G}$, all of whose t-strips are long, that is $\|\partial\Pi\| > \rho$ for any t-strip Π in Δ. Recall (see Section 4), that the constant ε bounds the length of the side arcs of contiguity subdiagrams and satisfies the inequality $\varepsilon > 2H + c_1$ (see Section 3). Recall also that every word of the set $\{x_i a^m x_j, y_i b^m y_j \mid i, j = 1, \ldots, n, \ m \in \mathbb{Z}\}$ is (λ, c)-quasigeodesic by the choice of the constants $\lambda > 0, c \ge 0$ in Section 3.

The following lemma is similar to Lemma 6.2 [2].

Lemma 7. *There exists a constant ρ, which depends on the group G only, such that the sum Σ_u of the lengths of all unbound arcs of t-strips in a minimal special diagram Δ over $\overline{G}$ is less than $m\sqrt{\rho}$, where m is the number of t-strips in Δ, provided all maximal subpaths of $\partial\Delta$ consisting of g-edges are (λ, c)-quasigeodesic.*

Proof We will use the notations introduced before Lemma 6. Let $\Delta_1, \ldots, \Delta_l$ be circular diagrams over G, then $\Sigma_u = \Sigma_1 + \ldots + \Sigma_l$, where Σ_i is the sum of the lengths of the arcs of type (1) in the boundary $\partial\Delta_i$, $i = 1, \ldots, l$.

Assume that $\Sigma_u \geq m\sqrt{\rho}$. Then $\Sigma_i \geq n_i\sqrt{\rho}/50$ for some i, since otherwise by Lemma 6, $\Sigma_u = \Sigma_1 + \ldots + \Sigma_l < (\sqrt{\rho}/50) \sum_{i=1}^{l} n_i \leq m\sqrt{\rho}$.

Consider a circular diagram Δ_i over G for which above inequality holds. After a series of elementary transformations of Δ_i one can find in it a path $\bar{p}$ such that $|\bar{p}| = \|\bar{p}\|$ for every arc p of types (1) and (2) in $\partial\Delta_i$, where $\bar{p}$ is homotopic to p in Δ_i. When considering the γ-image of the 1-skeleton of Δ_i in $C(G)$ (see Section 1), we will keep the same notation p, $\bar{p}$ for the images of the paths p and $\bar{p}$ in $C(G)$.

By Lemma 1, p and $\bar{p}$ are H-close to each other for some constant $H = H(\lambda, c, G)$. Below we will use the notation of Lemma 25 [8]. Consider the images of $\bar{p}$ for the arcs p of type (1) in $C(G)$ as elements of a set N_1 and the images of $\bar{p}$ for the arcs p of type (2) as elements of N_2. Finally, the images of arcs having type (3) will be treated as elements of N_3.

Using the notation of Lemma 2, we obtain $\sigma_3 \leq \varepsilon n_i$. If $\bar{p}_1, \ldots, \bar{p}_k$ belong to N_1, then $\sigma_1 = \sum_{j=1}^{k} |\bar{p}_j| > \sum_{j=1}^{k} (\lambda\|\bar{p}_j\| - c) \geq \lambda\Sigma_i - n_i c \geq \frac{1}{50}\lambda n_i\sqrt{\rho} - n_i c = n_i(\frac{1}{50}\lambda\sqrt{\rho} - c)$.

Therefore, if ρ is sufficiently large, one can apply Lemma 2 and find c_1-bound subsegments of length greater than $2 \cdot 10^{-4}\lambda\sqrt{\rho}/50$ in a segment $\bar{p}_1 \in N_1$ and another segment $\bar{p}_2 \in N_1 \cup N_2$. This means that p_1 and p_2 have $(2H + c_1)$-bound subsegments q_1 and q_2 of length greater than $10^{-4}\lambda\sqrt{\rho}/50$. There are two cases.

1. $\bar{p}_2 \in N_1$, i.e. q_1 and q_2 are subpaths of the boundary paths of t-strips in Δ. However, in this case, one can define a contiguity subdiagram of a t-strip to a t-strip with boundary $v_1 q_1 v_2 q_2$, where $\|v_1\|, \|v_2\| < \varepsilon$ since $\varepsilon > 2H + c_1$ contrary to the maximality of the choice of the distinguished system of contiguity subdiagrams in Δ, as p_2 is an unbound arc.

2. $\bar{p}_2 \in N_2$, that is q_2 is a subpath of the contour $\partial\Delta$, then a contradiction arises as above (here there is a new contiguity subdiagram between a t-strip and $\partial\Delta$).

So the inequality $\Sigma_u \geq m\sqrt{\rho}$ is false and the lemma is proved.

The following lemma gives a lower bound for the sum of the lengths of the outer arcs of long t-strips in a minimal special diagram Δ over $\overline{G}$.

Lemma 8. *For any minimal diagram Δ over $\overline{G}$, all of whose t-strips are long and all maximal subpaths of the contour $\partial\Delta$ consisting of g-edges are (λ, c)-quasigeodesic, the inequality $\Sigma_o > \frac{1}{4}\Sigma$ holds, where Σ_o is the sum of lengths of all outer arcs of long t-strips in Δ and Σ is the sum of the perimeters of all t-strips in Δ.*

Proof As $\Sigma_o = \Sigma - \Sigma_I - \Sigma_u$, where Σ_I is the sum of the lengths of all inner arcs of t-strips in Δ. Using Lemma 7, we will obtain the lower bound for Σ_I. Let us consider three different cases of contiguity subdiagrams between two t-strips in Δ.

1. Γ is a contiguity subdiagram between B-sections (see Section 5) of the t-strips Π_1 and Π_2 in Δ.

We put $\partial(\Pi_1, \Gamma, \Pi_2) = p_1 q_1 p_2 q_2$, where $\|p_1\|, \|p_2\| \leq \varepsilon$, q_1 (q_2) is a subpath of a B-section of the t-strip Π_1 (respectively of Π_2). Let

$$\phi(\partial\Pi_1) \equiv t^{-1} x_i^1 a^k x_j^1 t (y_j^1)^{-1} b^{-k} (y_i^1)^{-1}$$

and

$$\phi(\partial\Pi_2) \equiv t^{-1} x_i^2 a^l x_j^2 t (y_j^2)^{-1} b^{-l} (y_i^2)^{-1},$$

where $x_i^1, x_i^2, x_j^1, x_j^2$ are in $\{x_1, \ldots, x_n\}$ and $y_i^1, y_i^2, y_j^1, y_j^2$ are in $\{y_1, \ldots, y_n\}$ (Fig.9).

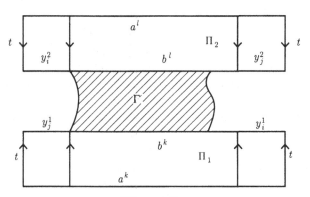

Figure 9.

Since the elements $\{a, b, x_i, y_i \ i = 1, \ldots, n\}$ are contained in our alphabet $\mathcal{A}$ the A-section and the B-section of Π_1 (and similarly of Π_2) have equal lengths. Therefore

$$\frac{\|q_1\|}{\|\partial\Pi_1\|}, \frac{\|q_2\|}{\|\partial\Pi_2\|} < \frac{1}{2}$$

2. Γ is a contiguity subdiagram between A-sections of t-strips Π_1 and Π_2 in Δ.

As above, we set $\partial(\Pi_1, \Gamma, \Pi_2) = p_1 q_1 p_2 q_2$, where q_1 and q_2 are subpaths of A-sections of the t-strips Π_1 and Π_2, respectively. We will use the same notation as in the previous case for the labels of the contours of Π_1 and Π_2 .

Let $\phi(q_1) \equiv c_1 a^{k_1} c_2$ and $\phi(q_2) \equiv d_1 a^{l_1} d_2$, where $\|c_i\|, \|d_i\| \leq 1$ $i = 1, 2$. Therefore Γ is a circular diagram over G with contour $p_1' q_1' p_2' q_2'$, where $\|p_1'\|, \|p_2'\| \leq \varepsilon + 2$ and $\phi(q_1') \equiv a^{k_1}$, $\phi(q_2') \equiv a^{l_1}$.

Hence, by Lemma 2.5 [2] there exists a constant $M = M(G) > 0$ such that either $|k_1|, |l_1| < M$ or the images of the paths q_1' and q_2' in $C(G)$ (see Section 1) are compatible, but as was shown in Section 5, the second case contradicts the minimality of Δ. So we have $|k_1|, |l_1| < M$.

Then $\|q_1\| < \|a\| M + 2 = M + 2$ as $a \in \mathcal{A}$ and similarly, $\|q_2\| < M + 2$. Therefore

$$\frac{\|q_1\|}{\|\partial\Pi_1\|}, \frac{\|q_2\|}{\|\partial\Pi_2\|} < \frac{M+2}{\rho}$$

since Π_1 and Π_2 are long t-strips, i.e. $\|\partial\Pi_1\|, \|\partial\Pi_2\| > \rho$.

3. Γ is a contiguity subdiagram between an A-section of t-strip Π_1 and a B-section of t-strip Π_2.

Let us set $\partial(\Pi_1, \Gamma, \Pi_2) = p_1 q_1 p_2 q_2$, $\|p_1\|, \|p_2\| \leq \varepsilon$ and
$\phi(\partial\Pi_1) \equiv t^{-1} x_i^1 a^k x_j^1 t(y_j^1)^{-1} b^{-k} (y_i^1)^{-1}$,
$\phi(\partial\Pi_2) \equiv t^{-1} x_i^2 a^l x_j^2 t(y_j^2)^{-1} b^{-l} (y_i^2)^{-1}$, $\phi(q_1) \equiv c_1 a^{k_1} c_2$, $\phi(q_2) \equiv d_1 b^{l_1} d_2$, where $\|c_i\|, \|d_i\| \leq 1$ $i = 1, 2$.

As in the previous case, we obtain that either $\|k_1\|, \|l_1\| < M$ or the paths q_1' and q_2' with labels a^{k_1}, b^{l_1} are compatible. The compatibility yields that there exists a path v in Γ joining some vertices of the paths q_1' and q_2' such that $\phi(v) b \phi(v)^{-1} \in E(A)$ (see Section 5) in G. Since the elements $\phi(v) b \phi(v)^{-1}$ and a have infinite order in the group G and belong to the same maximal elementary subgroup in G, we obtain, using Lemmas 1.16, 1.17 [2], that there are nonzero integers $\bar{k}, \bar{l}$ such that $\phi(v) b^{\bar{k}} \phi(v)^{-1} = a^{\bar{l}}$ in G. However, this contradicts the conditions of Theorem 3.

Therefore, we have $|k_1|, |l_1| < M$ and, similarly as in the previous case $\frac{\|q_1\|}{\|\partial\Pi_1\|}, \frac{\|q_2\|}{\|\partial\Pi_2\|} < \frac{M+2}{\rho}$.

An arbitrary t-strip in Δ may have several (a, b)- and (a, a)-contiguities. Nevertheless, in any case we have estimations $\frac{\|q_k\|}{\|\partial\Pi\|} < \frac{M+2}{\rho}$ for $k = 1, 2$, where Π is any long t-strip in Δ (not necessarily that, which has boundary subpaths q_1, q_2). In the case of (b, b)-contiguity, we have obtained the estimation $\frac{\|q\|}{\|\partial\Pi\|} < \frac{1}{2}$ for the length $\|q\|$ of the whole B-section of a t-strip Π (which may has several contiguity arcs).

Then, by Lemma 5, after adding up the lengths of all inner arcs of the t-strips in Δ, we obtain $\Sigma_I < (20\frac{M+2}{\rho} + \frac{1}{2})\Sigma$.

Finally by Lemma 7, $\Sigma_u < m\sqrt{\rho}$, where m is the number of t-strips in Δ. As all t-strips in Δ are long, $\Sigma_u < m\sqrt{\rho} = \frac{m\rho}{\sqrt{\rho}} \leq \frac{1}{\sqrt{\rho}}\Sigma$. Therefore

$$\Sigma_o = \Sigma - \Sigma_I - \Sigma_u > (1 - 20\frac{M+2}{\rho} - \frac{1}{2} - \frac{1}{\sqrt{\rho}})\Sigma > \frac{1}{4}\Sigma$$

since $1 - 20\frac{M+2}{\rho} - \frac{1}{2} - \frac{1}{\sqrt{\rho}} > \frac{1}{4}$ for sufficiently large ρ.

7. Linear isoperimetric inequality for the group $\overline{G}$

In this section we obtain a linear isoperimetric inequality for the group $\overline{G}$ given by the presentation (6), i.e. we prove the sufficiency of the conditions of Theorem 3.

Lemma 9. *There exists a constant $\theta_1 = \theta_1(G) > 0$ such that for any minimal diagram Δ over $\overline{G}$ the inequality $\|\partial\Delta\| \geq \theta_1 \Sigma_o$ holds.*

Proof Let us consider a fixed contiguity subdiagram Γ of a t-strip Π to $\partial\Delta$. We put $\partial(\Pi, \Gamma, \partial\Delta) = p_1 q_1 p_2 q_2$ (Fig.10). Then $\phi(q_1) = \phi(p_1)^{-1}\phi(q_2)^{-1}\phi(p_2)^{-1}$ in the group G and, using $\|p_1\|, \|p_2\| \leq \varepsilon$, we obtain $|q_2| \geq |q_1| - 2\varepsilon$.

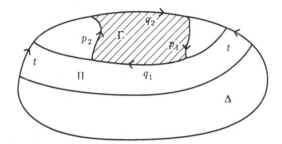

Figure 10.

As the contour of the t-strip Π is (λ, c)-quasigeodesic, we obtain $|q_1| \geq \lambda\|q_1\| - c$, and so $\|q_2\| \geq |q_2| \geq \lambda\|q_1\| - c - 2\varepsilon$, i.e. $\|q_1\| \leq \lambda^{-1}\|q_2\| + (c+2\varepsilon)\lambda^{-1}$. Finally, $\|q_1\| \leq \|q_2\|(\lambda^{-1}(2\varepsilon + c + 1))$ because $\|q_2\| \geq 1$ by definition of a contiguity subdiagram .

Let $\theta_1 = \lambda(2\varepsilon + c + 1)^{-1}$. Then $\|q_2\| \geq \theta_1\|q_1\|$. Using the fact that different contiguity subdiagrams are disjoint, we conclude that $\|\partial\Delta\| \geq \theta_1 \Sigma_o$, where θ_1 depends on the group G only.

Lemma 10. *There is a constant $\theta = \theta(G) > 0$ such that for any minimal special diagram Δ over $\overline{G}$, all of whose t-strips are long and all maximal subpaths of $\partial\Delta$ consisting of g-edges, are (λ, c)-quasigeodesic, we have $\|\partial\Delta\| \geq \theta\Sigma$.*

Proof This follows immediately from Lemmas 8 ,9 where one can set $\theta = \frac{1}{4}\theta_1$.

The following assertion establishes a linear isoperimetric inequality for diagrams with long t-strips.

Lemma 11. *There exists a constant $\alpha_1 = \alpha_1(G) > 0$ such that for any minimal diagram Δ over $\overline{G}$, all of whose t-strips are long, $n(\Delta) \leq \alpha_1\|\partial\Delta\|$ where $n(\Delta)$ is the number of faces in Δ.*

Proof Notice that one can consider all maximal subpaths of the contour $\partial\Delta$ consisting of g-edges to be $(\frac{1}{2}, 1)$-quasigeodesic and (λ, c)-quasigeodesic for $\lambda \leq \frac{1}{2}, c \geq 1$ as well (details can be found in [2]).

Excising all t-strips from Δ, one can obtain a set $\Delta_1, \ldots, \Delta_l$ of circular diagrams over G. Obviously, $\sum_{i=1}^{l} \|\partial \Delta_i\| \leq \Sigma + \|\partial \Delta\|$, where Σ is the sum of the perimeters of all t-strips in Δ.

Let n_i denotes the number of g-faces in Δ_i. Using the hyperbolicity of G, we can conclude that $n_i \leq \beta \|\partial \Delta_i\|$ $i = 1, \ldots, l$ and so $n_o = n_1 + \ldots + n_l \leq \beta (\sum_{i=1}^{l} \|\partial \Delta_i\|)$, where n_o is the number of g-faces in Δ. Therefore $n_o \leq \beta (\Sigma + \|\partial \Delta\|)$.

Let Π be a t-strip in Δ and $\phi(\partial \Pi) \equiv t^{-1} x_i a^k x_j t y_j^{-1} b^{-k} y_i^{-1}$. Then $\|\partial \Pi\| \geq |k|(\|a\| + \|b\|) = 2|k|$, i.e. the number of t-faces in Π is equal to $|k| + 2 \leq (\frac{1}{2} + 1)\|\partial \Pi\| = \frac{3}{2}\|\partial \Pi\|$. So the total number of t-faces n_t in Δ satisfies the inequality $n_t \leq \frac{3}{2}\Sigma$.

By Lemma 10, we obtain for the total number $n(\Delta)$ of faces in Δ the estimate

$$n(\Delta) = n_t + n_o \leq (\frac{3}{2} + \beta)\Sigma + \beta\|\partial \Delta\| \leq (\frac{3}{2} + \beta)\theta^{-1}\|\partial \Delta\| + \beta\|\partial \Delta\| = \alpha_1\|\partial \Delta\|,$$

where $\alpha_1 = \beta + (\frac{3}{2} + \beta)\theta^{-1}$ depends on the group G only.

Now we are able to prove the sufficiency of the conditions of Theorem 3.

According to van Kampen's lemma, to prove that the group $\overline{G}$ is hyperbolic, it suffices to find a constant $\alpha = \alpha(G) > 0$, which depends on the group G only, such that for any minimal special diagram Δ over $\overline{G}$ the linear inequality $n(\Delta) \leq \alpha\|\partial \Delta\|$ holds. Let us consider a minimal special diagram Δ over $\overline{G}$. Excising all short t-strips from Δ, we obtain a set $\Delta_1, \ldots, \Delta_d$ of circular diagrams over $\overline{G}$, all of whose t-strips are long.

Notice that the total number of t-strips in Δ is less than or equal to $\frac{1}{2}\|\partial \Delta\|$ as any distinguished t-strip has two boundary t-edges on the contour $\partial \Delta$ (see Section 2). A boundary $\partial \Delta_i$, $i = 1, \ldots, d$, of the diagram Δ_i consists of arcs of two types : 1)boundary arcs of the short t-strips of Δ; 2)subpaths of the contour $\partial \Delta$.

Therefore using the definition of short t-strips (see Section 5) and the remark above, we obtain $\sum_{i=1}^{d} \|\partial \Delta_i\| \leq \frac{1}{2}\rho\|\partial \Delta\| + \|\partial \Delta\| = (\frac{1}{2}\rho + 1)\|\partial \Delta\|$.

Since all t-strips of Δ_i $i = 1, \ldots, d$ are long, the number $n(\Delta_i)$ of faces in Δ_i satisfies the inequality $n(\Delta_i) \leq \alpha_1\|\partial \Delta_i\|$ by Lemma 11. Thus, for the total number n' of faces in the union of the diagrams Δ_i, $i = 1, \ldots, d$, we have

$$n' = n(\Delta_1) + \ldots + n(\Delta_d) \leq \alpha_1(\sum_{i=1}^{d} \|\partial \Delta_i\|) \leq \alpha_1(\frac{1}{2}\rho + 1)\|\partial \Delta\|.$$

Consider a short t-strip Π in Δ, i.e. $\|\partial \Pi\| \leq \rho$, and set

$$\phi(\partial \Pi) \equiv t^{-1} x_i a^k x_j t y_j^{-1} b^{-k} y_i^{-1}.$$

Then $\|\partial \Pi\| \geq |k|(\|a\| + \|b\|) = 2|k|$. Hence the number of t-faces in Π, which is equal to $|k| + 2$, satisfies the inequality $|k| + 2 \leq \frac{1}{2}\rho + 2$.

Therefore, for the total number of t-faces n'' in the short t-strips, one obtains
$n'' \leq (\frac{1}{2}\rho + 2) \cdot \frac{1}{2}\|\partial\Delta\|$.

Finaly, as the total number of faces in Δ is equal to $n(\Delta) = n' + n''$, we obtain , that $n(\Delta) \leq \alpha_1(\frac{1}{2}\rho + 1)\|\partial\Delta\| + \frac{1}{2}(\frac{1}{2}\rho + 2)\|\partial\Delta\| = \alpha\|\partial\Delta\|$, where $\alpha = \alpha_1(\frac{1}{2}\rho + 1) + \frac{1}{2}(\frac{1}{2}\rho + 2)$.

Since $\alpha = \alpha(G) > 0$, i.e. α depends on the group G only, this proves the hyperbolicity of the group $\overline{G}$.

8. Necessity of conditions of Theorem 3

1. Let us assume that neither of the subgroups A and B is maximal elementary in G. We will use the decompositions (4) and (5) of the subgroups A and B (see Section 1). Denote by $E(a)$, $E(b)$ the elementarizers of the elements a and b, respectively, in the group G, i.e. the maximal elementary subgroups of the hyperbolic group G containing a and b, respectively.

By Lemma 1.16 [2], $A \subset E(a)$, $B \subset E(b)$, where the the inclusions are proper. Therefore one can choose elements $g_1 \in E(a) \setminus A$ and $g_2 \in E(b) \setminus B$ in the group G. As we assume the group $\overline{G}$ to be hyperbolic, we can define the elementarizers $\overline{E}(a)$, $\overline{E}(b)$ of the elements a and b in $\overline{G}$.

Since $t^{-1}at = b$ in $\overline{G}$, we conclude that $t^{-1}\overline{E}(a)t = \overline{E}(b)$, by Lemma 1.16 [2]. But the group G is embedded into $\overline{G}$ by the natural map (see [6]), and so $E(a) \subset \overline{E}(a)$, $E(b) \subset \overline{E}(b)$ and $t^{-1}g_1t, g_2 \in \overline{E}(b)$, i.e. $\overline{g} = t^{-1}g_1tg_2 \in \overline{E}(b)$.

As the subgroup $\overline{E}(b)$ is maximal elementary in $\overline{G}$, there exists a nonzero integer m such that $x \in \overline{E}(b)$ if and only if $xb^mx^{-1} = b^{\pm m}$ in $\overline{G}$, by Lemmas 1.16, 1.17 [2]. As $\overline{g} \in \overline{E}(b)$, $\overline{g}b^m\overline{g}^{-1} = b^{-m}$ implies $\overline{g}^2b^m\overline{g}^{-2} = b^m$, i.e. $[\overline{g}^2, b^m] = 1$ in $\overline{G}$.

Notice that $g_1 \notin A$, $g_2 \notin B$, and so the element $\overline{g}^2 = (t^{-1}g_1tg_2)^2$ has infinite order in $\overline{G}$ by Britton's lemma [6]. Further, the order of b^m in $\overline{G}$ is infinite also (as the element b has infinite order in G) and $\langle \overline{g}^2 \rangle \cap \langle b^m \rangle = \{1\}$ in the group $\overline{G}$. Therefore $\overline{E}(b) \supset \mathbb{Z} \times \mathbb{Z}$, where the first factor of the direct product is generated by $\overline{g}^2$ and the second by b^m. But this contradicts the hyperbolicity of $\overline{G}$ [1,3], and so our assumption is false. Consequently at least one of subgroups A or B is maximal elementary in G.

2. Assume that for some $g \in G$ the subgroup $K = gAg^{-1} \cap B$ is infinite. Obviously, K is an infinite elementary subgroup of G. As $|A : \langle a \rangle|$, $|B : \langle b \rangle| < \infty$, one can choose nonzero integers u, v such that $ga^ug^{-1} = b^v$ in G. Since $t^{-1}at = b$ in $\overline{G}$, $t^{-1}a^vt = b^v$, we obtain the equality $ga^ug^{-1} = t^{-1}a^vt$ in $\overline{G}$, i.e. $(tg)a^u(tg)^{-1} = a^v$. By Lemmas 1.16, 1.17 [2], the above equality implies $tg \in \overline{E}(a)$ in $\overline{G}$, but the order of element tg is infinite in $\overline{G}$ (see [6]). Therefore as above, one can choose nonzero integers u_0, v_0 such that $(tg)^{u_0} = a^{v_0}$ in $\overline{G}$. But this is impossible, since the elements $(tg)^{u_0}$ and a^{v_0} have different normal

forms in $\overline{G}$ (see [6]). Hence our assumption is false. This completes the proof of Theorem 3. □

9. Proof of Corollary 7

Let G and H be hyperbolic groups, A and B be infinite elementary subgroups of G and H, respectively, and $\psi : A \to B$ a fixed isomorphism. Assume for definiteness that A is maximal elementary in G.

Define $P = G *_{A=B} H = \langle G * H \mid a = \psi(a), a \in A \rangle$. Below we prove the hyperbolicity of the group P, i.e. we establish the sufficiency of the conditions of Corollary 7.

Let $K = G * H$. It is easy to see that the hyperbolic group K and its isomorphic subgroups A and B satisfy the conditions of Theorem 3. Therefore the group $L = \langle K, t \mid t^{-1}at = \psi(a), a \in A \rangle$ is hyperbolic. Consider the subgroup M of L generated by the subgroups $t^{-1}Gt$ and H, i.e. $M = \langle t^{-1}Gt, H \rangle$ in L. Obviously, M is isomorphic to P via the map $\phi : P \to M$ given by $\phi(g) = t^{-1}gt$ for $g \in G$ and $\phi(h) = h$ for $h \in H$ (see also [6]).

Therefore we have to prove the hyperbolicity of the subgroup M of the hyperbolic group L. For this purpose, it is sufficient to establish that M is quasiconvex in L [1,3].

Let us choose sets of generators of the groups G and H, i.e. $G = \langle g_1, \ldots, g_m \rangle$, $H = \langle h_1, \ldots, h_l \rangle$. Then $L = \langle t, g_i, h_j \ i = 1, \ldots, m \ j = 1, \ldots, l \rangle$ and $M = \langle t^{-1}g_i t, h_j \ i = 1, \ldots, m \ j = 1, \ldots, l \rangle$.

One can consider two metrics on M: one is induced from L and another is the word metric in the generators $\{t^{-1}g_i t, h_j \ i = 1, \ldots, m \ j = 1, \ldots, l\}$ of the group M (for $W \in M$ we denote by $\|W\|_M$ the length of the word W in generators $\{t^{-1}g_i t, h_j \ i = 1, \ldots, m \ j = 1, \ldots, l\}$ and by $|W|_M$ we denote the minimal length of words $\overline{W}$ in this alphabet, representing W in M). It is clear that

$$|W|_M \le \|W\|_M \le \|W\| \le 3\|W\|_M$$

for $W \in M$. To prove the quasiconvexity of M it is sufficient [1,3] to show that there exists a constant $C = C(M) > 0$ such that any geodesic word $W \in M$ (i.e. a label of a geodesic path in the Cayley graph $C(M)$ of M) is $(C,0)$-quasigeodesic in L, i.e. for any $W \in M$ with the geodesic property $|W|_M = \|W\|_M$ we have to prove $\|W\| \le C|W|$, where $C = C(M)$ depends on the group M only.

Thus, let W be a geodesic element of M and let V be a geodesic element of L such that $W = V$ in L. Consider the Cayley graph $C(L)$ of the group L and two paths p and q in it with labels $\phi(p) = W$, $\phi(q) = V$ where $p_- = q_-$, $p_+ = q_+$. By van Kampen's lemma there exists a circular diagram Δ over L with contour $\partial\Delta = pq^{-1}$ (Fig.11).

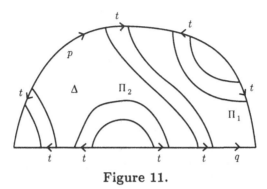

Figure 11.

The t-strips of Δ can be of three types: 1) call a t-strip Π_1 of Δ to be of the first type if the boundary t-edges of $\partial\Pi_1$ belong to the path p ; 2) call a t-strip Π_2 of Δ to be of the second type if the boundary t-edges of $\partial\Pi_2$ belong to the path q ; 3) the remaining t-strips will be called the third type strips.

Remove all t-strips of the second type from Δ in the following way. Let Π_2 be such a t-strip in Δ. Denote by $\overline{\Delta}$ a subdiagram of Δ with $\partial\overline{\Delta} = v_1 v_2^{-1}$, where v_1 is a subpath of the contour $\partial\Pi_2$, v_2 is a subpath of q and $\overline{\Delta}$ contains Π_2 (Fig.12). Let us put $q = q_1 v_2 q_2$.

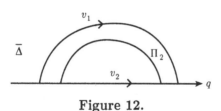

Figure 12.

Since the elementary subgroups A and B are quasiisometrically embedded in L (see [1,3]), there exists a constant $\overline{C} = \overline{C}(A, B)$ such that $\|v_1\| \leq \overline{C}\|v_2\|$. We can excise $\overline{\Delta}$ from Δ and replace the path q by $q' = q_1 v_1 q_2$. After removing all t-strips of the second type from Δ, we obtain a circular diagram Δ' with the contour $\partial\Delta' = p(q')^{-1}$, where $\|q'\| \leq \overline{C}\|q\|$.

Notice that the word $\phi(q')$ consists of syllables $t^{-1}ut$ where $u \in G$ and syllables $u' \in H$ since the symbols $t^{\pm 1}$ alternate in $\phi(q')$ as in $W = \phi(p)$. As W is a geodesic word in the alphabet $\{t^{-1}g_i t, h_j \ i = 1, \ldots, m \ j = 1, \ldots, l\}$, i.e. W has the shortest representation, we obtain $\|W\|_M \leq \|\phi(q')\|_M$. As was mentioned above $\|W\|_M \geq \frac{1}{3}\|W\|$ and $\|\phi(q')\|_M \leq \|\phi(q')\| \leq \overline{C}\|\phi(q)\| = \overline{C}\|V\|$. This implies that $\|W\| \leq 3\overline{C}\|V\|$, i.e. $\|W\| \leq C|W|$, where $C = 3\overline{C}$

depends on the group M only, and $|W| = \|V\|$ since $W = V$ in L and V is geodesic in L.

This completes the proof of the hyperbolicity for the group M as well as for $P = G *_{A=B} H$.

It remains to establish the necessity of the conditions of Corollary 7. Let us assume that the group $P = G *_{A=B} H$ is hyperbolic and neither of the subgroups A and B is maximal elementary in G and H, respectively. Let $\langle a \rangle$, $\langle b \rangle$ be infinite cyclic subgroups of finite index in A and B. By our assumption, we have the proper inclusions $A \subset E(a)$, $B \subset E(b)$. Therefore one can choose elements $g \in E(a) \setminus A$ and $h \in E(b) \setminus B$.

Denote by $\overline{E}(a)$ and $\overline{E}(b)$ the elementarizers of the elements a, b in the hyperbolic group P. Obviously, $\overline{E}(a) = \overline{E}(b)$ and $E(a) \subset \overline{E}(a)$, $E(b) \subset \overline{E}(b)$. Hence $u = gh \in \overline{E}(a)$ in the group P. Notice, that u has infinite order in the group P (see [6]) since $g,h \notin A = B$ in P, and since $u \in \overline{E}(a)$, there exist nonzero integers m,l such that $u^m = a^l$ in P, by Lemma 1.16 [2]. But this is impossible, since the elements u^m and a^l have different normal forms in the group P [6].

Therefore our assumption is false. This completes the proof of Corollary 7.

□

The following assertion is known and follows easily from the proof of Theorem 3 and Corollary 7 (in this case all t-strips of the minimal diagrams are short).

Proposition [1,3]. *HNN-extensions (amalgamated products) of two hyperbolic groups with finite associated (amalgamated) subgroups are hyperbolic.*

□

10. Proof of Theorem 1

We will use some results of the paper [2]. As the group G is torsion free, all elementary subgroups of G are cyclic.

The group G is finitely generated and so it is countable: $G \setminus \{1\} = \{g_1, g_2, \ldots\}$. Let $F(x_1, x_2, \ldots)$ be the free group with countable set of generators and $F(x_1, x_2, \ldots) \setminus \{1\} = \{v_1, v_2, \ldots\}$ be some enumeration of the non-trivial words in $F(x_1, x_2, \ldots)$. Let us order the countable set of pairs $\Omega = \{\omega_1, \omega_2, \ldots\} = \{ (g_i, v_j), i, j \in \mathbf{N} \}$ by the type of $\mathbf{N}$.

A pair $\omega_k = (g_i, v_j)$ will correspond to the operation of finding elements $\bar{g}_1, \ldots, \bar{g}_n$ of some quotient of G, such that $g_i = v_j(\bar{g}_1, \ldots, \bar{g}_n)$, where $v_j = v_j(x_1, \ldots, x_n)$.

Let G_i be a non-elementary torsion free hyperbolic quotient of the group G such that for any $\omega_j = (g_k, v_l)$ with the property $\omega_j \le \omega_i$ there exist elements $\bar{g}_1, \ldots, \bar{g}_n \in G_i$ such that $g'_k = v_l(\bar{g}_1, \ldots, \bar{g}_n)$ in G_i, where g'_k is an image of g_k

in G_i and $v_l = v_l(x_1, \ldots, x_n)$. Now we show how to construct a group G_{i+1}, where $\omega_{i+1} = (g_m, v_s)$ is a next element in Ω after ω_i.

For the image of an element g_m in G_i, we will keep the same notation. We can consider the word $v_s = v_s(x_1, \ldots, x_n)$ as an element of the free group $F(x_1, \ldots, x_n)$ for some n. Recall that $F(x_1, \ldots, x_n)$ is a hyperbolic group. Let $E(g_m)$ be the elemetarizer of the element g_m in G_i. Since the group G_i is torsion free, $E(g_m) = \langle h \rangle_\infty$ for some $h \in G_i$, i.e. $g_m = h^k$ for $k \in \mathbf{Z}$. There are two cases.

1. The word v_s is not a proper power in $F(x_1, \ldots, x_n)$, that is $E(v_s) = \langle v_s \rangle_\infty$ in the group $F(x_1, \ldots, x_n)$. Consider the group $G'_i = G_i * F(x_1, \ldots, x_n)$. Obviously, G'_i is a hyperbolic group as a free product of hyperbolic groups [1,3]. Let us denote by $A = \langle g_m \rangle_\infty$ and $B = \langle v_s \rangle_\infty$ the infinite cyclic subgroups of G'_i generated by g_m and v_s, respectively. It is easy to see that $g^{-1} A g \cap B = \{1\}$ in G'_i for any $g \in G'_i$, and as $E(v_s) = \langle v_s \rangle_\infty$ in G'_i, B is a maximal elementary subgroup of G'_i. Therefore, by Theorem 3, the HNN-extension $\overline{G}_i = \langle G'_i, t \mid t^{-1} g_m t = v_s \rangle$ of the group G'_i with associated subgroups A and B is a torsion free non-elementary hyperbolic group, since G'_i is a torsion free.

Since G_i is a non-elementary torsion free subgroup of the non-elementary hyperbolic group $\overline{G}_i$, by Theorem 2 [2] there is a non-elementary hyperbolic quotient G''_i of the group $\overline{G}_i$ such that: 1) the natural homomorphism ε of $\overline{G}_i$ onto G''_i is surjective on G_i, i.e. $\varepsilon(G_i) = G''_i$; 2) ε induces a bijective map on sets of conjugacy classes of elements having finite orders in $\overline{G}_i$ and G''_i respectively, i.e. the group G''_i is torsion free .

Therefore, the group G''_i is a non-elementary torsion free hyperbolic quotient of the group G since $\varepsilon(G_i) = G''_i$, and the group G_i is quotient of G. Let us keep the same notations for the images of the elements $t, g_m, v_s(x_1, \ldots, x_n)$ in G''_i (notice that t is the image of some element of G). Then we have $g_m = v_s(t x_1 t^{-1}, \ldots, t x_n t^{-1})$ in the group G''_i. So one can set $G_{i+1} = G''_i$.

2. The word v_s is a proper power in $F(x_1, \ldots, x_n)$, that is $E(v_s) = \langle \bar{v}_s \rangle_\infty$ in $F(x_1, \ldots, x_n)$ and $v_s = \bar{v}_s^u$ for some $u \in \mathbf{Z}$. If $k = \pm 1$, then the construction of the group G_{i+1} is similar to the one described above. Hence we are left with the case $|k| \geq 2$. As the group G_i is a non-elementary hyperbolic group, by Lemma 3.1 [2] there exists an element $b \in G_i$ such that $g \langle h \rangle_\infty g^{-1} \cap \langle b \rangle_\infty = \{1\}$ in G_i for any $g \in G_i$. Let us define the subgroups A and B of G_i as $A = \langle h \rangle_\infty$, $B = \langle b^u \rangle_\infty$. Since the subgroup A is maximal elementary in G_i and $g A g^{-1} \cap B = \{1\}$, the HNN-extension $\overline{G}_i = \langle G_i, t \mid t^{-1} h t = b^u \rangle$ of the group G_i with associated subgroups A and B is a torsion free hyperbolic group by Theorem 3

As G_i is a non-elementary torsion free subgroup of the non-elementary hyperbolic group $\overline{G}_i$, there exists a non-elementary torsion free hyperbolic quotient M of the group $\overline{G}_i$ such that the natural homomorphism ε of $\overline{G}_i$ onto M is surjective on G_i, i.e. $\varepsilon(G_i) = M$, by Theorem 2 [2]. Therefore M is a non-elementary torsion free hyperbolic quotient of the group G as G_i has the same property. Further, let us consider the group $M' = M * F(x_1, \ldots, x_n)$ and its two isomorphic subgroups $A = \langle b^k \rangle_\infty$, $B = \langle \bar{v}_s \rangle_\infty$. Obviously, $gAg^{-1} \cap B = \{1\}$ for any $g \in M'$, and B is a maximal elementary subgroup of M' (since $E(\bar{v}_s) = \langle \bar{v}_s \rangle_\infty$ in $F(x_1, \ldots, x_n)$). So by Theorem 3, the HNN-extension $\overline{M}' = \langle M', w \mid w^{-1} b^k w = \bar{v}_s \rangle$ is a torsion free (as M' is torsion free) non-elementary hyperbolic group. Similarly, as was described above (by Theorem 2 [2]), there is a non-elementary torsion free hyperbolic quotient L of the group $\overline{M}'$ such that the natural homomorphism ε of $\overline{M}'$ onto L is surjective on M. Thus L is a quotient of the group G since $\varepsilon(M) = L$ and M is a quotient of G.

Let us consider the images of the elements v_s, g_m, h, $\bar{v}_s$, t, w, b in L (we will keep the same notation for them). We have the following equations in L $g_m = h^k = (tb^u t^{-1})^k$, $v_s = \bar{v}_s^u = (w^{-1} b^k w)^u$. Therefore $g_m = (tw) v_s (tw)^{-1}$, and so $g_m = v_s((tw) x_1 (tw)^{-1}, \ldots, (tw) x_n (tw)^{-1})$ in the group L.

Now set $G_{i+1} = L$ and define $H = \lim_\to G_i$ as the inductive limit of the quotients G_i. Then H is a quotient of the group G. Clearly, H is non-trivial as all groups G_i are infinite (they are non-elementary hyperbolic groups and so infinite by definition). Also, H is a torsion free group since all G_i are forsion free. The group H is verbally complete by construction. So H is a required group.   □

11. Proof of Theorem 2

As in the proof of Theorem 1, we define an ordered set of pairs

$$\Omega = \{\omega_1, \omega_2, \ldots\} = \{(g_i, v_j), \ i, j \in \mathbf{N}\},$$

where $G \setminus \{1\} = \{g_1, g_2, \ldots\}$ and $F(x_1, x_2, \ldots) \setminus \{1\} = \{v_1, v_2, \ldots\}$.

Let G_i be a non-elementary hyperbolic quotient of the group G such that for any $\omega_j = (g_k, v_l)$ with the property $\omega_j \leq \omega_i$, the image g_k' of g_k in G_i has finite order in G_i and there exist elements $\bar{g}_1, \ldots, \bar{g}_n \in G_i$ such that $g_k' = v_l(\bar{g}_1, \ldots, \bar{g}_n)$ in G_i ,where $v_l = v_l(x_1, \ldots, x_n)$.

Let us show how to construct a group G_{i+1}, choosing $\omega_{i+1} = (g_m, v_s)$ as the next element of Ω after ω_i. For the image of the element g_m in G_i, we will keep the same notation. As $v_s \in F(x_1, x_2, \ldots)$, $v_s = v_s(x_1, \ldots, x_n)$ for some $n \in \mathbf{N}$, i.e. we can regard v_s to be an element of a free group $F(x_1, \ldots, x_n)$. If the order of g_m is infinite in G_i, we consider the elementarizer $E(g_m)$ of the element g_m in G_i and by Lemma 1.16 [2] we can find a nonzero integer l such that $\langle g_m^l \rangle_\infty$ is a normal subgroup of $E(g_m)$. Further, by Theorem 3 [2], there exsists a quotient $\overline{G}_i$ of the group G_i such that: 1) $\overline{G}_i$ is a non-elementary hyperbolic group; 2) the element g_m has finite order in $\overline{G}_i$. Then $\overline{G}_i$ is non-elementary hyperbolic quotient of the group G (as G_i has the same property).

Therefore the element g_m has finite order in G_i. Let k be the order of the element g_m in G_i.

Denote by $E(G_i)$ the unique maximal finite normal subgroup of the non-elementary hyperbolic group G_i (the elementarizer of G_i [2]). If $E(G_i)$ is non-trivial, then one can pass to the quotient $G_i/E(G_i)$. So we can assume that $E(G_i) = \{1\}$. Let us consider the group $H = \langle x_1, \ldots, x_n \mid v_s^k = 1 \rangle$, that is a quotient group of the free group $F(x_1, \ldots, x_n)$ by the normal closure of the word v_s^k. Obviously, one can assume v_s to be cyclically reduced in $F(x_1, \ldots, x_n)$. Then by Theorem 5.2 [6; ch.4] the order of the image of v_s in the group H, which will be denoted by the same letter, is equal to k. Further, H is a hyperbolic group by Newman's theorem 5.5 [6; ch.4] because there is Dehn's algorithm for it (see also [3]). Consider the subgroup $A = \langle g_m \rangle_k$ of G_i and the subgroup $B = \langle v_s \rangle_k$ of H. As the subgroups A and B are finite of the same order, the group $\overline{G}_i = G_i *_{A=B} H = \langle G_i, H \mid g_m = v_s \rangle$ is a non-elementary hyperbolic group by the above Proposition. Since $E(G_i) = \{1\}$ in G_i, obviously, $E(G_i) = \{1\}$ in $\overline{G}_i$ and so (by Theorem 2 [2]) there exists a non-elementary hyperbolic quotient G'_i of the group $\overline{G}_i$ such that the natural homomorphism ε of $\overline{G}_i$ onto G'_i is surjective on G_i, i.e. G'_i is a quotient of the group G_i and hence of G.

Notice that $g_m = v_s(x_1, \ldots, x_n)$ for the images of g_m, $v_s(x_1, \ldots, x_n)$ in the group G'_i (keeping the same notations). So one can put $G_{i+1} = G'_i$. Finally, define $\tilde{G} = \lim_\rightarrow G_i$ as the inductive limit of the quotients G_i. Clearly, $\tilde{G}$ is a quotient of the group G. The group $\tilde{G}$ is non-trivial as all groups G_i are infinite being non-elementary hyperbolic groups. Moreover, $\tilde{G}$ is a verbally complete torsion group by construction. Hence $\tilde{G}$ is a required group. □

Proof of Corollaries 1–6. Since hyperbolic groups are finitely generated, Corollary 1 and Corollary 2 follow immediately from Theorem 1, Theorem 2 respectively.

As any verbally complete group is divisible and since a finitely generated divisible abelian group is trivial, Corollary 3, Corollary 4 are just consequences of Theorem 1 and Theorem 2.

Then Corollary 5 follows from Corollary 3 and Corollary 6 follows from Corollary 4. □

References

[1]   M.Gromov, *Hyperbolic groups*, in: Essays in Group Theory, Publ., Math. Sci. Res. Inst. 8, (S.M.Gersten, ed.), (1987), Springer, New York pp. 75–263.

[2] A.Yu.Ol'shanskii, *On residualing homomorphisms and G-subgroups of hyperbolic groups,* Int.J.Algebra and Comput. **3 No. 4** (1993), pp. 365–409.

[3] E.Ghys and P. de la Harpe, *Espaces Metriques Hyperboliques sur les Groupes Hyperboliques d'apres Mikhael Gromov,* Birkhauser, (1991)

[4] M.Bestvina and M.Feighn, *A combimation theorem for negatively curved groups,* J. Differ. Geom. **35 No. 1** (1992), pp. 85–102.

[5] O.Kharlampovich, A.Myasnikov, *Hyperbolic groups and free constructions,* To appear.

[6] R.Lyndon and P.Schupp, *Combinatorial Group Theory,* Springer-Verlag, (1977).

[7] A.Yu.Ol'shanskii, *Geometry of defining relations in groups,* Math. and Its Applications (1991), (Russian) English translation.

[8] A.Yu.Ol'shanskii, *Periodic quotients of hyperbolic groups,* Mat. Zbornik **182 No. 4** (1991), pp. 543–567. In Russian, English translation in Math. USSR Sbornik **72 No. 2.**

[9] A.Yu.Ol'shanskii, *Infinite groups with cyclic subgroups,* In Russian. Dokl. Acad. Nauk SSSR **245** (1979), pp. 785–787.

[10] V.S.Guba, *A finitely generated divisible group,* Izv. Acad. Nauk SSSR **50** (1986), pp. 883–924.

Free Actions of Abelian Groups on Groups

Peter M. Neumann and Peter J. Rowley

Queen's College, Oxford, OX1 4AW.
UMIST, Manchester, M60 1QD.

The theorem

A group H of automorphisms of a group K is said to act freely if $a^h \neq a$ whenever $a \in K - \{1\}$ and $h \in H - \{1\}$. The main purpose of this note is to place the following theorem on record. Some commentary follows the proof.

Theorem. *Let K be a group that admits an infinite group H of automorphisms acting freely. If H is abelian and has only finitely many orbits in K then K is abelian.*

Proof. For each $h \in H$ there is a commutator map $\gamma_h : x \mapsto x^{-1}x^h$ on K to itself. Since H acts freely, if $h \neq 1$ then γ_h is injective. Because H is commutative γ_h commutes with the action of H, and therefore it maps H-orbits to H-orbits. Since there are only finitely many H-orbits, if $h \neq 1$ then γ_h must be surjective.

Suppose now that $k \in K - \{1\}$, $h \in H$ and k^h is conjugate to k in K, that is, $k^h = y^{-1}ky$ for some $y \in K$. If $h \neq 1$ then there would exist $x \in K$ for which $y = x\gamma_h$, and so

$$k^h = x^{-h}xkx^{-1}x^h,$$

whence xkx^{-1} would be fixed by h. Since the action of H is free we conclude that $h = 1$. Thus, for any $k \in K - \{1\}$ the only element of the H-orbit of k that is conjugate to k is k itself. Equivalently, each conjugacy class of K meets each H-orbit in at most one element. It follows that if the number of H-orbits in $K - \{1\}$ is m then all conjugacy classes of K have size $\leq m$. By a well-known old theorem of B.H. Neumann (see [**10**], Theorem 3.1) the commutator subgroup K' is finite. But K' is H-invariant and non-trivial H-orbits are infinite. Therefore $K' = \{1\}$, that is, K is abelian, as the theorem states.

Commentary

Let G be a permutation group on a set Ω. Following Wielandt we write G_α for the stabiliser of the point α of Ω, and $G_{\alpha,\beta}$ for the subgroup fixing both α and β. Recall that the *rank* of G is the number of orbits of G acting on Ω^2. If G is transitive on Ω it is the number of orbits of G_α. In his symposium lecture [Durham, 13 July 1994] the first-named author explained that, in order to make both a historical point and a mathematical point, he would use the following (nowadays non-standard) terminology: if

 (1) for any two distinct points α, β, we have $G_{\alpha,\beta} = \{1\}$, and
 (2) there is a normal subgroup K of G that acts regularly on Ω

(that is, K is transitive and $K_\alpha = \{1\}$), then G would be said to be a *Frobenius group*. In standard terminology K is then known as a Frobenius kernel and a stabiliser G_α is known as a Frobenius complement for G. Since K acts regularly, once a base-point α is chosen, Ω may be identified with K (a point ω is identified with the unique element x of K such that $\alpha x = \omega$) and it is well-known (and not hard to see) that the action of G_α on Ω is then identified with its action by conjugation on K. An immediate consequence of the theorem is

Corollary 1. *Let G be a Frobenius group with kernel K and complement H. If H is abelian and G has finite rank then either G is finite or K is abelian.*

Typical examples are groups of affine transformations $z \mapsto az + b$ of a field F, where b ranges over F and a ranges over a subgroup of finite index in the multiplicative group $F^\times$.

Corollary 2. *If G is a Frobenius group of finite rank whose complement H is cyclic (or finitely generated abelian), then G is finite.*

We believe it to be a matter of folk-lore that Corollary 2 follows from Corollary 1. Here is one line of argument (for the case where H is cyclic). Let a be a generator of H, let K be the Frobenius kernel, let $x \in K - \{1\}$, and let $x_i := x^{a^i}$ for $i \in \mathbb{Z}$. The elements x_0, $x_0 x_1$, $x_0 x_1 x_2$, ... cannot all be in different H-orbits and it follows easily that there exist r, s, t such that $0 < s < t$ and $x_0 x_1 \cdots x_r = x_s x_{s+1} \cdots x_t$. If $r = t$ then $x_0 x_1 \cdots x_{s-1} = 1$ (from which it follows that $s > 1$) and we set $m := s - 1$; otherwise we set $m := \max(r, t)$. The equation yields that $x_m \in \langle x_0, \ldots, x_{m-1} \rangle$, and from this, by applying a as many times as necessary, we find that $x_i \in \langle x_0, \ldots, x_{m-1} \rangle$ for $i \geq m$. Similarly, applying a^{-1}, a^{-2}, ... successively, we find that $x_i \in \langle x_0, \ldots, x_{m-1} \rangle$ for $i < 0$. Thus $\langle x_0, \ldots, x_{m-1} \rangle$ is an H-invariant subgroup L of K. By Corollary 1 we may assume that K is abelian. Then L is a finitely generated abelian group that has only finitely many characteristic subgroups (since H has only finitely many orbits in L), and it follows easily that L is finite. Letting x range through a set of representatives for the H-orbits in K we see that K

may be expressed as the product of finitely many such finite groups L, and so K is finite. Therefore G is finite, as asserted.

Corollary 2 ought to be a significant step towards the solution of a more general problem about permutation groups (see [11], or [5, §3] for some background). This concerns groups which, in the symposium lecture mentioned above, were called *Maillet groups*, namely, transitive permutation groups (G, Ω) satisfying condition (1). Edmond Maillet introduced and studied such groups in 1892 in his doctoral thesis [6] and some later papers (in particular [7, 8]). He showed that if $|\Omega| \leq 200$ then they have regular normal subgroups. William Burnside took the matter further and, using results which we discuss below, extended Maillet's theorem to all cases where $|\Omega| \leq 81\,000\,000$. Ultimately, a famous theorem proved by Frobenius in 1901 (see [3]) completely solved Maillet's problem. In our language it asserts that every finite Maillet group has a regular normal subgroup; that is, for finite groups there is no distinction between Maillet groups and Frobenius groups. For infinite groups, however, the former constitute a very much larger class—for example, a free group of rank 2 has $2^{\aleph_0}$ different faithful representations as a Maillet group, but only countably many faithful representations as a Frobenius group. In this language (which has been changed from that used in [5] and [11]) the problem is this:

Problem A. *Does there exist an infinite Maillet group (G, Ω) of finite rank in which a stabiliser G_α is cyclic? (Or finitely generated abelian?)*

By analogy with Burnside's partial results for the finite case it is tempting to conjecture that the answer is *No*. Before Frobenius proved his theorem Burnside had already given quite simple proofs in the special cases

    (i) where $|G_\alpha| > |\Omega|^{\frac{1}{2}}$ (see [1, pp. 142–143]), and

    (ii) where the stabilisers are abelian (even soluble; see [2]).

The condition in (i) can be rewritten as the inequality $|G_\alpha| > r$, where r is the rank of (G, Ω). One might hope therefore that Burnside's argument for (i) could be extended to the case of a group where the stabilisers are infinite and the rank is finite, but it uses Sylow's theorems and combinatorial and arithmetical arguments depending essentially on the finiteness of G. A more promising possibility is his proof of (ii), which was, in effect, the first application of the transfer homomorphism (in a character-theoretic form). It is possible that some such technique might be used to show that a Maillet group of finite rank with cyclic stabilisers (perhaps even with abelian stabilisers) has a regular normal subgroup. If so, then our theorem, or its Corollary 2, would solve the problem.

As was noted in [5], the special case of the conjecture in which the rank of G is 2 (that is, G is doubly transitive) is known to be true. It was proved by Károlyi, Kovács and Pálfy in [4] and independently by Mazurov in [9] that if the stabilisers in a doubly transitive permutation group G are abelian (in which case the group must be a Maillet group) then G has an abelian regular normal subgroup; in particular, if the stabilisers are cyclic then G must be finite.

In a different direction, the theorem suggests the following further problems:

Problem B. *Does there exist an infinite group K that has a cyclic (or finitely generated abelian) group of automorphisms with only finitely many orbits?*

Problem C. *What can be said about a group K that admits an abelian group H of automorphisms having only finitely many orbits?*

A natural conjecture is that such a group must be finite-by-abelian-by-finite, but again, we have made little progress towards a proof.

Acknowledgements

We offer warm thanks to the organisers and to the backers (LMS, EPSRC) of the LMS Durham Symposium at which the main part of this work was done. We are also grateful to an anonymous, conscientious and kindly referee for helpful suggestions.

Notes added in proof, 1 March 1997

(1) The argument presented in the first five sentences of the second paragraph of the proof of the theorem is to be found in W. Burnside, *Theory of groups of finite order*, Second Edition, Cambridge 1911, p.90; it appears also in B. Huppert, *Endliche Gruppen I*, Springer, Berlin 1967, p.500.

(2) The first named author acknowledges that the terminology he introduced in his lecture is not standard and that it sould be better to conform with convention so that what are here called 'Maillet groups' should be called 'Frobenius groups', and what are here called 'Frobenius groups' should be called 'split Frobenius groups'.

References

[1] W. Burnside, *Theory of groups of finite order*, Cambridge University Press, Cambridge (1897).

[2] W. Burnside, *On transitive groups of degree n and class n − 1*, Proc. London Math. Soc. **32** (1900), pp. 240–246.

[3]   G. Frobenius, *Über auflösbare Gruppen IV*, Sitzungsber. Kön. Preuss.
      Akad. Wiss. Berlin (1901), pp. 1216–1230.= Gesammelte Abhandlungen
      III pp. 189–203.

[4]   Gy. Károlyi, S. J. Kovács and P. P. Pálfy, *Doubly transitive permutation
      groups with abelian stabilizers*, Aequationes Math. **39** (1990), pp. 161–
      166.

[5]   Dugald Macpherson,   *Permutation groups whose subgroups have just
      finitely many orbits*, Preprint (April, 1994).

[6]   E. Maillet, *Recherches sur les substitutions et en particulier sur les groupes
      transitifs*, Thèse de Doctorat, Gauthier-Villars, Paris (1892).

[7]   E. Maillet, *Des groupes primitifs de classe $N - 1$ et de degré N*, Bull.
      Soc. Math. de France **25** (1897), pp. 16–32.

[8]   E. Maillet, *Des groupes transitifs de substitutions de degré N et de classe
      $N - 1$*, Bull. Soc. Math. de France **26** (1898), pp. 249–259.

[9]   V. D. Mazurov, *Doubly transitive permutation groups*, (Russian), Sibirsk.
      Mat. Zh. **31** (1990), pp. 102–104. (Translated in Siberian Math. J. **31**
      (1990), pp. 615–617.

[10] B. H. Neumann, *Groups covered by permutable subsets*, J. London Math.
      Soc. **29** (1954), pp. 236–248. (Also in: Selected works of B. H. Neumann
      and Hanna Neumann, Charles Babbage Research Centre, Winnipeg 1988,
      pp. 940–943.)

[11] Peter M. Neumann, *Problem 12.62*, in: Unsolved problems in group
      theory—the Kourovka notebook. Twelfth edition. (V. D. Mazurov and
      E. I. Khukhro, eds.), Novosibirsk (1992).

Finitely Presented Soluble Groups

John S. Wilson

Lecture 1. The Golod–Shafarevich Theorem for finitely presented groups

In the course of their work on the class field tower problem, Golod and Shafarevich [15] obtained an important result concerning presentations of finite p-groups. This result, as improved by Vinberg [24] and Gaschütz, asserts that if G is a finite p-group which can be generated by d and no fewer elements, then, in any presentation of G with d generators, the number r of relations satisfies $r > \frac{1}{4}d^2$. It follows easily that for a presentation of G with n generators and r relations the inequality

$$r > n - d + \tfrac{1}{4}d^2$$

must hold. As a by-product of this work, Golod [14] was able to give a construction demonstrating for the first time the existence of infinite finitely generated p-torsion groups. In [25] a result was proved which applies to all finitely presented groups and tightens the link between the above inequality and Golod's construction. The proof in [25] was given in the context of pro-p groups. In this lecture we shall give a direct proof, and afterwards discuss some extensions and some related results for Lie algebras and pro-p groups. We write $d(G)$ for the smallest number of elements that can generate a finitely generated group G, and G^{ab} for the abelianization of G; thus $G^{\mathrm{ab}} = G/G'$, where G' is the derived group of G.

Theorem 1.1. *Let G be a group having a presentation with n generators and r relations, and let $d = d(G^{\mathrm{ab}}) \geqslant 2$. Then either*

(i) *the inequality*

$$r \geqslant n - d + \tfrac{1}{4}d^2 \tag{1}$$

holds, or

(ii) *there exist a prime p and normal subgroups $N \triangleleft G$ such that G/N is a (finitely generated) infinite residually finite p-torsion group.*

Theorem 1.1 demonstrates that infinite finitely generated p-torsion groups exist in abundance. If G is a finite p-group then $d(G^{ab}) = d(G)$, and Theorem 1.1 applied for presentations on d generators gives $r \geqslant \frac{1}{4}d^2$; therefore we recover the Golod–Shafarevich Theorem in a slightly weakened form. The restriction $d \geqslant 2$ in the hypothesis is necessary, since the inequality (1) fails, for example, for the presentation of $\mathbb{Z}$ with one generator and no relations. Consideration of abelianizations yields the inequality $r \geqslant n - d$ for any finitely presented group, and for the cases $d = 0, 1$, the method of proof of Theorem 1.1 gives no more information than this.

Since finitely generated soluble torsion groups are finite, Theorem 1.1 has the following result as a special case:

Corollary 1.2. *If G is a soluble group having a presentation with n generators and r relators and if $d = d(G^{ab}) \geqslant 2$ then (1) holds.*

We have $n - d + \frac{1}{4}d^2 = n + \frac{1}{4}(d-2)^2 - 1$, and so (1) implies $r \geqslant n - 1$, with equality only if $d = 2$. Thus Theorem 1.1 has the following consequence:

Corollary 1.3. *Suppose that G is a group having a presentation with n generators and r relations. If $n - r \geqslant 1$ and if $d(G^{ab}) > 2$, then G has an infinite residually finite p-torsion image for some prime p.*

It was shown by B. Baumslag and S. Pride [3] that if G is a group having a presentation with n generators and r relations with $n - r \geqslant 2$, then G has a subgroup of finite index which maps onto a free group of rank 2.

Our third corollary gives information about the subgroups of finite index in finitely presented groups.

Corollary 1.4. *If G is a finitely presented group which has no infinite torsion quotient groups then there is a constant $\kappa > 0$ such that*

$$d(H^{ab}) \leqslant \kappa |G : H|^{\frac{1}{2}}$$

for all subgroups H of finite index.

The significance of this result lies in the index $\frac{1}{2}$; a similar inequality without this index (and with $\kappa = d(G)$) holds for any finitely generated group G, by the Reidermeister–Schreier theorem.

Proof. Suppose that G has a presentation with n generators and r relations. Set $h = |G : H|$ and $d = d(H^{ab})$. We only need to consider the case when $d \geqslant 2$. By the Reidermeister–Schreier theorem, H has a presentation with $nh - (h - 1)$ generators and rh relations, and it is easy to see that H can have no infinite torsion quotient groups. Therefore Theorem 1.1 applied for H gives

$$rh \geqslant nh - h + 1 - d + \tfrac{1}{4}d^2,$$

and hence

$$(\tfrac{1}{2}d - 1)^2 \leqslant (r - n + 1)h.$$

The result follows on taking square roots.

Notation

The following notation will be used in the proof of Theorem 1.1. Let F be the free group on a set X with n elements and let $\mathfrak{k}$ be a field. Write R for the group algebra $\mathfrak{k}F$ of F over $\mathfrak{k}$ and I for the augmentation ideal of R. As a left ideal, I is generated freely by the set $Y = \{x - 1 \mid x \in X\}$. Moreover, if $k \in \mathbb{N}$ and $u \in I^k$, then u can be written uniquely in the form $\sum_{y \in Y} v_y y$ with each v_y in I^{k-1}. We have $\bigcap_k I^k = \{0\}$; if $u \in I^k \backslash I^{k+1}$ write $\delta(u) = k$, and write $\delta(0) = \infty$.

Lemma 1.5. *(Vinberg, [24]) Let $S \subseteq I$. Define $S_k = \{s \in S \mid \delta(s) = k\}$, suppose that each S_k is finite and write $s_k := |S_k|$. Let J be the ideal of R generated by S.*
(a) *Set $c_k = \dim (R/(J + I^{k+1}))$ for $k \geqslant 0$. Then*

$$c_k - 1 \geqslant nc_{k-1} - \sum_{j=1}^{k} s_j c_{k-j} \qquad (2)$$

for $k \geqslant 1$.
(b) *Define the power series $\sigma_S(t) = \sum_{i \geqslant 1} s_i t^i$ and let t be an element of $[0, 1]$ such that $\sigma_S(t)$ converges. If the sequence $(\dim R/(J + I^k))$ of integers is eventually constant then $1 - nt + \sigma_S(t) > 0$.*

Proof. (a) Fix k, write $\bigcup_{j \leqslant k} S_j = \{z_1, \cdots, z_m\}$ where $\delta(z_{i+1}) \geqslant \delta(z_i)$ for each appropriate i, and set $k_i = \delta(z_i)$ for each i. Define

$$\overline{A} = \bigoplus_{i=1}^{m} R/(J + I^{k-k_i+1}),$$

and let $\overline{B}$ be the direct sum of n copies of $R/(J + I^k)$. Thus

$$\sum_{j=1}^{k} s_j c_{k-j} = \sum_{i=1}^{m} c_{k-k_i} := \dim \overline{A},$$

and we must prove that

$$c_k - 1 \geqslant nc_{k-1} - \dim \overline{A}.$$

This will follow if there is a exact sequence

$$\overline{A}\xrightarrow{\varphi}\overline{B}\xrightarrow{\psi}I/(J+I^{k+1})\longrightarrow 0,$$

since then

$$nc_{k-1} = \dim \overline{B} = \dim \operatorname{im}\psi + \dim \ker\psi$$

$$= c_k - 1 + \dim \operatorname{im}\varphi \leqslant c_k - 1 + \dim \overline{A}.$$

Set $Y = \{y_1, \cdots, y_n\}$ and write A, B respectively for the direct sums of m, n copies of R. Define

$$\Psi : B \to I \qquad \text{by} \qquad (t_1, \cdots t_n) \mapsto \sum t_i y_i$$

and

$$\Phi : A \to B \qquad \text{by} \qquad (v_1, \cdots, v_m) \mapsto (u_1, \cdots u_n),$$

where $u_1, \cdots, u_n$ are the elements of R uniquely determined by $\sum v_j z_j = \sum u_i y_i$. It is not difficult to verify that Φ, Ψ induce maps φ, ψ with the required properties.

(b) Set $\gamma(t) = \sum_0^\infty c_k t^k$. Multiplying the inequality (2) by t^k and summing over k, we have

$$\gamma(t) - (1-t)^{-1} \geqslant nt\gamma(t) - \sigma_S(t)\gamma(t),$$

and hence

$$\gamma(t)(1 - nt + \sigma_S(t)) \geqslant (1-t)^{-1}, \tag{3}$$

provided that $\gamma(t)$ and $\sigma_S(t)$ are convergent.

Now set $b_0 = c_0$, define $b_k = c_k - c_{k-1}$ for $k \geqslant 1$ and write $\beta(t) = \sum_0^\infty b_k t^k$. Since the sequence (c_k) is eventually constant, $\beta(t)$ is a polynomial. We have $c_k = \sum_{i=0}^k b_i$, and so $\gamma(t) = \beta(t)/(1-t)$. Therefore $\gamma(t)$ converges in $[0,1)$, and if $\sigma_S(t)$ is convergent at $t \in [0,1)$ we conclude from (3) that $1 - nt + \sigma_S(t) > 0$. If $\sigma_S(1)$ is convergent then $\sigma_S(1) = \lim_{t \to 1-} \sigma_S(t)$ by Abel's theorem on power series, and so $1 - n + \sigma_S(1) > 0$ as required. This completes the proof of Lemma 1.5.

Let S be a subset of R and let J be the ideal generated by S. We define the *closed ideal* $\overline{J}$ generated by S to be $\bigcap_{k \geqslant 1}(J + I^k)$. Clearly we have $\overline{J} + I^k = J + I^k$ for every integer k; and $R/\overline{J}$ is finite-dimensional if and only if the sequence $(\dim(R/(J+I^{k+1})))$ is eventually constant.

Lemma 1.6. *As in Lemma 1.5 let $S \subseteq I$, suppose $s_k = |\{s \in S \mid \delta(s) = k\}|$ finite for each k, and write $\sigma_S(t) = \sum_{i \geqslant 1} s_i t^i$. Let $E = \{e_i \mid i \in \mathbb{N}\} \subseteq I$ and let $t \in [0,1)$. If $\sigma_S(t)$ converges and $1 - nt + \sigma_S(t) < 0$, then there is a set S^+ with $S \subseteq S^+ \subseteq I$ such that*

(i) $|\{s \in S^+ \mid \delta(s) = k\}|$ *is finite for each* k,
(ii) $1 - nt + \sigma_{S^+}(t) < 0$, *so that* $\dim R/K$ *is infinite, where* K *is the closed ideal generated by* S^+, *and*
(iii) *every element of* E *has nilpotent image in* R/K.

For example, we could take $E = \{u \mid 1 + u \in F\}$, or, if $\mathfrak{k}$ is countable, we could take $E = I$; in the latter case we would conclude that the image of I in R/K is a nil ideal.

Proof. We find a positive integer q large enough to ensure that

$$1 - nt + \sigma_S(t) + q^{-1}(1 - t)^{-1} < 0.$$

Thus

$$1 - nt + \sigma_S(t) + \sum_1^\infty t^{qi} < 0,$$

since

$$t^{qi} < q^{-1}t^{q(i-1)}(1 + \cdots + t^{q-1})$$

for each $i \geq 1$. Let $S^+ = S \cup \{e_i^{qi} \mid i \in \mathbf{N}\}$. Clearly we have $\delta(e_i^{qi}) \geq qi$ for each i. It follows that

$$\sigma_{S^+}(t) = \sigma_S(t) + \sum_1^\infty t^{\delta(e_i^{qi})} \leq \sigma_S(t) + \sum_1^\infty t^{qi},$$

and

$$1 - nt + \sigma_{S^+}(t) < 0.$$

This completes the proof of Lemma 1.6.

We now specialize to the case when $\mathfrak{k} = \mathbf{F}_p$ and, in the notation of Lemma 1.6, choose $E = \{u \mid 1 + u \in F\}$. If $u \in E$ and $u^m \in S^+$ then for $p^j \geq m$ we have

$$(1 + u)^{p^j} = 1 + u^{p^j} \equiv 1 \pmod{K}.$$

Therefore the image $\overline{G}$ of F in R/K is a p-torsion group. Each of the rings $R/(K + I^k)$ is finite, and the kernels of the maps from $\overline{G}$ to the groups of units of these rings have trivial intersection, so that $\overline{G}$ is residually finite. Finally, $\overline{G}$ spans R/K as a vector space over $\mathbf{F}_p$, so that if $\dim R/K$ is infinite then so is $\overline{G}$.

Lemma 1.7. *If G has a presentation $N \rightarrowtail F \twoheadrightarrow G$ with n generators and r relations, and if $d = d(G/G'G^p)$, then G has such a presentation with just $n - d$ relations not in $F'F^p$.*

Proof. The group $F/F'F^p$ may be regarded as an n-dimensional $\mathbf{F}_p$-vector space, and since $F/N \cong G$ we have $F/F'F^pN \cong G/G'G^p$ so that the image of N in $F/F'F^p$ has dimension $n - d$. We take relators $w_1, \cdots, w_{n-d}$ from N mapping to a basis of this subspace, and multiply each remaining relator by an element of $\langle w_1, \cdots, w_{n-r} \rangle$ such that the product is in $F'F^p$.

Proof of Theorem 1.1 We choose p with $d = d(G/G'G^p)$, set $\mathfrak{k} = \mathbb{F}_p$, and take a presentation $N \rightarrowtail F \twoheadrightarrow G$ with the property given in Lemma 1.7; say $N = \langle w_1^F, \cdots, w_r^F \rangle$. Suppose that (i) does not hold. Then $r < n - d + \frac{1}{4}d^2$, so that $r_1 < \frac{1}{4}d^2$ where $r_1 = r - (n - d)$. Hence, since $d \geqslant 2$, we can find $t \in (0, 1)$ with $1 - dt + r_1t^2 < 0$, i.e. with

$$1 - nt + ((n - d)t + r_1t^2) < 0.$$

Now r_1 is the number of relators in $F'F^p$, and it is well known and easy to check that $F'F^p = F \cap (1 + I^2)$. Thus we certainly have

$$1 - nt + \sigma_S(t) < 0$$

where $S = \{w_1 - 1, \cdots, w_r - 1\}$.

Choose $E = \{u \mid 1 + u \in F\}$, and construct S^+, K and $\overline{G}$ as above. The group $\overline{G}$ is an image of G which is an infinite residually finite p-group, and Theorem 1.1 follows.

Some related results

A result analogous to Theorem 1.1 holds for presentations (in the category of pro-p groups and continuous homomorphisms) of pro-p groups.

Theorem 1.8. *Let G be a pro-p group having a presentation with n generators and r relations. If $d = d(G) \geqslant 2$, then either the inequality*

$$r \geqslant n - d + \tfrac{1}{4}d^2 \tag{1}$$

appearing in Theorem 1.1 holds, or each finitely generated dense abstract subgroup of G has an infinite residually finite p-torsion quotient group.

Theorem 1.1 can be deduced from the above result by considering pro-p completions. Theorem 1.8 has the following consequence, which has no direct counterpart for abstract groups:

Corollary 1.9. *If G is a finitely generated soluble pro-p group and N is a closed normal subgroup such that G/N is isomorphic to the group of p-adic integers, then N is a finitely generated pro-p group.*

For more details about the above results, see [25].

Next we consider Lie algebras. What follows is work of my research student, Jeremy King. Before describing King's result we need a definition. An element y of a Lie algebra L is called ad-nilpotent if the map $\mathrm{ad}(y) : l \mapsto ly$ is nilpotent, and L is said to be weakly ad-nilpotent with respect to a generating set Y if every commutator of weight at least 1 in the elements of Y is ad-nilpotent. It follows from work in Gruenberg [18] that a finitely generated soluble weakly ad-nilpotent Lie algebra is nilpotent.

Theorem 1.10. *(King, [19]) Let L be a Lie algebra having a presentation with n generators and r relations and let $d = \dim{}_{\ell}(L/[L, L]) \geqslant 2$. Then either the inequality (1) holds or L has a quotient which is infinite-dimensional, residually nilpotent and weakly ad-nilpotent. In particular, (1) holds if L is soluble.*

As in the proof of Theorem 1.1, let F be the free group on a finite set X, let $Y = \{x - 1 \mid x \in X\}$, and let $R = \ell F$. The subalgebra A of R generated by Y is the free associative ℓ-algebra on Y. Theorems corresponding to Lemmas 1.5, 1.6 hold for A, and in fact they follow easily from Lemmas 1.5, 1.6. The Lie algebra $L(Y)$ in A generated by Y is the free Lie algebra on Y. Thus some of the proof of Theorem 1.10 can be modelled on the proof of Theorem 1.1. However the easy argument used in the proof of Theorem 1.1 to show that the group $\overline{G}$ is infinite is not available for Lie algebras. In the proof of the final assertion, the result of Gruenberg plays an analogous role to the finiteness of finitely generated soluble torsion groups in Corollary 1.2.

Finally, we describe some results which depend on nilpotence criteria of Zelmanov [28], according to which every finitely generated residually finite p-torsion group satisfying a law is finite and every finitely generated weakly ad-nilpotent Lie algebra satisfying a polynomial identity is nilpotent. Combining these assertions with Theorems 1.1 and 1.10, we conclude that the inequality (1) must hold for finitely presented groups with laws, and for Lie algebras with polynomial identities.

A closer analysis yields the following result.

Theorem 1.11. *(Wilson and Zelmanov, [27]) If G is a finitely presented group having a presentation for which the inequality (1) does not hold, then, for some prime p, the pro-p completion of G has a free abstract subgroup of rank 2.*

This suggests the following problem, which seems likely to be difficult.

Problem. *Suppose that G is a group having a presentation for which (1) does not hold. Must G have a free subgroup of rank 2?*

Lecture 2. Criteria for finitely presentability and a case study

Since the class of finitely presented groups is extension-closed and contains the cyclic groups, all polycyclic groups are finitely presented; and, in particular, all finitely generated nilpotent groups are finitely presented. However, the significance of finite presentability for the structure of more general soluble groups seems hard to determine. Even for metabelian groups the matter is quite delicate. While there are many finitely generated metabelian groups which are not finitely presented, it was shown by Baumslag [4] and Remeslennikov [21] in 1973 that every finitely generated metabelian group can be embedded in a finitely presented metabelian group. For example, the wreath product $\mathbb{Z} \operatorname{wr} \mathbb{Z}$, which is isomorphic to

$$\langle a, x \mid [a, a^w] = 1 \text{ for all words } w \rangle,$$

can be embedded in the metabelian group

$$\langle a, x, y \mid a^y = aa^x, [x, y] = 1, [a, a^y] = 1 \rangle.$$

In 1980, Bieri and Strebel [10] succeeded in finding a geometrical interpretation of the the property of being finitely presented for metabelian groups. We shall discuss this interpretation later in the lecture.

Further examples of finitely presented soluble groups are provided by certain groups of matrices over fields (especially arithmetic and S-arithmetic groups). Every such group is virtually nilpotent-by-abelian, i.e., it has a nilpotent-by-abelian normal subgroup of finite index. The group G with presentation

$$\langle x, y, a \mid 1 = [x, y, y] = [x, y, x] = [a^x, a^y], a^{x^2} = a^x a^y, a^{[x,y]} = a^2 \rangle$$

is an example of a finitely presented group which is soluble but not virtually nilpotent-by-abelian: Robinson and Strebel showed in [22] that the normal subgroup A generated by a is abelian and is the largest nilpotent normal subgroup, and clearly the quotient group G/A is a free nilpotent group of class 2 on 2 generators.

I know of no examples of finitely presented soluble groups, or, indeed, of finitely presented residually finite groups having no free subgroups of rank 2, which are not virtually nilpotent-by-(nilpotent of class at most 2).

Necessary conditions for finite presentability

We shall now recall briefly some necessary conditions for a soluble group to be finitely presented, and then we shall make a case study using the last of these, which is the condition given by Corollary 1.4.

(1) The homological criterion

It is well known that the finite presentability of a group G implies that its multiplicator $M(G) = H_2(G, \mathbb{Z})$ is finitely generated. Using the fact that finitely generated modules for polycyclic groups are noetherian, one can easily strengthen this to the following statement:

Lemma 2.1. *If G is finitely presented and K is a normal subgroup such that G/K is polycyclic, then the multiplicator $M(K)$ is finitely generated, regarded as a G/K-module.*

(2) The HNN criterion

In [9], Bieri and Strebel proved the following result:

Lemma 2.2. *Suppose that G is finitely presented and has no free subgroup of rank 2. Then for each N such that $G/N \cong \mathbb{Z}$, the group G is an ascending HNN extension over a finitely generated base group contained in N.*

This result allowed Bieri and Strebel to give a rather precise description of finitely presented nilpotent-by-(infinite cyclic) groups:

Theorem 2.3. *Suppose that N is a nilpotent normal subgroup of a finitely generated group G and that G/N is an infinite cyclic group. The following conditions are equivalent:*
  (i) *G is finitely presented;*
 (ii) *G/N' is finitely presented;*
(iii) *G is an ascending HNN extension over a finitely generated subgroup of N;*
 (iv) *there is an element $t \in G$ such that $G = \langle N, t \rangle$ and such that the characteristic polynomial of the endomorphism of $N^{\mathrm{ab}} \otimes \mathbb{Q}$ induced by t has all coefficients in $\mathbb{Z}$.*

(3) The sphere criterion

Let Q be a finitely generated abelian group and let $V(Q) = \mathrm{Hom}\,(Q, \mathbb{R})$. Thus $V(Q)$ is a real vector space of dimension n, where $n = \dim_{\mathbb{Q}}(Q \otimes \mathbb{Q})$. Fix a norm $\| \ \|$ in $V(Q)$ and write $S(Q)$ for the $(n-1)$-sphere $\{ v \in V(Q) \mid \|v\| = 1 \}$. For $v \in S(Q)$ let Q_v be the submonoid $\{ q \in Q \mid v(q) \geqslant 0 \}$. In [10], Bieri and Strebel associated with each $\mathbb{Z}Q$-module M the set

$$\Sigma_M = \{ v \in S(Q) \mid M \text{ is a finitely generated module for the ring } \mathbb{Z}Q_v \}.$$

It transpires that Σ_M is an open subset of $S(Q)$.

Now let G be a finitely generated group, write $K = G'$ and set $Q_G = G^{\mathrm{ab}}$ and $M_G = K^{\mathrm{ab}}$. So M_G can be regarded as a Q_G-module and the set Σ_{M_G} may be defined; it is an open subset of the sphere $S(Q_G)$ and is an invariant of G. Consider the following condition:

(SC) $\Sigma_{M_G} \cup -\Sigma_{M_G} = S(Q)$.

Bieri and Strebel proved the following remarkable result:

Theorem 2.3. *If G is a finitely presented group having no free subgroups of rank 2 then G satisfies (SC). Conversely, if G is a finitely generated metabelian group satisfying (SC) then G is finitely presented.*

Thus (SC) provides a geometrical characterization of finitely presented metabelian groups. But the utility of the invariant Σ_{M_G} and the condition (SC) is not restricted to metabelian groups. Abels has shown in [1] that the finitely presented S-arithmetic groups G can be characterized using the set Σ_{M_G}: in particular, his results give the following theorem.

Theorem 2.4. *If G is a finitely generated S-arithmetic group and G is nilpotent-by-abelian then G is finitely presented if and only if* (i) (SC) *holds and* (ii) $H_2(G', \mathbb{Z})$ *is a finitely generated $\mathbb{Z}G^{\mathrm{ab}}$-module.*

Of course, the necessity of (i) and (ii) above follows from Theorem 2.3 and Lemma 2.1. Theorem 2.3 also implies that if G is any group satisfying (SC), in particular, if G is finitely presented and has no free subgroups of rank 2, then G/G'' is finitely presented. A result of Bieri, Neumann and Strebel [8] whose proof relies on the property (SC) is the following.

Theorem 2.5. *If G is a finitely presented group having no free subgroups of rank 2, then for all $L \geqslant G'$ with $G/L \cong \mathbb{Z}^2$ there is a a finitely generated subgroup $N \geqslant L$ with $G/N \cong \mathbb{Z}$.*

The invariant Σ_{M_G} encodes important structural information for an arbitrary finitely generated group G and it can be used, for example, in the study of the automorphism group of G. However, when investigating finitely presented soluble groups which are not nilpotent-by-abelian, one encounters modules for non-abelian groups, and so far all attempts to find a natural extension of the definition of Σ_M when M is a module for a non-abelian group have proved elusive.

Moreover, if G does not have $\mathbb{Z}^2$ as a homomorphic image, then the sphere on which Σ_M lies contains at most two points, and so facts about open subsets are of small utility. We note, for instance, that Theorem 2.5, whose proof depends on properties of Σ_{M_G}, is vacuous in this case. If there is a subgroup G_0 of finite index in G which maps surjectively to $\mathbb{Z}^2$, then for many purposes one can replace G by G_0. So the groups for which methods using invariants on spheres are ineffective are those having no subgroup of finite index which maps surjectively to $\mathbb{Z}^2$. We shall return to these groups at the end of the lecture.

(4) The growth criterion

We saw in Lecture 1 that finitely presented soluble groups G satisfy the following condition.

(GC) There is a constant $\kappa > 0$ such that $d(H^{\mathrm{ab}}) \leqslant \kappa |G : H|^{\frac{1}{2}}$ for all subgroups H of finite index in G.

We shall illustrate how (GC) can be used to give information about finitely presented soluble groups. First we establish a module-theoretic consequence of (GC).

Suppose that $G = M \rtimes Q$, where M is an elementary abelian p-group for some prime p. Let H be a subgroup of finite index in Q and write $K = [M, H]$. Thus $K \triangleleft MH$ and we have $MH/K = M/K \times HK/K$, so that

$$d((MH)^{\mathrm{ab}}) \geqslant d(M/K) = \dim_{\mathbb{F}_p}(M/[M, H]).$$

Combining this with (GC) we obtain useful information about the structure of M; and this information is sometimes available for extensions which are not split (see [26] or [16]):

Lemma 2.6. *Suppose that G is a group which satisfies (GC) and which is an extension $M \rightarrowtail G \twoheadrightarrow Q$, where M is an elementary abelian p-group. If either G splits over M or Q is a virtually torsion-free soluble minimax group, then there is a constant $k > 0$ such that $\dim_{\mathbb{F}_p}(M/[M, H]) \leqslant k|Q : H|^{\frac{1}{2}}$ for each subgroup H of finite index in Q.*

We shall now test the power of the criterion (GC) on some specific groups. Suppose that G satisfies (GC) and has the following structure:

(a) $G = M \rtimes Q$, with M abelian and $pM = 0$;
(b) $Q = A \rtimes T$, with $A \neq 1$;
(c) M, regarded as an $\mathbb{F}_p A$-module, is free of finite rank $r > 0$.

We take a subgroup D of finite index in A such that $D \triangleleft Q$, and write $V = (D - 1)\mathbb{F}_p A$. We have

$$M/MV \cong (\mathbb{F}_p A/V)^r \cong (\mathbb{F}_p(A/D))^r,$$

so that $\dim_{\mathbb{F}_p}(M/MV) = r|A/D|$. Moreover $MV = [M, D]$ is normal in G. Let $H_D = C_Q(M/MV)$, so that $[M, H_D] = MV$, and let $T_D = C_T(M/MV)$. Thus

$$|Q : H_D| \leqslant |Q : DT_D| = |A/D||T/T_D|.$$

Writing k for the constant given by Lemma 2.6, we conclude that

$$r|A/D| \leqslant k|A/D|^{\frac{1}{2}}|T/T_D|^{\frac{1}{2}},$$

and so

$$|A/D| \leqslant k_1|T/T_D| \qquad \text{where } k_1 = k^2/r^2. \tag{4}$$

Let $a \in A$ and $u \in T_D$, and choose a free generator m for the $\mathbb{F}_p A$-module M. Modulo MV we have

$$m[a, u] = ma^{-1}u^{-1}au \equiv ma^{-1}a = m,$$

so that $[a, u] \in D$, since the image of $m\mathbb{F}_p A$ in M/MV is a free $\mathbb{F}_p(A/D)$-module. Thus $T_D \leqslant C_T(A/D)$. Because of (4), a bound independent of D for the order of $C_T(A/D)/T_D$ would clearly have important implications for the structure of Q, but of course in general no such bound will exist. In order to make further progress, we add some extra conditions:

(d) A is a torsion-free abelian group of finite rank;

(e) $T = \langle z \rangle$ is infinite cyclic;

(f) A is cyclic as a $\mathbb{Z}T$-module, and the characteristic polynomial f of the endomorphism of $A \otimes \mathbb{Q}$ induced by z has integer coefficients;

(g) every non-trivial subgroup of T acts irreducibly on $A \otimes \mathbb{Q}$.

(For example, let $A = \mathbb{Z}[1/p]$ and $f(t) = t - s$ with $s \in \mathbb{Z}\backslash\{0\}$.) The group G is clearly soluble but not virtually metanilpotent. We suspect that most, if not all, finitely presented soluble groups are virtually metanilpotent, and so we would like the hypothesis that G satisfies (GC) to lead to a contradiction. We shall make careful estimates of indices for a range of choices of D, and at this stage number-theoretic considerations will turn out to be relevant. For $n \in \mathbf{N}$ let $\varepsilon(n)$ be the exponent of $\mathrm{Aut}\,(\mathbb{Z}/n\mathbb{Z})$ and $\delta(n)$ the number of distinct primes which divide n. The following result is elementary.

Lemma 2.7. *Let $S \subseteq \mathbf{N}$. If the set $\{n/\varepsilon(n) \mid n \in S\}$ is bounded above then so is the set $\{\delta(n) \mid n \in S\}$.*

Take $l, m \in \mathbf{N}$, and define $D_{lm} = \langle a^z a^{-p^l}, a^{z^m} a^{-p} \mid a \in A \rangle$. We have

$$A/D_{lm} \cong \mathbb{Z}\langle t \rangle / (f(t), t - p^l, t^m - p) \cong \mathbb{Z}/(f(p^l), p^{lm} - p),$$

and this has order $a_{lm} = (f(p^l), p^{lm} - p)$. It can be shown that there is an upper bound k_2 for all of the indices $|C_T(A/D_{lm})/T_{D_{lm}}|$. Therefore from (4) we have for all l, m

$$a_{lm} \leqslant k_1 k_2 \,|T/C_T(A/D_{lm})| \leqslant k_1 k_2 \,\varepsilon(a_{lm}),$$

so that the set $\{a_{lm}/\varepsilon(a_{lm}) \mid l, m \in \mathbf{N}\}$ is bounded. It follows from Lemma 2.7 that $\{\delta(n) \mid n \in S\}$ is bounded, where $S = \{(f(p^l), p^{lm} - p) \mid l, m \in \mathbf{N}\}$. We note that by condition (g) the polynomial f is irreducible. The following easy result gives a different interpretation of the above property.

Lemma 2.8. *Let $n \in \mathbb{Z}$, let $f(t)$ be an irreducible monic polynomial in $\mathbb{Z}[t]$, and let c be a root of f in a splitting field. The following are equivalent:*

(i) *for all integers $b \in \mathbf{N}$ there are integers $l, m \in \mathbf{N}$ such $(f(n^l), n^{lm} - n)$ is divisible by at least b distinct primes;*

(ii) *the ring R generated by c has infinitely many maximal ideals I such that the images of c, n in R/I are non-zero and generate the same multiplicative group.*

Thus we have our contradiction if a root of f has the property expressed in condition (ii) above, with $n = p$. In the case when $f(t) = t - s$ with

$s \in \mathbb{Z}\backslash\{0\}$, we need to know whether there are infinitely many primes q such that the images of s, p in $\mathbb{Z}/q\mathbb{Z}$ generate the same multiplicative group.

It is natural to conjecture that if c, d are non-zero algebraic numbers which are not roots of 1, then the ring R generated by c, d has infinitely many maximal ideals I such that the images of c, d in R/I are non-zero and generate the same multiplicative group. Unfortunately this seems hard to establish. Even in the case when $c, d \in \mathbb{Q}$ the matter is non-trivial. For the case when $c, d \in \mathbb{Q}$ and $cd > 0$, there is a short argument (based on the proof of Lemma 4.2 of [26]) using Dirichlet's theorem on primes in arithmetic progression and quadratic reciprocity. A proof for the case when $c, d \in \mathbb{Q}$ and $cd < 0$ was given by Schinzel and Wójcik [23]; it occupies six journal pages. A few more partial results are known in the special case when one of the algebraic numbers is a rational prime.

Our attempts above to arrive at a contradiction seem therefore to have met with rather qualified success. For the specific examples that we have discussed, there may be other ways of obtaining a contradiction. However the point about the above calculations is that they can be made to apply in a more general context. We have already noted that if G is a group none of whose subgroups of finite index has $\mathbb{Z}^2$ as a homomorphic image, then the information available from the geometric methods of Bieri and Strebel [10] is of limited use. In [26] the following result was proved:

Theorem 2.9. *Let G be a finitely presented soluble group and suppose that no subgroup of finite index in G has $\mathbb{Z}^2$ as a homomorphic image. Then G has a nilpotent-by-cyclic subgroup of finite index.*

The finitely presented nilpotent-by-cyclic groups are reasonably well understood, by Theorem 2.3. In the proof of Theorem 2.9, deep results in the representation theory of soluble minimax groups are used to reduce to some groups resembling those discussed above, and these are shown not to satisfy (GC), using calculations like those above, coupled with partial solutions to the conjecture on pairs of algebraic numbers.

Lecture 3. Hilbert–Serre dimension

The condition (GC) discussed in the last lecture gives an inequality for each subgroup of finite index, and we have seen how some of the inequalities can be combined to elucidate the structure of finitely presented groups in a context where other techniques (such as those associated with spheres) are unavailable. The arguments led rapidly to detailed calculations and an open problem in number theory; and so it is clear that we can only expect (GC) to give decisive results in very special circumstances.

In this lecture I will describe a rather less *ad hoc* way of reassembling information from the inequalities given by (GC). Though more information is lost in this procedure, it seems natural and it can be carried out for a large class of finitely presented groups. It involves ideas similar to those in the classical dimension theory of commutative algebra.

Let R be a ring. We fix an additive function λ on R-modules of finite composition length and a right ideal I of R with $\lambda(R/I)$ finite.

Let M be a right R-module. We define the *Hilbert–Serre dimension of M with respect to I*, denoted $\mathrm{hs}\,(M, I)$, as follows:

$$\mathrm{hs}\,(M, I) = \infty \quad \text{if } \lambda(M/MI^n) \text{ is infinite for some } n \in \mathbf{N}$$

and

$$\mathrm{hs}\,(M, I) = \limsup_{n \to \infty} \frac{\log \lambda(M/MI^n)}{\log n} \quad \text{otherwise.}$$

Thus if $\mathrm{hs}\,(M, I) = d$ is finite, then for all $\varepsilon > 0$ one has

$$\lambda(M/MI^n) \leqslant n^{d+\varepsilon} \quad \text{for } n \text{ large enough;}$$

and d is the smallest number with this property. Equivalently, $\mathrm{hs}\,(M, I)$ may be defined to be the Gelfand–Kirillov dimension of the associated graded module $\bigoplus_{n \geqslant 0}(MI^n/MI^{n+1})$.

It is clear that $\mathrm{hs}\,(N, I) \leqslant \mathrm{hs}\,(M, I)$ if N is an R-module image of M, and that $\mathrm{hs}\,(M_1 \oplus M_2, I) = \max\{\mathrm{hs}\,(M_1, I), \mathrm{hs}\,(M_2, I)\}$ for any R-modules M_1, M_2. It follows that if M is finitely generated then $\mathrm{hs}\,(M, I) \leqslant \mathrm{hs}\,(R, I)$. It is very easy to prove

Lemma 3.1. *Let I be an ideal of R with $\lambda(R/I)$ finite, and suppose that $M_1 \rightarrowtail M \twoheadrightarrow M_2$ is a short exact sequence of R-modules. Then*
(a) $\mathrm{hs}\,(M, I) \leqslant \max\{\mathrm{hs}\,(M_1, I), \mathrm{hs}\,(M_2, I)\}$.
(b) *If in addition I has the strong Artin–Rees property and M, M_1 are finitely generated, then $\mathrm{hs}\,(M, I) = \max\{\mathrm{hs}\,(M_1, I), \mathrm{hs}\,(M_2, I)\}$.*

We recall that an ideal I of a ring R is said to have the *strong Artin–Rees property* if for each finitely generated R-module A and finitely generated submodule B there is an $n_0 \in \mathbf{N}$ such that $AI^n \cap B = (AI^{n_0} \cap B)I^{n-n_0}$ for all $n \geqslant n_0$.

In the case of interest to us, R is the group algebra $\mathbb{F}_p Q$, where Q is a polycyclic group, and $\lambda(N) = \dim_{\mathbb{F}_p}(N)$ for each finite-dimensional R-module N. Most of the results we describe below remain true if Q is a finitely generated soluble minimax group which is virtually torsion-free (with the torsion-free rank of Q playing the role of the Hirsch number), with slightly different proofs. For the remainder of this section, the prime p and the polycyclic group Q will remain fixed and R will denote $\mathbb{F}_p Q$.

Lemma 3.2. *If M is a finitely generated R-module and I is an ideal of finite codimension in R, then* $\mathrm{hs}\,(M, I) \leqslant h(Q)$.

Proof. Write $r = h(Q)$. By the remarks preceding Lemma 3.1, it is sufficient to prove the lemma in the case when $M = R$.

Let $Q_1 = Q \cap (1 + I)$. Thus Q_1 is the centralizer of R/I in Q and $|Q : Q_1|$ is finite. Since Q_1 is polycyclic, there is a constant c such that $|Q_1/Q_1^k| \leqslant ck^r$ for all $k \in \mathbb{N}$, where r is the Hirsch number of Q.

Let $n \in \mathbb{N}$ and let $j \in \mathbb{N}$ satisfy $p^{j-1} < n \leqslant p^j$. Since R has characteristic p we have

$$Q_1^{p^j} \leqslant 1 + I^{p^j} \leqslant 1 + I^n,$$

so that the image of Q in R/I^n has order bounded by

$$|Q/Q_1^{p^j}| \leqslant c|Q : Q_1|(p^j)^r \leqslant c'n^r \quad \text{where} \quad c' = c|Q : Q_1|p^r.$$

Since $\dim_{\mathbb{F}_p}(R/I^n)$ is at most the dimension of the group ring of the image of Q in R/I^n, we conclude that

$$\dim_{\mathbb{F}_p}(R/I^n) \leqslant c'n^r.$$

The result follows.

It does not seem at all clear in this degree of generality whether or not the statement $\mathrm{hs}\,(M, I) \in \mathbb{N} \cup \{0\}$ must hold for all finitely generated modules M. However the following result, which is an easy extension of the Hilbert–Serre theorem of commutative algebra, shows that this is so if Q is nilpotent and I is the augmentation ideal of R. It can be proved by imitating the standard proof of the Hilbert–Serre theorem (as described, for example, in [2], Chapter 11), and using the fact that the augmentation ideal of the group algebra of a finitely generated nilpotent group has the strong Artin–Rees property (by Theorem 5 of [20]) and has a polycentral generating set.

Lemma 3.3. *Suppose that $R = \mathbb{F}_p Q$, where Q is a finitely generated nilpotent group, and I is the augmentation ideal of R, and let M be a finitely generated R-module. Then the Poincaré series*

$$P(M, I, t) = \sum_{k \geqslant 0} \dim_{\mathbb{F}_p}(MI^k/MI^{k+1})t^k$$

may be written in the form

$$\frac{f(t)}{\Pi_{i=1}^s (1 - t^{l_i})},$$

where $f(t) \in \mathbb{Z}[t]$ and each l_i is a positive integer. Moreover $\mathrm{hs}\,(M, I)$ is the degree of the pole of $P(M, I, t)$ at 1.

We now return to the general case in which Q is polycyclic and I has finite codimension in R. We need information about modules M which is stable under passage from Q to subgroups of finite index. We define the (*total*) *Hilbert–Serre dimension* of an R-module M to be the supremum of the numbers hs (M, I) as I runs through all ideals of R of finite codimension, and we denote this by hs (M). Thus hs $(M) \leqslant h(Q)$ if M is finitely generated, by Lemma 3.2. If Q_1 is a subgroup of Q we write hs (M_{Q_1}) for the total Hilbert–Serre dimension of M regarded as an $\mathbb{F}_p Q_1$-module. The next result gives some information about the relationship between hs (M_Q) and hs (M_{Q_1}).

Lemma 3.4. *Let Q_1 be a subgroup of Q and write $d =$ hs (M_Q) and $d_1 =$ hs (M_{Q_1}). The following assertions hold:*
(a) $d \leqslant d_1$;
(b) *if $|Q : Q_1|$ is finite then $d = d_1$;*
(c) *if Q_1 is a subnormal subgroup of Q and d_1 is finite, then $d = d_1$.*

If Q is nilpotent, the function hs (M) defined above coincides with a dimension function that is better known. We recall that the *Krull dimension* $k(M)$ of a non-zero module M is the greatest ordinal d such that M has a descending chain of submodules of order-type ω^d. We have

Lemma 3.5. *(Groves and Wilson, [16]) If Q is nilpotent, then hs $(M) = k(M)$ for every non-zero finitely generated $\mathbb{F}_p Q$-module M.*

The relevance of the above results to the condition (GC) comes from the next result.

Lemma 3.6. *Let M be an R-module with the property that some extension G of M by Q satisfies (GC), and write C for the centralizer of M in Q. Then hs $(M) \leqslant \frac{1}{2} h(Q/C)$.*

Proof. We fix an ideal I of finite codimension in R. Set $Q_1 = Q \cap (1 + I)$. Thus $|Q : Q_1|$ is finite. Write $b = |Q : CQ_1|$ and $r = h(Q/C)$, and let c be a constant satisfying

$$|CQ_1/C : (CQ_1/C)^k| \leqslant ck^r$$

for all $k \in \mathbf{N}$.

Let $n \in \mathbf{N}$ and define $j \in \mathbf{N}$ by $p^{j-1} < n \leqslant p^j$. We have $Q_1^{p^j} \leqslant 1 + I^{p^j}$, and so $[M, CQ_1^{p^j}] \leqslant MI^{p^j}$. Thus we have

$$\dim_{\mathbb{F}_p}(M/MI^{p^j}) \leqslant k|Q : CQ_1^{p^j}|^{\frac{1}{2}} \leqslant kb^{\frac{1}{2}}|CQ_1/C : (CQ_1/C)^{p^j}|^{\frac{1}{2}},$$

where k is the constant given by Lemma 2.6, and hence

$$\dim_{\mathbb{F}_p}(M/MI^n) \leqslant \dim_{\mathbb{F}_p}(M/MI^{p^j}) \leqslant kb^{\frac{1}{2}}cp^{j(\frac{1}{2}r)} \leqslant kb^{\frac{1}{2}}c(pn)^{\frac{1}{2}r}.$$

We conclude that hs $(M, I) \leqslant \frac{1}{2} h(Q/C)$, and since this holds for all ideals I of finite codimension in R the result follows.

Now we describe three applications of these ideas to finitely presented groups. The results which follow are all deeper and more difficult than the results discussed so far in this lecture.

1. Extensions by abelian groups

Let Q be an abelian group with $n = h(Q)$, and let the $(n-1)$-sphere $S(Q)$ be defined as in Lecture 2.

Definitions. . A rational hemisphere is a subset of $S(Q)$ of the form

$$\{v \mid v(q) \geqslant 0\}$$

with $q \in Q$. A rational spherical polyhedron (RSP) is a finite union of sets, each of which is an intersection of finitely many rational hemispheres. The dimension of an RSP is the dimension of a neighbourhood of an interior point.

The next two results explain our interest in rational spherical polyhedra.

Theorem 3.7. *(Bieri and Groves, [7]) If M is a finitely generated $\mathbb{Z}Q$-module then the complement Σ_M^c of Σ_M in $S(Q)$ is a RSP.*

Lemma 3.8. *(Groves and Wilson, [16]) Suppose that M is a finitely generated $\mathbb{Z}Q$-module and that $k(M) = m$. Then*
(a) $\dim \Sigma_M^c \leqslant m - 1$, and
(b) *there is a subgroup Q_1 of Q with $h(Q_1) = m$ such that M is finitely generated as a $\mathbb{Z}Q_1$-module.*

The deduction of assertion (b) above from the previous results is geometrical in character. For each subgroup Q_1 with $h(Q_1) = m$ define

$$S(Q, Q_1) = \{v \in S(Q) \mid v(Q_1) = 0\}.$$

Thus $S(Q, Q_1)$ is a subsphere of dimension $n - m - 1$. Moreover Bieri and Strebel showed in [11] that if M is a finitely generated $\mathbb{Z}Q$-module then M is finitely generated for $\mathbb{Z}Q_1$ if and only if $S(Q, Q_1) \cap \Sigma_M^c = \emptyset$. So our task is to find a subsphere of the form $S(Q, Q_1)$ which is disjoint from a RSP of dimension at most m, and it is fairly plausible on geometrical grounds that this can be done.

It follows immediately from the above four results that if some extension of the $\mathbb{F}_p Q$-module M by Q satisfies (GC), then Σ_M^c is a RSP of dimension at most $\frac{1}{2} h(Q) - 1$, and moreover M is finitely generated as an $\mathbb{F}_p Q_1$-module for some subgroup Q_1 with $h(Q_1) \leqslant \frac{1}{2} h(Q)$. This leads to the following result on finitely presented nilpotent-by-abelian groups.

Corollary 3.9. *(Groves and Wilson, [16]) Let G be a finitely generated nilpotent-by-abelian group satisfying* (GC). *Then there is a finitely generated subgroup G_1 satisfying $G' \leqslant G_1 \leqslant G$ such that*

$$h(G/G_1) \geqslant \begin{cases} \frac{1}{2}h(G^{\mathrm{ab}}) & \text{if } G' \text{ is a torsion group,} \\ \\ \frac{1}{2}h(G^{\mathrm{ab}}) - 1 & \text{in general.} \end{cases}$$

This should be contrasted with the result Theorem 2.5, proved by Bieri, Neumann and Strebel, which implies that if G is a finitely presented group without free subgroups of rank 2, and if $h(G^{\mathrm{ab}}) \geqslant 2$, then there is a finitely generated subgroup $G_1 \geqslant G'$ with $G/G_1 \cong \mathbb{Z}$.

2. Extensions by nilpotent groups

The assertions collected in Theorem 3.10 are obtained using techniques and results in the representation theory of finitely generated nilpotent groups. The proofs are difficult, and yet one suspects that the results are far from best possible.

Theorem 3.10. *(Groves and Wilson, [16], [17]) Let Q be a torsion-free nilpotent group and suppose that there is an $\mathbb{F}_p Q$-module M on which Q acts faithfully and satisfying* $\mathrm{hs}\,(M) \leqslant \frac{1}{2}h(Q)$.
 (a) *If $d(Q) \leqslant 2$ then Q is abelian.*
 (b) *If $d(Q) \leqslant 5$ then Q is nilpotent of class at most two.*
 (c) *The centre of Q cannot be equal to Q' and be infinite cyclic.*
 (d) *The group Q cannot be free nilpotent of class 2 and rank 3.*

We would like to thank Aeroflot for providing us with the unexpected opportunity to work together with Hermann Heineken on the case $d(Q) = 5$ of assertion (b), in the Special Delegates Lounge at Krasnoyarsk Airport.

The assertions above yield structural information about finitely presented metanilpotent groups, in the form of restrictions on the structure of the quotients of these groups by their Fitting subgroups. For example, assertion (a) leads to the following result.

Theorem 3.11. *(Groves and Wilson, [16]) If G is a finitely presented meta-nilpotent group and $d(G) \leqslant 2$ then G is virtually nilpotent-by-(nilpotent of class at most two).*

A result valid for arbitrary finitely presented metanilpotent groups but with a weaker conclusion has been proved by Brookes and Groves [12]: if G is such a group then G is virtually nilpotent-by-(nilpotent of class at most $d(G)$). Indeed, somewhat stronger results can be obtained using their techniques.

3. Extensions by polycyclic groups

The behaviour and significance of hs(M) when Q is an arbitrary polycyclic group seem a little obscure. We shall illustrate that hs(M) and $k(M)$ do not in general coincide. We begin by describing some important examples of polycyclic groups.

Let A be a non-cyclic free abelian group of finite rank and let T be a free abelian subgroup of the automorphism group of A such that $A \otimes_{\mathbb{Z}} \mathbb{Q}$ is irreducible as a $\mathbb{Q}T_1$-module for all subgroups T_1 of T of finite index. Thus the ring F of endomorphisms of $A \otimes_{\mathbb{Z}} \mathbb{Q}$ as a $\mathbb{Z}T$-module is an algebraic number field, and T is a subgroup of the group of units of the ring of algebraic integers of F. Set $Q = A \rtimes T$. It is apparent that some assertions from number theory have ramifications for the structure of these groups. For example, Dirichlet's unit theorem implies that $h(T) < h(A)$.

With Q as above, define $M = \mathbb{F}_p A$. Then M becomes an $\mathbb{F}_p Q$-module with A acting on M by multiplication and T acting by conjugation. From the commutative Hilbert–Serre theory we have hs$(M_A) = h(A)$ (see for example Atiyah and Macdonald [2], Chapter 11). Lemma 3.4 shows that hs$(M_A) =$ hs(M_Q). Therefore

$$\text{hs}(M_Q) = h(A) > 1.$$

The $\mathbb{F}_p Q$-submodules of M are just the ideals of $\mathbb{F}_p A$ which are invariant under the action of Q, and, by a theorem of Bergman [6], all non-zero such ideals have finite codimension in $\mathbb{F}_p A$. It follows that

$$k(M) = 1.$$

Now suppose that the group Q is as above, but that M is any $\mathbb{F}_p Q$-module which, regarded as a module for the integral domain $\mathbb{F}_p A$, is torsion-free of finite rank. Considerations like those in the above paragraph, together with the inequality $h(T) < h(A)$ which comes from Dirichlet's unit theorem, give

$$\text{hs}(M_Q) = \text{hs}(M_A) = h(A) > \tfrac{1}{2}(h(A) + h(T)) = \tfrac{1}{2}h(G).$$

Thus no extension of M by Q can satisfy (GC); and so no such extension can be finitely presented. This result can be extended.

Theorem 3.12. *(Brookes, Roseblade and Wilson, [13]) Suppose that Q is a polycyclic group and M is an $\mathbb{F}_p Q$-module. If some extension of M by Q is finitely presented, then Q has a normal subgroup L such that Q/L is virtually nilpotent and such that L acts nilpotently on M. Finitely presented abelian-by-polycyclic groups are virtually metanilpotent.*

There are many examples of finitely generated abelian-by-polycyclic groups which are not virtually metanilpotent (for example, standard wreath products of $\mathbb{Z}$ with polycyclic groups which are not virtually nilpotent), and the above result shows that such groups certainly cannot be embedded in finitely presented abelian-by-polycyclic groups. This gives a negative answer to a question raised in 1973 by G. Baumslag [5].

An important part of the proof of this difficult result is a reduction to the case when Q is a split extension $A \rtimes T$ of the type described above. The hypothesis of the theorem has consequences not dissimilar to the assertion $\mathrm{hs}\,(M) \leqslant \frac{1}{2}h(Q)$, and the strategy is to use the unit theorem just as above. The notion of applying the unit theorem to show that groups are not finitely presented seems somewhat curious (and the strategy would break down for extensions of nilpotent groups by soluble minimax groups), but it is the only way known at the moment to tackle such problems.

References

[1] H. Abels, *Finite Presentability of S-arithmetic Groups. Compact Presentability of Soluble Groups*, Lecture Notes in Maths., **1261** Springer-Verlag (1987).

[2] M. F. Atiyah and I. G. Macdonald, *Introduction to Commutative Algebra*, Addison Wesley, (1969).

[3] B. Baumslag and S. Pride, *Groups with two more generators than relators*, J. London Math. Soc. (2) **17** (1978), pp. 425–426.

[4] G. Baumslag, *Subgroups of finitely presented metabelian groups*, J. Austral. Math. Soc. **16** (1973), pp. 98–110.

[5] G. Baumslag, *Finitely presented metabelian groups*, in: Proceedings of the Second International Conference on the Theory of Groups, Lecture Notes in Mathematics, **372** Springer-Verlag, (1974) pp. 65–74.

[6] G. M. Bergman, *The logarithmic limit-set of an algebraic variety*, Trans. Amer. Math. Soc. **157** (1971), pp. 459–469.

[7] R. Bieri and J. R. J. Groves, *On the geometry of the set of characters induced by valuations*, J. Reine Angew. Math. **347** (1984), pp. 168–195.

[8] R. Bieri, W. D. Neumann AND R. Strebel, *A geometric invariant of discrete groups*, Invent. Math. **90** (1987), pp. 451–477.

[9] R. Bieri and R. Strebel, *Almost finitely presented soluble groups*, Comment. Math. Helv. **53** (1978), pp. 258–278.

[10] R. Bieri and R. Strebel, *Valuations and finitely presented metabelian groups*, Proc. London Math. Soc. (3) **41** (1980), pp. 439–464.

[11] R. Bieri and R. Strebel, *A geometric invariant for modules over an abelian group*, J. Reine Angew. Math. **322** (1981), pp. 170–189.

[12] C. J. B. Brookes and J. R. J. Groves, *Modules over nilpotent group rings*, J. London Math. Soc. **52** (1995), pp. 467–481.

[13] C. J. B. Brookes, J. E. Roseblade and J. S. Wilson, Modules for group rings of polycyclic groups, to appear in J. London Math. Soc.

[14] E. S. Golod, *On nil-algebras and residually finite p-groups*, Izv. Akad. Nauk. SSSR Ser. Mat. **28** (1964), pp. 273–276.

[15] E. S. Golod and I. R. Shafarevich, *On the class field tower*, Izv. Akad. Nauk SSSR Ser. Mat. **28** (1964), pp. 261–272.

[16] J. R. J. Groves and J. S. Wilson, *Finitely presented metanilpotent groups*, J. London Math. Soc. (2) **50** (1994), pp. 87–104.

[17] J. R. J. Groves and J. S. Wilson, Unpublished notes.

[18] K. W. Gruenberg, *Two theorems on Engel groups*, Proc. Cambridge Philos. Soc. **49** (1953), pp. 377–380.

[19] J. D. King, *Finite presentations of Lie algebras and restricted Lie algebras*, Bull. London Math. Soc. **28** (1996), pp. 249–254.

[20] P. F. Pickel, *Rational cohomology of nilpotent groups and Lie algebras*, Comm. Algebra **6** (1978), pp. 409–419.

[21] V. N. Remeslennikov, *On finitely presented groups*, in: Proceedings of the Fourth All-Union Symposium on the Theory of Groups, Novosibirsk, (1973), pp. 164–169.

[22] D. J. S. Robinson and R. Strebel, *Some finitely presented soluble groups which are not nilpotent by abelian by finite*, J. London Math. Soc. (2) **26** (1982), pp. 435-440.

[23] A. Schinzel and J. Wójcik, *On a problem in elementary number theory*, Math. Proc. Cambridge Philos. Soc. **112** (1992), pp. 225–232.

[24] E. B. Vinberg, *On the theorem concerning the infinite dimension of an associative algebra*, Izv. Akad. Nauk SSSR Ser. Mat. **29** (1965), pp. 209–214.

[25] J. S. Wilson, *Finite presentations of pro-p groups and discrete groups*, Invent. Math. **105** (1991), pp. 177–183.

[26] J. S. Wilson, *On finitely presented soluble groups with small abelian-by-finite images*, J. London Math. Soc. (2) **48** (1993), pp. 229–248.

[27] J. S. Wilson and E. I. Zelmanov, *Identities for Lie algebras of pro-p groups*, J. Pure Appl. Algebra **81** (1992), pp. 103–109.

[28] E. I. Zelmanov, *On the restricted Burnside problem*, in: Proceedings of the International Congress of Mathematicians, Kyoto 1990, Springer-Verlag, (1992), pp. 395–402.

Printed in the United States
By Bookmasters